The Essence of Oil & Gas Depletion

Collected Papers and Excerpts
Compiled by

C. J. Campbell

©2002
Multi-Science Publishing Company

Table of Contents

Abbreviations and Conversions

T	=	trillion (10^{12})	d	=	day
G	=	billion (10^9)	b	=	barrel
M	=	million (10^6)	t	=	tonne
k	=	thousand (10^3)	cf	=	cubic foot
a	=	year	Sm^3	=	cubic metre

Conversion

Barrel	Sm³	Tonne*	cf gas
1	0.159	0.136	–
6.29	1	0.855	35.3
7.33	1.17	1	–

(*varies with density: Arabian Light 33.5°API quoted)
1 b/d =50 tonnes/a
1 kb/d x 0.000365 = 1 Gb/a

Density
°API (American Petroleum Institute scale) = (141.5/sp.gr) – 131.5
(eg. Density of water = 10° API)

Oil-Gas Equivalence (oe)
1 boe = 6 kcf (calorific) = 10 kcf (approximate value)

Natural Gas Liquids (NGL)
Condensate condenses from gas naturally at surface temperature and pressure
NGL(P) = liquids extracted from gas in plants
LNG = liquefied natural gas
LPG = liquefied petroleum gas

Categories
Bitumen (tar) defined by viscosity (>10,000 mPa.s)
Extra-Heavy – <10° API density
Heavy – 10–17.5° API density (<22°API in Venezuela; <25°API in Canada)
Polar – >66.6° Latitude
Deepwater – >500m
(Above items here treated as *Non-Conventional*, but no standard definition)

For

Anne, Patrick, Emma, Clara and Oscar,

Jack and Julia, Simon and Oddny

and

Bobbins

Acknowledgments

Many people have contributed to this study in many different ways. In particular the help and support of the following is acknowledged with gratitude.

Kjell Aleklett, Carlos Alvarez, Inger Anda, Sarah Astor, Ali Bakhtiari, Jean-Marie Baudaire, Pierre-René Bauquis, Roger Bentley, Wolfgang Blendinger, Richard Douthwaite, David Fleming, Peter Gerling, Nick von Glahn, Jan Hagland, Buzz. Ivanhoe, Richard Hardman, Klaus Illum, Jens Junghans, Jean Laherrère, Maarten van Mourik, Alain Perrodon, Susanne Peters, Rui Rosa, Richard Shepherd, Chris Skrebowski, David Strahan, Ron Swenson, Randy Udall, Frank de Winter, Walter Youngquist, Walter Ziegler, Werner Zittel.

Thanks too go to Bill Hughes and his team at Multi-Science for bringing the effort into print.

Aspiration

That knowledge of oil and gas depletion may help people and their governments understand the limits imposed by Nature such that they may be able to plan their future so as to live in better harmony with themselves, their environment and each other.

The General Depletion Picture

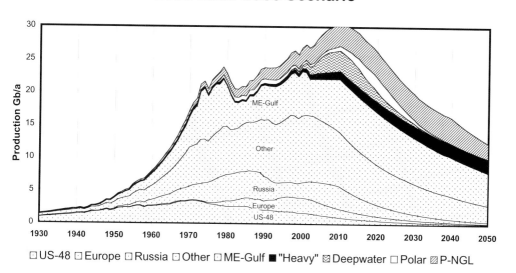

**Oil & Natural Gas Liquids
2003 Base Case Scenario**

Tables of essential data

Table 1. Estimated *regular oil* produced to 2075

Past		Future		Total
	Known Fields		New Fields	
896		871	133	1900
		All liquids		
986		1714		2700

In billion barrels (Gb)   *Status: end 2002*

Table 2. Production rate forecast Mb/d

	2005	2010	2020	2050	Total to 2075
Regular Oil	60	60	47	22	1900
US-48	3.5	2.6	1.4	0.2	195
Europe	5.1	3.7	1.9	0.3	76
Russia	8.6	9.4	4.9	0.7	200
M. East Gulf	17	22	22	13	749
Other	26	23	17	8	680
Heavy, bitumen etc	2.8	4	5	6	300
Deepwater (>500m)	6.6	9	4	0	63
Polar	1.2	2	6	0	30
Natural Gas Liquids	8.2	9	11	6	400
Total	**78**	**83**	**72**	**33**	**2700**

Base Case Scenario: flat demand for conventional oil due to recession; M. East swing role ending in 2010
Regular Oil includes *Condensate* but not liquids produced from gasfields by processing — PNGL

Status: end 2002

Part 1

Introduction

The Association for the Study of Peak Oil ("ASPO") is a network of concerned scientists in European universities and institutions who are committed to study the issue of the peak of world oil and gas supply, and evaluate its impact. Its declared mission is

1   To evaluate the world's endowment of oil and gas;
2   To model depletion, taking due account of economics, technology and politics;
3   To raise awareness of the serious consequences for Mankind.

The World's oil was formed in the geological past during a few brief epochs of intense global warming when algal growths proliferated to provide the organic material that was later converted to oil under now well-understood processes. Gas was derived both from vegetal material in a similar fashion and from the break down of ordinary oil at high temperature on deep burial. It needs a good seal to hold it in the reservoir, and much has been lost over geological time.

If we admit that oil and gas were formed in the geological past, we cannot deny that they are fossil fuels subject to depletion. It follows that we started running out when the first barrel was produced. As every beer drinker knows: the glass starts full and ends empty.

The issue is not so much to determine when we will finally run out, which is far in the future, but rather to concentrate on when production will reach a peak and begin to decline due to resource constraints. This threatens to be a discontinuity of historic proportions, given that oil provides 40% of traded energy and 90% of transport fuel. The World, which runs on oil, will certainly notice the impact of the onset of shortage. The Stone Age did not end because we ran out of stone, but because we found out how to make better implements of bronze and, later, iron and steel. It was an upward voluntary progression that led to more complex societies and an explosion of population. The Oil Age will however end because we ran out of oil, meaning that this time we will have to stumble down the other side of the pyramid of progress, not because we have found a better alternative, but out of raw necessity. As of to-day, no substitute fuel is in sight that comes close to oil and gas in terms of convenience, efficiency or cost. In short, by any reasonable reckoning, *Hydrocarbon Man* will have become extinct by the end of the Century, without as yet a sign of a successor.

The importance of the subject can hardly be exaggerated. It may be asked therefore why it is not better understood and placed at the head of the agenda when planning for the future. There are several possible explanations:

1   It is counter-intuitive: we have never run out of essential supplies before
    – and, with reason, we have come to have a blind faith that technology
    will come to the rescue as in the past.
2   Public data on reserves are grossly unreliable, so it is difficult for anyone
    to see the trend of past discovery with which to estimate realistic future
    possibilities. We live in a world of poor accounting.
3   It runs counter to classical economics, which, having been built on the
    experience of the Industrial Revolution, relies on the ineluctable laws of
    supply and demand, perceiving Man to be Master of his Environment.
4   There are vested interests in industry and government with motives to
    deny and confuse the issue. Democratic governments find it easier to
    react to crises than prepare for them.

But there is a now a new awakening, as ASPO and other concerned entities around the world succeed in drawing attention to these critical issues. The evidence is building. The clouds of obfuscation are being swept away. Conferences are being held. Television programmes are being made. Governments are being alerted. This awakening is itself a fascinating subject in its own right. Those, who only a few years ago were lone voices in the wilderness, now find themselves being taken seriously. New opinions, attitudes and instincts are being formed, although there remain many uncertain points of detail.

In short, world oil production will start to decline from around 2010, with gas following not long afterwards.

This book has six parts:

1   A summary of the situation as submitted to the United Kingdom
    Government in its consultative process on energy policy
2   Edited and selected extracts from the ASPO Newsletters, which have
    helped influence the change of attitude

It can be seen that some of this material has already been published; however, bringing it together in this way was felt to create a new whole greater than the sum of its parts.

These papers contain inconsistencies and repetitions because the subject is still in a state of flux due to varying definitions, data-sets and evolving understandings. The economist with a reluctance to face resource depletion may latch onto the inconsistencies as evidence of a fundamental weakness in the analysis, while the scientist will see them as evidence of progress. A good response to the detractors is to refer to the famous quotation from Maynard Keynes, one of the foremost economists, who, on being accused of inconsistency, replied:

> *"When I get new information, I change my conclusions. What do you do, Sir?".*

Submission to H. M. Government Consultation on Energy Policy
www.dti.gov.uk/energy/energyep/index.htm

by C. J. Campbell, The Association for the Study of Peak Oil (ASPO)

Forecasting Global Oil Supply

Summary

1 *Economists* treat the oil and gas resource as near infinite, seeing supply being driven by market forces, whereas *Natural Scientists* evaluate the occurrence in Nature, seeing supply being controlled by immutable physics.

2 The *Discovery Trend* is all-important because production mirrors it after a time-lag.

3 *Public Reserve Data* are very unreliable and definitions ambiguous – vested interests in industry and government indulge in *Denial and Obfuscation*

4 A *Realistic Forecast*, readily confirm-able from the industry database and robust modelling, shows that oil production will peak around 2010, with oil and gas combined following five years later.

5 *Peak Oil* production will be a *Watershed* for the world economy, with many political, social and geopolitical consequences. The transition to decline will be difficult, but much could be done in terms of energy saving and the provision of alternative fuels, if action is taken early enough.

6 There is time to change direction: so it is *Not a Hopeless Mission*.

The Conflict

Few would deny that the world runs on oil. By describing oil as a fossil fuel, everyone admits that it was formed in the past, which means that we started running out when we consumed the first barrel. That much can surely be agreed, but opinions differ about how far along the depletion curve we are.

On one side of the debate are the classical economists of the flat-earth sect, whose views are summed up by Professor M. A. Adelman:

> *"Minerals are inexhaustible and will never be depleted. A stream of investment creates additions to proved reserves from a very large in-ground inventory. The reserves are constantly being renewed as they are extracted... How much was in the ground at the start and how much will be left at the end are unknown and irrelevant"*

On the other side, stand the natural scientists, who have been trained to observe Nature and understand its immutable physical laws. When they look at the issue, they ask three simple questions:

* *What was found?*
* *How much was found?* and
* *When was it found?*

They want this information to extrapolate the past discovery trends to show what is likely to be found in the future. They recognise that oil has to be found before it can be produced, meaning that the production trend has to reflect, in some manner, an earlier discovery trend. They know that an oilfield contains what it contains, because it was filled in the geological past, even if the amount it holds is not at first known accurately.

The amounts involved are commonly described in terms of *Reserves* and *Resources*, which are, however, often used in different and confusing senses. In plain language, we need to know the following parameters:

* How much has been produced to-date (*Cumulative Production*);
* Estimates of how much remains to be produced from known fields (*Reserves*); and
* Estimates of how much will be produced from new fields (*Yet-to-Find*);
* The total endowment, being the sum of these elements (*Ultimate Recovery*)

It is also expedient to set a time limit of, say, 2075, to avoid having to worry about tail end production that can drag on for a very long time, and is irrelevant to the determination of peak production, the main issue.

Denial and Obfuscation

The dispute between the rival disciplines is clouded by statements from vested interests with motives to confuse. The major oil companies issue bland scenarios and imagery, avoiding the word *depletion* like the plague for fear that it would smell like a depleting asset to the investment community. Their actions, as they merge, downsize, outsource staff and desist from investment in new refining capacity, speak louder than their words.

Even so, the chief executive of one major company has recently admitted that less than half the production needed to meet a 2% annual increase in demand to 2010 can be delivered from present fields. He suggests that more than $100 billion dollars a year would have to be invested to secure the supply, which is far more than currently planned. The statement implies that demand will not be met. Others speak of an

Note: Some of the referenced figures and tables are produced elsewhere in the book

imminent peak of production, when free to do so on leaving office. No criticism is implied. The directors of oil companies have a fiduciary duty to make money for their shareholders. It is simply not their job to do more than meet mandatory financial reporting requirements. Such requirements, like many accounting practices, are however archaic and need to be thoroughly revised to provide answers to the three cardinal questions posed above.

The implications of the decline of the world's premier energy source are so pervasive that in political terms it is easier for governments and responsible agencies "not to know". This is certainly the position of the International Energy Agency on which many governments rely. It resolutely declines to make any serious study of the resource base or access the industry database. Its implausible forecasts change from year to year without explanation. For example, in 1998, it estimated that as much as 17% of world supply in 2020 would have to come from "unidentified unconventional" sources, but, without explanation, it removed the item altogether in its subsequent forecast, when it was able to hide behind a flawed study by the United States Geological Survey, which is reviewed below.

The critical issue is not so much when oil will eventually run out, but rather when production will reach a peak and begin to decline. That will surely represent a major watershed for the world's economy, with far reaching social and political consequences.

Unreliable Public Data and Confusing Definitions

Confusing definitions, which have been inherited from the early days of the oil industry, largely explain the extremely unreliable nature of public data. It is worth briefly reviewing the industry practices of estimation and reporting.

Explorers search for promising prospects and determine their potential. They measure the size of the prospect from seismic surveys, now giving extremely high resolution, and assess the likely reservoir conditions from regional knowledge and nearby boreholes. Based on this information they make best estimates of the volume of oil in the ground and the likely percentage recovery. While sound scientific estimates can be made, the explorers are in practice normally under pressure to exaggerate in order to secure internal funding to test the prospect. The application of subjective probability theory is one means for doing so by the simple expedient of exaggerating the unquantifiable high case (P_5), which lifts the *Mean* value that is normally used for economic evaluation and management decision. These estimates, in any event, do not enter the public domain, being confidential to the company.

If the prospect is drilled successfully, engineers move in to plan the development. Their task is to design the facilities to balance investment against production rate so as to maximise profit. A large field normally has several phases of development with the reserves attributable to each being duly reported as it is completed, and termed *Proved* for financial purposes. The overall reserve estimates are accordingly revised upwards with each successive phase of development, such that they commonly end up approximating with the original volumetric estimates by the explorers. Only in small fields with a single short development plan do reported *Proved Reserves* reflect the actual size of the field. Furthermore, companies have had good commercial incentives to be able to draw on inventories of under-reported reserves, reducing tax and providing a better image by smoothing the wild fluctuations in their assets that otherwise result from periodic discoveries separated by lean years. In this way, they are able to report high levels of reserve replacement to impress the stockmarket, which do not properly reflect the actual results of discovery. They are wont to explain the upward revisions in terms of technological progress when they are in fact primarily a reporting phenomenon. The growth of reported *Proved Reserves* has misled many analysts into thinking that more is being found than is the case.

No one denies the huge technological advances, but the impact has been to hold production higher for longer without adding to the reserves themselves, save in certain exceptional cases. Thus, technological progress has served to improve profit, especially under the principles of discount economics, at the cost of accelerating depletion.

It is clearly vital to backdate any reserve revisions to the discovery of the field concerned to obtain a genuine discovery trend. A field is found by the first successful borehole drilled upon it, so that all the oil, ever to be produced from it under whatever circumstances, is logically attributable to the date of that borehole. Backdating, however, calls for access to the industry database of individual fields, and cannot be achieved on the basis of published national statistics.

The Flawed US Geological Survey Report

This debate has been much influenced by a report published by the USGS in 2000. Some serious questions should be asked about the methodology employed. In brief, it made an academic assessment of the *subjective* probability of new discovery in each of the world's basins, tacitly assuming an infinite number of wildcats to find it. For example, in an unknown, untested, basin in East Greenland, it concluded that there was a 95% probability (F_{95}) of finding more than zero, namely at least one barrel, and a 5% probability (F_5) of finding more than 112 billion barrels. A *Mean* value of 47 billion was computed from this range. Since the estimates were quoted to three decimal places, the reader could be forgiven for assuming them to be accurate. But a moment's reflection would question the very concept of a *subjective* 5% probability of the actual result being greater than the stated value. In plain language, it was a guess that could as well be the half or the double, yet it entered the calculations distorting the critical *Mean* value. We are now seven years into the study period and can compare the forecast with what has been found in the real world. The USGS forecast, as a *Mean* estimate, that 732 Gb (billion barrels) are to be found between 1995 and 2025, which gives an average of 24 Gb a year. So far, the average has been only 15 Gb, when above average performance should be expected because the larger fields are usually found first because they are the biggest targets. Its estimates furthermore fail to indicate the amount of exploration drilling required to deliver the indicated amounts, so that it is not possible to see if they are plausible in relation to past performance. What the USGS failed to do was to properly model discovery over time and against wildcat drilling, and evaluate field size distributions, which would have improved the abstract geological assessment by taking into account the experience of the past.

The USGS did not forecast production as such, but Figure 1 compares past actual discovery with three hypothetical trends to deliver the USGS range of estimates. It is obvious that only the low (F_{95}) case bears any remotely reasonable relationship with the past actual trend, which, it is stressed, resulted from the diligent efforts of the industry in a worldwide quest for the biggest and best prospects. The industry had the benefit of all the much-vaunted advances of technology and geological knowledge. If more could have been found, it would have been found, especially recognising that the international industry operates under extraordinarily favourable economic terms whereby the cost of exploration is offset against taxable income under high marginal tax rates. In many countries, it was effectively spending 10c dollars on exploration.

Not content with exaggerating the *Undiscovered* potential, the USGS went on to add *Reserve Growth*, based on US onshore experience on which it had data. It failed to understand the nature of reserve reporting, as outlined above. Stated in plain language, *Proved Reserves*, as reported for financial purposes, refer to what the existing wells of the current stage of an oilfield's development are expected to deliver: in other words, they are *Proved So Far*, saying little about the eventual size of the field as a whole. In the case of the large old fields, initially reported *Proved Reserves* understated the ultimate field size by as much as one-third, as is amply documented by mature North Sea fields or Prudhoe Bay in Alaska. However, most of the offshore finds over the past decade are too small to have more than one development phase, so their initially reported *Proved Reserves* may indeed represent what the fields can deliver, leaving no scope for later growth. Accordingly, the USGS made the double mistake of assuming that the experience of the old onshore fields of the Lower-48 United States was even remotely representative of the offshore or international arena, and secondly of applying the same growth factor to fields of differing size.

It is relevant to note that the USGS now faces legal proceedings under Public Law 106-554; H.R. 5658 Section 515, which is concerned with the obligation of government departments to provide sound information.

A more realistic approach

In trying to explain the real position, the first step is to define clearly the different categories of oil, distinguishing the easy and cheap, commonly called *Conventional*, from the expensive and difficult. As explained above, public data on reserves are extremely unreliable, but if a detective were hired to piece together all the evidence and clues, he would likely report the following rounded estimates for *Conventional* oil, as defined herein.

	World	**Persian Gulf***	**Russia**
Produced-to-date	875	225 (26%)	120 (14%) Gb (billion barrels)
To be produced from			
known fields	900	500 (53%)	70 (8%)
new fields	150	40 (29%)	15 (10%)

*(Iran, Iraq, Kuwait, Saudi Arabia and the United Arab Emirates)

World demand naturally influences the rate of depletion. A Base Case Scenario might conclude that a near absence of spare capacity in late 2000 was forcing up the price of oil. It could have risen higher had the economy

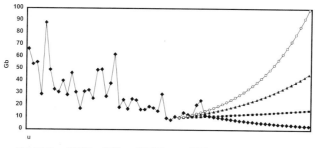

Discovery (Crude Oil & Condensate)
Past Actual & USGS Forecasts

Figure 1

not moved into recession, reducing the demand for oil, and thereby, the pressure on price. It follows that if the economy were to recover, oil demand would rise in parallel until it again hit the falling ceiling of capacity, when prices would soar, re-imposing recession in a vicious circle. For these reasons, the scenario assumes that demand will be on average about flat, giving a plateau of conventional oil production until the five Swing countries of the Persian Gulf are no longer able to offset the natural decline of the rest of the world. This threshold is expected around 2010, when they would have to produce over 20 Mb/d (million barrels a day), despite the natural decline of their ageing fields. It amounts to about 37% of world demand for conventional oil. World production is then assumed to commence its long-term decline at the then depletion rate.

Although described as a production "plateau", it is likely to be anything but flat. It will more likely be a period of recurring price surges, recessions, international tensions, and growing conflicts for access to critical oil supplies, as indigenous production in the United States and Europe declines. The calls, as modelled, on the individual swing countries are shown in Figure 2. They are considerable, even with flat world demand, especially if their reserves are still exaggerated as suspected. In the event that supply should be higher than anticipated, then peak would clearly come sooner, to be followed by a steeper decline.

Non-Conventional Oil

In addition to *Conventional* oil, which, generally speaking, is the easy, cheap stuff that has supplied most oil to-date and will dominate all supply far into the future, are other more difficult and expensive sources. The economists tend to claim that they can be progressively tapped as the need arises, but the sad truth is that most of them can come on only slowly for the very reason that they are expensive and difficult to extract. Some carry environmental costs too. They cannot accordingly have much impact on global peak although they will ameliorate the subsequent decline. They are subject to their own depletion profiles, which are less readily modelled.

Heavy Oil

Oils heavier than 17.5° API, including tarsands, are here treated together, with production being controlled by extraction rate rather than the resource base. The assessment shows production rising gradually to 4.5 Mb/d by 2020.

Deepwater Oil (>500m water depth)

The deepwater domain is characterised by special geological conditions. Prolific oil generation occurred only in certain divergent plate-tectonic settings having early rifts, in which source rocks were deposited and preserved. The right conditions are probably confined to the Gulf of Mexico and margins of the South Atlantic. Elsewhere, deltas, which do locally extend into deep water, are mainly gas-bearing because they have to rely on the source-rocks within the delta itself.

It is evident that deepwater operations test technology and management to the limit, which means in turn that only the larger prospects or clusters of prospects are likely to be viable. A further constraint is the availability of floating production equipment. It is concluded that, from an endowment of about 60 billion barrels, deepwater production might rise with heroic effort to a peak of about 8 Mb/d by 2010. It is likely that the international oil companies, which control this production, will produce it at maximum rate, irrespective of world demand or the supply coming from Russia and the Middle East.

Polar Oil

Antarctica has very limited geological prospects, being in any case closed to exploration by agreement. The Arctic regions of Alaska, Canada, Greenland, Norway and Russia are more promising, having some huge sedimentary basins. However, the evidence to-date suggests that they are mainly gas-prone, save for parts of Alaska and Siberia, because vertical movements of the crust under the weight of fluctuating ice-caps in the geological past depressed the source-rocks into the high temperature gas-generating window. The development of these remote and hostile areas calls for substantial high-risk investment, which is unlikely to bear fruit for many years.

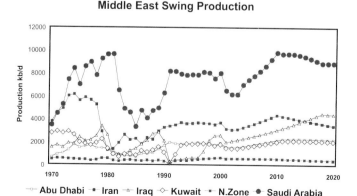

Figure 2

Natural Gas

Natural gas is commonly treated as a potential substitute for oil, although it is far less convenient as a transport fuel. It was more widely generated in nature than was oil, but required a better seal to hold it in the reservoir, much having been lost over geological time.

The higher molecular mobility of gas means that it depletes very differently from oil. An uncontrolled well would exhaust a gas accumulation very quickly. Accordingly in practice, production is generally capped at far below capacity to provide a long plateau, set by pipeline infrastructure, with most fluctuation being seasonal. The capped production provides in-built spare capacity, which is progressively drawn down, while costs and prices generally fall under competitive market pressures as the original investments are written off. The end of the plateau comes abruptly when this in-built spare capacity has been exhausted, and it comes without warning market signals.

It appears that the United States is now close to the end of its plateau, such that new gas wells have to be produced at maximum rate, being depleted in a matter of months. It is increasingly forced to rely on Canadian imports, which themselves are progressively subject to the same depletion pressures.

It is difficult to model gas supply since so much depends on uncertain market forces and the construction of new pipelines. Based on the consensus world endowment of 10000 Tcf (trillion cubic feet), production is here modelled to rise to a plateau from 2015 to 2040, before declining sharply. This may be optimistic because much gas is stranded in remote areas too far from markets for pipeline links. While liquefaction and transport in specialised tankers is possible, it is both costly and constrained by the sheer scale of the undertaking. To move only 2 Tcf a year would take almost 100 tankers with a monthly turnaround.

The overall depletion profile with its abrupt end carries grave risks to supply unless properly evaluated and anticipated, as Europe is likely to discover to its cost.

Natural Gas Liquids (NGL)

The production of gas in turn gives a substantial supply of NGL, especially as the percentage extracted is likely to increase. NGL will form an important additional supply around oil peak, but will eventually decline in parallel with the natural gas.

Non-Conventional Gases

Non-conventional gases are important too. The two most promising are Arctic gas, much from Siberia, and coalbed methane from the world's coal basins, although it is slow and difficult to extract. Gas hydrates from the ocean depths can be confidently dismissed. They occur mainly as disseminated ice-like granules and laminae, from which the methane cannot migrate to accumulate in commercial quantities. A great deal of misplaced research funding is wasted on this chimera.

Future Global Oil and Gas Production

The following tables and Figure 3 provide an assessment of past and future production of oil and gas from all sources, based on available information and robust modelling techniques is given in the Introduction.

The watershed for oil comes around 2010, followed five years later by the peak of oil and natural gas combined. The indicated supply constraints will inevitably curb demand in one way or another. It implies that the economic growth of the past, which was largely driven by an abundant supply of cheap oil-based energy, cannot continue on the same trajectory.

Oil discovery in the United States peaked in 1930 with the discovery of the East Texas field. Peak production inexorably followed forty years later, but the US consumer barely noticed as cheap imports made up the difference. Since then, the same pattern of peak and decline has been repeated from one country to another, and the time-lag from peak discovery to peak production has fallen thanks to modern technology. Given that peak world discovery was passed in 1964, it should be no surprise that the corresponding peak of global production is now close. Exactly when it will come depends on many short-term factors, not least of which are military intervention in the Middle East and the degree to which Russia steps up its exports. The base-case scenario points to 2010, but it could come sooner if economic recovery should drive up the demand for oil.

Geopolitical Consequences

The uneven distribution of conventional oil in Nature is illustrated in Figure 4. Approximately half of what is left to produce lies in the swing countries of the Middle East. It is evident that the United States, whose indigenous production has been in decline for thirty years without hope of reprieve, has a desperate need to secure control of foreign oil. Indeed, access to such oil has long been officially classed as a vital national interest, justifying military intervention where necessary. Whereas in the past such a policy was designed to meet the threat of short politically-inspired interruptions, now it faces the iron grip of depletion as the nations of the world vie with each other for critical supplies. Under the principles of globalism, resources anywhere are supposed to be accessible to the highest bidder, but as shortages appear, producing countries may find it expedient to conserve their resources for themselves, further exacerbating the competition for imports by others in need. The United Kingdom itself is not immune. With flat demand, consumption will exceed production by around 2007, with the percentage of imports set to rise to 20% in 2010, 50% in 2020, and 95% in 2050.

Russian oil production is rising to a second peak around 2010, partly making good the anomalous fall occasioned by the fall of the Soviets, and partly from the application of modern methods by newly privatised companies. There is evidently scope for new discovery, although earlier hopes that the Caspian, one of the world's earliest oil provinces, might counter Middle East dominance fade in the face of the disappointing results.

As North Sea production declines, Europe will find itself increasingly dependent on Russian supplies of oil and particularly gas, carrying heavy geopolitical implications. Meanwhile, the pressures and tensions upon the Middle East, that are growing to the point of a threatened war, are readily explainable by the resource disparities depicted in Figure 4.

The Time to Act is Now

The inevitable conclusion from a realistic assessment of the situation is that world oil production, which provides about 40% of global energy needs and about 90% of transport fuel, will start to decline within about ten years. It is evident that the World will have to learn to use less – much less. It will be difficult, but not impossible given the current monumental level of waste. Every effort is also called for to step up the production of alternative sources of energy to make up as much of the shortfall as possible. There is a great deal at stake as solutions have long lead times and call for difficult adjustments, but much could be done if governments could be alerted in time to act. It is not a hopeless mission as there is a little breathing space in

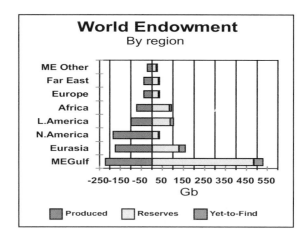

Figure 4

which to make appropriate plans since the production of all liquid hydrocarbons, as forecast and illustrated, need not fall below present levels for some twenty years.

While a Depletion Protocol to manage natural depletion under an equitable international agreement has been proposed, it is unrealistic to expect much progress as powerful nations will likely continue to be motivated by perceived self-interest. Accordingly, it will rather be for each country and each community to find individual solutions within their own resources and capabilities. Those that do plan and prepare will clearly have great advantage over those that simply react to the crisis when it hits them.

Part 2

Extracts from the ASPO Newsletters

Newsletter No. 1, December 2000

1. From small beginnings
This first humble and somewhat egocentric edition is intended to do no more than plant a seed that hopefully will take root and grow. It is egocentric for the simple reason that since it is the first edition, no other contributors have had reason to report in on their work, but that too will change as we progress.

2. Lecture in the Danish Parliament
On December 5th C. J. Campbell gave a lecture in the Danish Parliament building to an audience of about 80 in a meeting of the Society of Green Technology of Denmark. It was followed by a Panel Discussion in which a Director of Shell, the Head of the Danish energy commission and two politicians of rival parties were asked to react to the presentation. The important point was that all accepted that we do face a *Transition* as oil production reaches peak and goes into decline. Opinions might differ as to the date and scale of the transition, as well as what might be done about it, but that it was a serious matter of concern was not disputed. This indeed is progress.

3. Clausthal Lecture
On December 7th C. J. Campbell gave a similar public lecture at Clausthal Technical University at the invitation of Professor Wolfgang Blendinger, whose prior experiences with Shell had made him convinced of the reality of depletion. Approximately fifty attended. Interviews were given to the press and to a radio correspondent, which have apparently resulted in good

articles in important German papers
[See video on http://www.rz.tu-clausthal.de/realvideo/event/peak-oil.ram]

4. The Formation of ASPO

These meetings and discussions with interested parties led to the idea to create this Association, which was then discussed with Dr Wellmer and his colleagues at the BGR in Hannover. Drs Kehrer and Rempel provided several publications on the subject, which have yet to be filed and entered into the database.

Later in December, meetings were held with Mr Jan Hagland of the Norwegian Petroleum Directorate and Mr Jan-Erik Karlsen and his colleagues of Rogaland Research to discuss their participation. Additional publications were obtained from these sources.

5. Data Sources

The December issue of the *Oil & Gas Journal* was obtained with the reserve updates for 2000. It is noteworthy that 71 countries announce no change in their estimates (in many cases not having done so for many years), which demonstrates the unreliable nature of this public data. It will no doubt be reproduced in the *BP Statistical Review of World Energy*, which many analysts mistakenly regard as an authoritative source.

6. Solar Hydrogen Energy Group

Mr von Glahn, the Chairman, reported on a meeting in Brussels on December 5th, also attended by Dr Bentley, in which Europe's new policy of encouraging renewable energy was debated. It appears that the Commission is belatedly becoming aware of the depletion of oil and the imminent peak, despite the flawed advice to the contrary it has been receiving from many quarters. Papers explaining the true position were placed in the hands of the officials.

Later von Glahn requested a review of the flawed study by the US Geological Survey study, which he succeeded in placing on the official website of the G8 Renewable Energy Task Force, chaired by the Shell Chairman. It remains to be seen if it will be suppressed.
(www.renewabletaskforce.org/viewscomments.asp?id=46)

7. TV Programmes

Interviews have been granted to the following: Dutch TV, CBC, Toronto, and the BBC. The latter produced an excellent 30 minute programme for the Money Programme, also re-broadcast on BBC World. It is being made available to other TV stations, currently being shown on Dutch TV. It is noteworthy that Shell in the programme confessed to new discovery of no more than 250 Gb, and an equal amount from the correction of under-reported reserves (so-called "reserve growth"). Given that it presumably relates to all liquids, it is not far out of line from our own estimates.

8. Weak Oil Price
It is difficult to explain the recent fall in oil price in resource or spare capacity terms. Three explanations have been voiced:

- Financial institutions, which no doubt made substantial profit from speculating in oil futures over the past year, liquidated their positions for year-end financial reporting purposes.
- The winter has already ended for the refinery business due to the time lag for product to pass through the system. They just squeezed through, and may now expect a slight lessening of pressure as springtime demand lessens.
- It is a reaction to Saddam's failure to withhold Iraq production in the face of the release of strategic stocks in the USA, and offers from the IEA to do the same if necessary. This comfort, plus offers by Saudi Arabia to make good any shortfall (whether or not that is feasible in practice) gave the market confidence that OPEC would be able to hold price in the $22–28 band.

The wide fluctuations in price underline suggestions that it is an ignorant market, much subject to sentiment and facile commentary.

Newsletter No. 2, February 2000

9. Reading University
Dr Bentley gave a keynote address on oil depletion and its significance for India at the Petrotech Conference in New Delhi on January 9th. He found that there was already a serious interest in this subject by the Indian officials, who desire to secure better knowledge of the world position in connection with their national planning.

Dr Bentley made a visit to the Indian research establishment for consultations.

By way of introduction, Dr Bentley contributes the following record of Reading's growing interest in this subject:

> A group of people at The University of Reading in the UK has been working on the question of oil depletion since they first held a seminar on the topic in 1995, to which, among others, Dr. Campbell and Professor Odell were invited.
>
> Reading has seen its role as largely being that of checking the validity and suitability of the Petroconsultants' data, and the mid-point peaking model, for predicting the future of regional oil production in broad terms.

Reading has had no external funds for this research, but estimate that they have spent a total of about three man-years of internally-funded research time on the problem in the period since 1995.

Reading's publications to-date on this topic include:

2000

'Global Oil and Gas Depletion' Prepared as a hand-out paper for the EU Conference: 'What energy options for Europe in 2020?', held in Brussels, 4th & 5th December, 2000.

'Were those past oil forecasts really so wrong?' R.W. Bentley. Tomorrow's Oil, November, 2000. (Subscription newsletter, OneOffshore Ltd., PetroData Group.)

'World Oil Supply: Near & Medium-Term.' R.W. Bentley, R.H. Booth, J.D. Burton, M.L. Coleman, B.W. Sellwood, and G.R. Whitfield. Report submitted to the UK's Dept. for Trade & Industry 'Energy Advisory Panel', Sept 5th, 2000.

'Perspectives on the Future of Oil.' R.W. Bentley, R.H. Booth, J.D. Burton, M.L. Coleman, B.W. Sellwood, and G.R. Whitfield. EnergyExploration and Exploitation, Vol. 18, Nos. 2 & 3, pp 147-206, Multi-Science, 2000.

1999

'The Oil Future – A Very Different View.' R.W. Bentley, R.H. Booth, J.D. Burton, B.W. Sellwood, G.R. Whitfield. Newsletter, Int'l. Assocn. forEnergy Economics, 4th Quarter, 1999.

'The Future of Oil.' R.W. Bentley. Proc. 1999 European Environment Confr., Sept. 13/14, 1999, LSE, London, pp 29-38. ERP Environment, Shipley, W. Yorks, ISBN 1 872677 27 4.

1998

'UK Energy – The Next 5 to 10 Years.' R.W. Bentley. Report submitted to the UK Dept. of Trade & Industry, and circulated among UK and overseas energy analysts, Jan. 1998. Updated in 1999 as: 'Oil – The Next 5 to 10 Years', and subsequently expanded as a report: 'Perspectives on the Future of Oil'.

1997

'Oil Shock 'Imminent' if Heavies are Slow or Expensive to Produce.' R.W. Bentley. Energy World, June 1997, pp 20-22. The Institute of Energy, London.

1996

'The Conventional Oil Production Peak – 5 or 20 Years Away?' R.W. Bentley. Energy World, Dec. 1996, pp 11-12. The Institute of Energy, London.

10. Cork University

Dr Campbell gave a seminar to a group of M.A. students in the Geography Department of University College, Cork. The subject was met with interest and enthusiasm, so much so that it has been decided to try to establish a new unit within the Geography Department to cover the subject. A request for research funding has been submitted to support this initiative, proposing a three-phase project

a) an assessment of the resource base and depletion in order to forecast the date of peak production;
b) an evaluation of the global consequences of peak oil; and
c) an evaluation of the impact on Ireland.

Ireland's energy needs have doubled since 1980, with 80% coming from oil and gas. A long-term national plan has been formulated for the construction of roads, industrial infrastructure, transport and so forth, but it assumed a "business as usual" energy supply, which is now in question. Ireland is indeed vulnerable, being very much at the end of the line as North Sea production falls, and as Europe has to tap distant gas in Siberia and North Africa. So the new initiative at Cork can make an important national contribution.

Dr Campbell gave a lecture to the Irish Institute of Petroleum on February 7th.

11. L-B Systemstechnik

In January, Dr Werner Zittel completed a short study on oil the depletion of UK oil fields, showing the increasing number of new (small) fields needed to compensate for the accelerating decline rate of already producing fields. Though the first thirteen fields from the 1970s still contribute 15 % of today's oil production, about seventy new fields were needed in the period 1995 to 1999 to compensate just for the decline of the old fields. Plotting each field's production in historical order and adding all fields to the total output shows convincingly the growing speed of the "treadmill". In fact, it looks as if the foreseeable decline already started late in 1999. According to monthly data from the Royal Bank of Scotland, oil output in 2000 was down by about 7 % on the year, but production in October 2000 (the latest available data) shows a 16% decline relative to October 1999.

A short version (without individual field data) is available on the web in German at
http://www.energiekrise.de/oel/artikel/docs/oelproduktion_uk.pdf

12. Germany/Austria

On 23rd January 2001, a public discussion on the recent oil price increase was held at the Technical Museum in Vienna, Austria. Werner Zittel contributed for ASPO with a short presentation entitled "The evening before the global energy crisis". The presentation (in German) is available on the net at http://www.eva.ac.at/energiegespraeche. Other speakers were J. Benigni (PVM Oil Associates), Franz Wirl (Lehrstuhl für Energie, Industrie und Umwelt, University Vienna), and Nadir Guerer (OPEC Secretariat). All statements are available at the above internet address. It was common ground among the speakers that today's oil production system has shifted from being a supply driven system to "just-in-time" production.

Other speakers traced this back to the economics of the industry, but Zittel pointed out that it might be also due to supply constraints, reflecting the tight production situation of Non – OPEC countries, which are at peak. He added that it might be very susceptible to small fluctuations that in turn might produce chaotic consequences. The situation is very similar to 1973 where the largest producer and consumer (USA) had passed its production maximum and changed to a just-in-time supply-system without any spare capacity. But this time it hits the world, thirty years later on the depletion curve, at a stage of much higher oil dependency, both in absolute terms and on a global scale. It is only a matter of time before the spark ignites. It remains to be seen whether that spark will be Saddam Hussein or the US electricity and gas crises.

13. Gas

The US gas price has been in the range of between $ 2 – 4 /Mcf for more than two decades, both in nominal and real values. But within the last two months, it suddenly rose to over $10/Mcf, reflecting the onset of a somewhat chaotic behaviour in the US gas market. However, prices have weakened again over the past weeks, falling to about $6 – 7/Mcf. This decrease is widely attributed to a reduced draw on US natural gas storage. However, what has not been spotted by most observers is that there was a corresponding draw on Canadian storage, which was greater than expected. The recently completed trans-border "Alliance" pipeline, thought to feed the Californian thirst for energy, exports not new gas, but taps the older "Trans Canada Pipeline System" and storage. If present trends continue, it looks as if Canadian storage will be reduced to close to zero by late March. An informative collection of data on US and Canadian gas storage levels and analysis is given at http://www.raymondjames.com/ecm/energy/stat.htm.

The US energy crisis is just beginning, and is very dangerous. It may easily lead to a global disruption of the energy supply system.

14. Depletion Protocol

Mr Simmons, a Houston banker, who has been in touch for some years, has joined President Bush's Energy Committee. He has taken a keen interest in the idea of a Depletion Protocol and is to bring it to the attention of the US cabinet. It was first proposed at the 1998 Centre for Global Energy Studies conference (see below).

Campbell C.J., 1998, The enigma of oil prices in times of pending oil shortage; *World Oil Prices: oil supply/demand dynamics to the year 2020; Centre for Global Energy Studies, 208–225*

15. Interest in Canada and the USA

The Rt. Hon. Edward Schreyer, former Governor General of Canada, has been in touch with a view to consultation in relation to what he sees as "The Imminent Energy Crisis", a matter of long concern to him.

Mr Rogers, formerly with the Canadian National Energy Board, has also been in touch, and may be able to help the difficult problem of resolving Canada's differing definition of heavy oil (at 25° API).

Stewart Udall, President Kennedy's Secretary of the Interior, writes again of his growing concern and desire to alert his country's leaders to the situation.

16. Oil Discovery in 2000

Petroconsultants has made a press release giving new discoveries in 2000 as follows:

	Oil Mb	Gas bcf	Cond Mb
L.America	547	5508	92
Europe	311	1996	6
N.Africa	128	5086	122
C&S Africa	2785	3260	0
M.East	232	27655	751
Far East	813	16957	175
FSU	6400	12000	0
Total	**11216**	**72462**	**1146**

Probably most of the discovery in C & S Africa is non-conventional deepwater oil, as is some of Latin America's. Much of the FSU discovery relates to the single large find at Kashagan East in the north Caspian. So the general trend is probably around 6 Gb/a, one-quarter of what is consumed. It may be noted in passing that 1 Tcf of gas yields on average about 16 Mb of Condensate, which is a useful relationship when trying to identify the condensate component of total liquids which are commonly reported.

Newsletter No. 3, March 2000

17. New U.S. Government position

The new US Secretary of Energy has made an important speech, finally admitting that his country faces a long and intractable energy crisis. Whereas the previous administration adopted a policy of denial, pretending that the problems were short term issues related to OPEC politics, the new administration, having no responsibility for the failure of past policies, can now address the issue more directly. As is always the case, politicians find it easier to react to situations rather than to anticipate them. It is in any event an extraordinary admission of the gravity of the situation, highlighted by the stock market, which is collapsing for not unrelated reasons.

Quote

Energy Secretary Warns of Shortage, Recession
By H. JOSEF HEBERT, The Associated Press

WASHINGTON (March 19) – Energy Secretary Spencer Abraham said Monday the country is facing the most serious energy shortages since the 1970s. *Without a solution, he said, the energy crisis will threaten prosperity and national security and change the way Americans live.*

Abraham, in a speech hours before a special energy task force was to make recommendations to President Bush, *said California's power problems are neither isolated nor temporary and could spread to other parts of the country, including the Northeast.*

"The power crisis isn't just pinching our wallets, it's changing the way we live our lives," Abraham said in a speech described by aides as his most comprehensive assessment of the country's energy concerns since becoming energy secretary.

Abraham said the demand for energy "is rising across the board" but particularly for natural gas and electricity. As he spoke, the Energy Department also was raising concern of possible gasoline price spikes this summer.

The department said both crude-oil inventories and gasoline stocks are 6 percent to 7 percent lower that what they traditionally have been this time of year. The estimates came even before calculation of the potential impact of a decision last week by the Organization of Petroleum Exporting Countries to lower production by 1 million barrels a day.

At the White House, spokesman Ari Fleischer said Abraham had communicated with OPEC members before the production cut was made.

Fleischer said Bush is focused on long-term strategies. "While we can encourage and should encourage conservation, the most realistic approach our nation can take, in the president's opinion, is to increase supplies of energy."

As for tapping the nation's strategic petroleum reserve, Fleischer said, "The president has never ruled that out, but the president believes that is an emergency action that should be taken not as a consequence of supply-and-demand imbalances but in terms of national security emergencies."

John Cook, chief petroleum analyst for the department's Energy Information Administration, called the current inventory levels a disturbing sign for this summer's heavy driving season.

While stocks likely will increase somewhat, "We are beginning the driving season with very little stock cushion," Cook said in remarks prepared for an oil refiners' conference in New Orleans.

Abraham, addressing an energy conference sponsored by the U.S. Chamber of Commerce, said the administration is determined to pursue "*a rational balance between our energy needs and our environmental concerns.*"

He gave no details about the proposals expected to be presented to Bush by the energy task force headed by Vice President Dick Cheney. He said the administration's policy "*will stress the need to diversify America's energy supply.*"

Abraham said it is a myth that oil companies and power companies are "engaged in a massive conspiracy to gouge consumers."

"*There is no magic source of supply, no hidden pool of energy that can be turned on and off like a faucet,*" he declared.

Energy analysts are concerned that this summer will bring not only another round of power blackouts in California and possibly other states, as well as increases in the cost of not only electricity, but also gasoline during the peak summer driving season.

Gasoline prices nationwide for all brands average $1.41 a gallon, slightly lower than in recent weeks, the Energy Department said.

While the DOE earlier this month predicted prices at the pump this summer would increase modestly to about $1.49 on average, analysts said they would not rule out severe price spikes beyond that should supply problems develop.

Last year gasoline soared past $2 a gallon in some parts of the country, especially in the Midwest.

In a presentation to the refiners conference, Cook said gasoline stocks in the Gulf Coast area are nearly 13 percent below the normal five-year average, and 9 percent lower than at this time last year.

"For crude oil the situation is not much better," said Cook.

Nationwide, he said, crude stocks are 7 percent below the low end of the normal range for this time of year.

While the impact of OPEC's latest production cuts is not expected to show up in the U.S. market for six to eight weeks, it is likely to aggravate stock supplies even more, analysts said.

Newsletter No. 4, April 2000

18. Letter to EU Commissioner

The following letter was sent to Loyola de Palacio, the EU Commissioner for Energy requesting financial support to update and improve a 1995 study of depletion based on the industry database. A reply is awaited. If it succeeds, ASPO members will be able to access the data needed to support their own studies. Members may wish to use their own contacts in Brussels to support this initiative.

> 11 April 2001
> Sra Loyolla de Palacio
> Commissioner for Energy
> European Commission
> Brussels
> Dear Commissioner,
>
> I have been in touch with you previously in connection with Europe's future oil supply, and write now because there have been certain new developments, which strengthen the arguments for the Commission to sponsor a quantitative analysis of world oil supply, based on the depletion of the resource. This would involve evaluating the true oil reserves and discovery trends so as to forecast realistic production by country on a sound technical basis.
>
> Since, as you no doubt know, public data on oil reserves are grossly unreliable, it is necessary to use the commercial database, compiled from information supplied by both national and international oil companies. While by far the most reliable source of information, it itself requires a degree of further evaluation, because it contains various anomalies that have to be identified and resolved. There are also issues of definition to be addressed.
>
> J.H.Laherrère and I prepared such a study from this database in 1995, which led to conclusions that have been broadly confirmed by subsequent events. I enclose a CD copy of this report, so that you may evaluate its contents. But I would ask you to return it as soon as you have finished with it, and also to treat it as confidential, since it is subject to copyright.
>
> The study needs now to be updated and improved with the help of new analytical techniques, definitions and understandings that have evolved since the first report was written. Given the seriousness of the situation, I believe that the Commission is now amply justified in providing the necessary financial support to undertake the work.

In this connection, I should mention that a new institute, the Oil Depletion Analysis ("ODAC"), is being established in London to address these issues and also act as the secretariat for a wider group, known as The Association for the Study of Peak Oil ("ASPO"), currently comprising the following institutions:

BGR (Germany)

Clausthal University (Germany)

L-B Systemstechnik (Germany)

Norwegian Petroleum Directorate

Reading University (UK)

Rogaland Research, Norway

University College, Cork, Ireland

Uppsala University, Sweden

It is proposed that the study should be undertaken as a cooperative effort by ODAC-ASPO, so that the member institutions may contribute their expertise in order to check the accuracy and validity of the assessment. I will be very glad to assist in every way possible.

You are no doubt aware of a recent report by the US Geological Survey, which proposes that the world's undiscovered potential lies within a wide statistical probability range of 239 to 1376 billion barrels; and that reserves may be under-reported or under-estimated by between about 20% and 100%. The USGS is now exerting an influence on the International Energy Agency, which is not itself in a position to evaluate the USGS report, since it lacks the data and has not thoroughly studied supply from a resource standpoint.

The purpose of the proposed new study is to narrow this excessive range to provide a sound basis for calculating the impact of depletion on oil supply, an issue that is not itself covered by the USGS. It will be based on the actual discovery trends achieved by the industry in the real world rather than abstract geological hypothesis and unspecified technological aspirations.

Few would deny the importance of the subject for Europe, given that North Sea production is now at peak and set to decline sharply in the years ahead. The geopolitical consequences of growing reliance on gas from Russia, North Africa and eventually the Middle East must also be a matter for concern.

There are lessons, too, to be learned from the energy crisis that now hits the United States due to its failure to properly evaluate the consequences of depletion, which in turn threatens to push the World into recession

My colleagues from ODAC-ASPO and I would be very pleased

to come to Brussels to discuss this matter further, if you should so wish.

Yours sincerely

C. J. Campbell

19. Presentation by Chairman of Yukos Oil Company

Dr Khordakovsky, the Chairman of Yukos, the important Russian oil and gas company, has been briefed and supplied with slides for a keynote speech to the International Petroleum Forum in Paris. It is interesting that a Russian company, being less subject to stock market imagery than its western counterparts, is willing to address this sensitive subject. Russia is evidently beginning to realise its geopolitical strength as Europe's future premier energy supplier. An article in *The Times* already comments that Gazprom, the Russian gas company, now wields more influence in especially Eastern Europe than did the Red Army in the Cold War. The link with Yukos may provide useful insights into Russian oil and gas. It has already undertaken to investigate past production and reserves in the Former Soviet Union.

20. Uppsala Initiative

Professor Aleklett has raised Swedish awareness of oil depletion in an important article in the Svenska Dagenblat, which is being followed up by lectures and contacts in political circles. This may be of particular significance while Sweden holds the EU presidency.

21. Denmark

Dr Illum, who is keeping the flame alight in Denmark, had an article in the Danish newspaper *Information* entitled Når oliekrisen kommer (When the Oil Crisis Comes). The Danish Energy Agency copied the article and circulated it at a meeting in the Danish Energy & Environment Council, at which he presented a paper on future energy perspectives, focusing on transportation in particular. On, April 30, the same paper will be presented at public conference in Copenhagen.

22. Reading University responded to the EU paper with thirteen points as follows:

Contribution to the debate on the Green Paper: *Towards a European strategy for the security of energy supply*

1. Can the European Union accept an increase in its dependence on external energy sources without compromising its security of supply and European competitiveness? For which sources of energy would it be appropriate, if this were the case, to foresee a framework policy for imports? In this context, is it appropriate to favour an economic approach:

energy cost; or geopolitical approach: risk of disruption?

Answer: Europe has no choice. Her production of indigenous oil and gas supplies reaches peak in a very few years, and then declines rapidly. Whatever she does, Europe will become increasingly dependent on external sources of hydrocarbon fuel, and her security of energy supply and competitiveness will be increasingly compromised.

For oil, it is virtually certain the Middle East will be unable to meet international demand in the near term, due to internal investment constraints.

In any event, within a decade or so, the Middle East will hit its own resource-limited peak oil production levels.

A *global* oil policy framework is therefore vital.

This must include:

- Methods to correctly evaluate the resource base in the exporter countries.
- Methods to prioritise markets, and to ration supplies fairly.
- Methods to adequately compensate producer countries for their depleting resource, while at the same time setting oil prices that do not unnecessarily damage the economies of the world.

The *Protocol* suggested by Dr. Campbell could provide a suitable starting point for such a framework.

Attempts to set up such a framework *after* the oil crisis hits are likely to fail, as national self-interest and international enmity will discourage co-operation. The time to work on an international framework for global oil depletion is now.

The proper body for this is the IEA, perhaps under UN auspices (cf. the IPCC). But the IEA do not yet understand the problem. It is for the European Commission to:

- Pressure the IEA to recruit a cohort of experienced petroleum geologists, reservoir, and production engineers seconded from the petroleum industry to carry out, within six months, a competent appraisal of the situation.
- Hold a sequence of workshops to bring all relevant parties (EU, IEA, OPEC, USGS, oil majors, energy users, and key academics) to formulate an international strategy.

For gas, Europe will become an increasing importer from Russia, the Middle East, and N. Africa. But even here supplies will not last for long and increasing demands from other markets (China, Japan, and the US) will put Europe's supplies in jeopardy. An *initial* framework is therefore also required for gas imports to the EU.

Europe needs to be rapidly weaned off her addiction to oil and gas if she is not to face GDP collapse. Gas in particular, comes to a shuddering end with virtually no warning (and hence none of the price signals so beloved of economists). The example of the US should be in Europe's mind: oil peaking in 1970, gas about now.

To address the coming hydrocarbon shortages major European initiatives will be required. These should be started as early as possible. They include extending research into carbon-sequestered coal; and legislating social and technical measures to reduce hydrocarbon use. The main solution open to consumer governments facing a shortage of hydrocarbons will be *demand reduction*.

Industry will be good at providing some of the answers to the coming hydrocarbon supply problems, and they respond well and imaginatively to price signals. But the price signals will be severe and come with little warning. The current energy industry across Europe has absolutely no appreciation of the approaching problems: their time horizons are short, their focus is on their own business (not their nation's), and they must answer to the stockmarket so can invest little in advance of problems. Therefore, Europe's government must be involved to:

- Research and understand the situation;
- Explain this to her industry, and the wider population;
- Lead in those tasks (such as setting taxes and fuel duties, and legislating energy efficiency measures) that are outside the purview of industry.

2. Does not Europe's increasingly integrated internal market, where decisions taken in one country have on an impact on the others, call for a consistent and co-ordinated policy at Community level?

Answer: Yes it does (see Question 1). Europe needs an energy policy. It needs to be harmonised with her environment policy.

3. Are tax and state aid policies in the energy sector an obstacle to competitiveness in the European Union or not? Given the failure of attempts to harmonise indirect taxation, should not the whole issue of energy taxation be re-examined taking account of energy and environmental objectives?

Answer: Yes. Energy taxation should be re-examined to prevent market distortions.

4. In the framework of an ongoing dialogue with producer countries, what should supply and investment promotion agreements contain? Given the importance of a partnership with Russia in particular, how can stable quantities, prices and investments be guaranteed?

Answer: Given that the producer countries own the resources, stable quantities and prices cannot be guaranteed.

The best that can be done with producer regions is:

- Sign diplomatic 'Memoranda of Understanding'.
- Encourage the countries to join the international arrangements envisaged (see Q.1) to face the supply difficulties.
- Foster economic linkages, trading, and mutual understanding.
- Make all oil and gas import agreements contingent on detailed resource analysis assessments (as is currently done before financing a specific gas pipeline, for example).
- Learn from both the successes and mistakes of the 1973 and 1978 oil price shocks. In those periods, international rationing was largely successful through the essentially a-political administration of the global oil companies (resisting, for example, special pleading by the UK's Prime Minister, Heath). On the other hand, oil producers found oil prices denominated in dollars soon gave rise to problems, as inflation rapidly eroded the value of the dollar. This erosion possibly softened the oil shocks' blows on the world economy, but so annoyed the producer nations that a sequence of price hikes were initiated that undermined international attempts to stabilise the situation. Oil may have to be denominated in terms of other tradeable commodities.

5. Should more reserves be stockpiled – as already done for oil – and should other energy sources be included, such as gas or coal? Should the Community take on a greater role in stock management and, if so, what should the objectives and modalities be? Does the risk of physical disruption to energy supplies justify more onerous measures for access to resources?

Answer: Short-term security stocks are a vital protection against short term technical or political problems in one or a few producer countries.

For this reason they need to be up-to-date and well thought out. For oil, supplies need to be of good quality (not 'old' oil), real (not just in pipelines and tank bottoms), accessible, and their rationing well thought out *in advance* (for example, we are told that at least one EU country's existing stock of fuel ration-tokens can be easily counterfeited with modern colour photocopiers).

For gas, we do not know about the scope for storage over and above existing systems. In the US, large gas stocks are held in underground storage facilities for commercial reasons to balance the supply chains over the heating and cooling seasons.

For coal, given that a proportion of Europe's coal for power stations is carried by diesel-driven freight trains, above-ground-stocks of coal at the power station sites (*pace* Mrs. Thatcher's policy during the UK Miners' strike) seem warranted.

However the inevitable coming supply problems are long-term, and no amount of short-term stocks can help here. Attempts by Europe to 'manage' supplies to outwit speculators, as was suggested during last year's 'Fuel Crisis', will definitely fail, as do all commodity-management programmes unless instituted as part of effective and strict control of both demand and supply. The latter controls are difficult to envisage in the situation anticipated.

6. How can we develop and ensure better operation of energy transport networks in the European Union and neighbouring countries so as to enable the internal market to function properly and guarantee security of supply?

Answer: Continue present plans for well-thought out inter-regional links, accepting some cost of network 'over-design' as valuable insurance against regional shortages.

7. The development of some renewable energy sources calls for major efforts in terms of research and technological development, investment aid and operational aid. Should co-financing of this aid include a contribution from sectors which received substantial initial development aid and which are now highly profitable (gas, oil, nuclear)?

Answer: Yes, definitely. The time-scales that Europe faces conventional fuel supply difficulties are very short.

8. Seeing that nuclear energy is one of the elements in the debate on tackling climate change and energy autonomy, how can the Community find a solution to the problem of nuclear waste, reinforcing nuclear safety and developing research into reactors of the future, in particular fusion technology?

Answer: Research and development are the only answers. Potential, at least partial, solutions exist in terms of intrinsically safe non-breeder fission reactors; and energy beam sources that transmute more dangerous radio-nucleotides to safer daughter products.

In the near term, vitrification of the high-level waste products with clearly thought-through geological storage needs to be pursued.

Despite their high energy potential, breeder reactors probably should not be pursued. The potential waste problems, and their very fast response times to criticality (well outside human-mediated control) mean they should only be developed if these problems can be overcome.

Tokamak fusion has made considerable progress of recent years, and has good prospects; though in use it will need international supervision to prevent the illicit production of weapons-grade material. This research should be given more assistance than at present.

A variety of other routes to fusion exist, and these should be pursued.

The whole 'Nuclear' vs. 'Renewables and Energy saving' debate needs to be carried to a much higher level.

A world with a population of in excess of 7 or 8 billion, facing high levels of water stress (see WMO reports); increasing soil loss; and ever-increasing fertiliser and transport difficulties due to the decline of oil and gas; needs to think ahead very carefully. Can population management, plus renewables and extensive energy saving, bring us to the 'sunny uplands'; or do we have to go the fission / fusion route; or was *The Limits to Growth* right all along, and increasing production is just unsustainable in the face of resource and investment limits?

Mankind can let this mega, unplanned experiment work itself out; or try to do some calculations, backed by small-scale experiments (Crete, Sweden?), to see what futures might be feasible.

9. *Which policies should permit the European Union to fulfil its obligations under the Kyoto Protocol? What measures could be taken in order to exploit fully potential energy savings which would help to reduce both our external dependence and CO_2 emissions?*

Answer: As is well-known, the IEA's *World Energy Outlook, 2000*, calculates that OECD Europe, under a reference scenario, will increase CO_2 emissions over 1990 levels by 15% in 2010, and by 23% in 2020. This scenario *includes* measures anticipated by member countries as part of their Kyoto commitment. Significant new actions are clearly necessary.

Some of the recent EU-funded studies, and those of the IEA themselves, have the best data on what policies would have what effects. Deciding which of these polices *can* be implemented will be difficult. Previous experience with a proposed EU carbon tax is salutary, but it is almost certain that something similar will have to be implemented in the end.

Other required measures include:

- Large-scale investment in energy-efficient equipment, products and services. Policies related to efficiency of energy use are outlined in a proposed 'Campaign for Take-Off' currently under consideration by the EU. Exploitation of potential energy savings requires a large-scale publicity campaign at citizen level. The focus should be on exploitation of existing energy-efficient technology over a substantial time horizon. All stakeholders will need to be empowered for such a 'Campaign for Take-Off' to fully succeed.

- Substantial increases in funding for research into CO_2 sequestration and alternative energies; via Eureka, SAVE and ALTENER. Framework 6 should include a budget line on improving transport efficiency.

10. Can an ambitious programme to promote biofuels and other substitute fuels, including hydrogen, geared to 20% of total fuel consumption by 2020, continue to be implemented via national initiatives, or are co-ordinated decisions required on taxation, distribution and prospects for agricultural production?

Answer: The longer-term sustainability of biofuels needs to look hard-headedly at the issues of: the EU's future agricultural requirements, soil and water availabilities, and at the various fuels' energy inputs.

Hydrogen, except from renewables, is questionable against a backdrop of declining hydrocarbon provision; and also from coal, via gasification with water, unless the CO_2 sequestration problem is solved.

11. Should energy saving in buildings (40% of energy consumption), whether public or private, new or under renovation, be promoted through incentives such as tax breaks, or are regulatory measures required along the lines of those adopted for major industrial installations?

Answer: Tax breaks: VAT should reflect the energy efficiency of the equipment installed, with the most efficient equipment zero-rated. For commerce and industry, accelerated depreciation should be standard for energy efficient equipment.

Regulatory measures: Given the time-scale for turn-over in buildings, and the proximity of energy difficulties, enhancing current legislation on buildings efficiency to those of the best in Europe (subject to climate zone adjustment) is warranted.

12. Energy saving in the transport sector (32% of energy consumption) depends on redressing the growing imbalance between road and rail. Is this imbalance inevitable, or could corrective action be taken, however unpopular, notably to encourage lower use of cars in urban areas? How can the aims of opening up the sector to competition, investment in infrastructure to remove bottlenecks and intermodality be reconciled?

Answer: The imbalance is not inevitable. Actions required include:

- Better use of planning procedures to reduce the need for travel; for example, multi-purpose buildings, and co-location of work and living.
- Re-allocation of road space away from passenger cars to public transport, walking and cycling.
- Retention of transport infrastructure in public ownership, with competition encouraged through transport usage charges, where these include the full social, environmental and economic costs.
- Encouragement of inter-modality by reduced access charges or changes in axle loads. Much better physical facilities for both freight and passenger transport to encourage inter-modal switching are technically possible and relatively cheap.

13. How can we develop more collaborative visions and integrate the long-term dimension into deliberations and actions undertaken by public authorities and other involved parties in order to evolve a sustainable system of energy supply. How are we to prepare the energy options for the future?

Answer: To prepare the energy options for the future, Europe needs:

- Good data.
- Good calculations (modelling).
- EU-wide co-ordinated research.
- EU co-ordination of policy; Workshops of key participants.
- New and harmonised tax codes.
- New legislation.

Overall, sustainable use of energy requires change to a sustainable lifestyle. This transformation requires making energy users aware that their life-style contributes to global warming, and to resource depletion. Information, education and training are key.

23. Shocking Response by UK Minister

Copies have been received of correspondence from the former UK Minister of Energy to a Member of Parliament, who had been concerned by the issues raised by the BBC film of November 8th "*The Next Oil Crisis*". It demonstrates abject official ignorance of the subject. The Minister claims to "keep abreast" of expert opinion, which seems to imply that the government itself is not studying the matter. The Minister expresses the view that oil consumption will rise from 28 Gb to 41 Gb by 2020, but claims that reserves will be adequate to meet demand for 25 to 30 years (evidently misunderstanding Reserve to Production ratios). She also makes the bizarre statement "As with all commodities that are finite in supply, substitutes take over long before supply runs out". She relies on the flawed USGS study to claim that new discovery will match reserves. She has fortunately now left office, but if this calibre of advice being given by the responsible department, there is little reason to expect her successor to be better informed.

The tone of the Minister's letter suggests flat ignorance rather than a deliberate attempt to mislead, making it perhaps even more serious. It highlights the important mission of ASPO in providing sound advice to governments to counter the gross failings of the officials.

24. US Situation

President Bush's decision to junk Kyoto has rocked the world, but may be seen as a pragmatic response by a nation facing an intractable energy crisis, due to a failure to properly analyse the depletion of oil and gas and plan for the consequences. No doubt the country will now increase coal production, reverse the decline of nuclear power and open various areas, closed for environmental reasons, to oil and gas drilling. While is makes eminent sense to determine what resources these areas hold, it would be unwise to assume that they will deliver sufficient to make any material difference to the country's dire energy situation. It will be interesting to see how long Canada will allow its declining oil and gas resources to be exported.

Meanwhile, the Federal Reserve Bank has reduced interest rates in the hope of propping up the flagging economy in the belief that it is just another business cycle, susceptible to such stimulus.

25. UK Gas

The Times comments of soaring UK gas prices, pointing out that several of the privatised energy suppliers are in financial difficulties, and that anticipated demand will far exceed supply by 2005. Failing the grasp the nature of depletion, it attributes the difficulty to under-investment.

It remains to be seen if Europe will learn the lessons from US experience.

26. Hubbert Center Newsletter
The current issue of the newsletter carries an article *"Peak Oil – A Turning Point for Mankind"* by C. J. Campbell, similar to that attached to an earlier ASPO Newsletter. J.Zagar also contributes an article explaining oil how production in ageing fields in Saudi Arabia and the USA calls for ever increasing work and effort, which touches on the critical issue of spare capacity.

27. Caspian Exploration
Great hopes have been expressed that the Caspian would deliver much needed new oil. Particular interest has focussed on the giant Kashagan structure in the north. The first well, Kashagan East, was completed some time ago, finding about 7 Gb of high sulphur oil at great depth in a poor reservoir. Two major companies in the project withdrew from the venture in the light of these results. The second well, Kashagan West, is now reported to be testing. Operating conditions are extreme and the costs are very high.

The results, while promising, appear to dispel the notion that the Caspian may be "a second Middle East", as had been irresponsibly claimed.

28. Deepwater Setbacks
An explosion, leading to the sinking of the world's largest floating production platform off Brasil, underlines the extreme technical challenges of deepwater operations. Earlier interpretations suggesting that deepwater production would be characterised by high short-lived peaks, as suggested by geological circumstances, are due for revision, as in practice production seems to follow a more conventional "plateau" pattern, being subject to severe operational constraints.

29. Website
Attention is drawn to an interesting website covering the subject, including Jean Laherrère's graphs and commentary. http://mwhodges.home.att.net/energy/energy

Newsletter No. 5, May 2000

30. IEA shifts ground again
The International Energy Agency has yet again shifted ground. It will be recalled that in its *1998 World Energy Outlook*, it did succeed in delivering a coded message, explaining the gravity of the oil supply situation, albeit in bland diplomatic terms. Then in the Edition of 2000 it reverted to its earlier position of there not being a cloud on the horizon. A new policy statement has now been issued which does express "concern" about future oil supply

and emphasises the need for new study of the matter.

This may reflect the changing position of the United States government. It previously seemed oblivious to resource constraints, possibly being influenced by a thoroughly flawed study of the US Geological Survey that claims the world to have a huge potential for new discovery and "reserve growth" with a miraculous reversal of the recorded trend of the past forty years.

A.M.S. Bahktiari in a telling article in the *Oil & Gas Journal* of April 30th 2001demonstrates the absurdity of the IEA forecasts by comparing the 1998 and 2000 Editions of the World Energy Outlook as follows

	1998 Forecast	2000 Forecast
Total supply in 2020	112 Gb	115 Gb
Non-OPEC	27	46
OPEC	60	62
Processing gains	2.5	2.6
Identified Unconventional	2.4	4.2
Unidentified Unconventional	19	0

It is absolutely inconceivable that any serious analysis, taking into account the resource base, could, in the span of only two years, almost double the estimate of non-OPEC supply in 2020, when most of the news between the two reports has been negative. No new major province has been discovered: the Caspian continues to disappoint; and the North Sea is seen to peak.

In 1998, the mythical balancing item of *Unidentified Unconventional* was expected to deliver 17% of the supply by 2020, implying the discovery of huge new, albeit unspecified, resources, but two years later this expectation had vanished without explanation.

It is a shocking commentary on a world organisation with prime responsibility for advising the OECD governments and the European Union on oil supply. If nothing else, it demonstrates in very clear terms the importance of the ASPO mission to unmask this situation, and come forth with well reasoned, logical and consistent estimates upon which sound and rationale government planning may be made.

31. American policy

The Bush administration confesses to a serious energy crisis, with the Secretary of Energy admitting that it will have colossal economic impact, even changing the "American way of life". The government proposes to ameliorate it with the following provisions:

- open new federal lands to oil and gas exploration, including the

Arctic Wildlife Reserve;
- ease regulations on pipelines and refineries;
- facilitate new power line construction;
- streamline regulations on new power stations;
- provide federal aid to nuclear power stations;
- expand coal-fired power generation;
- apply new energy efficiency standards on electrical appliances;
- provide tax incentives for smaller vehicles.

They sound sensible, but the drilling in sensitive areas will no doubt face opposition from environmentalists. It is worth remembering in this connection that the Prudhoe Bay Field in Alaska is a "king", head and shoulders larger than any other field in the areas, due to a unique set of geological circumstances. It is accordingly doubtful that the Wildlife Reserve will be found to contain anything comparable despite its promising address. A few key boreholes drilled by the government as "research" would resolve the issue without risking the ire of the environmental lobby by opening it to indiscriminate exploration by competing companies. The government has also junked Kyoto, no doubt to facilitate expansion by the coal companies, in an action, which has attracted worldwide condemnation.

Meanwhile the influential Center for Strategic and International Studies in Washington has issued a report about energy supply, recommending that the US should act as the world's policeman using military means to ensure that the "sea-lanes remain open". This has been seen as a veiled threat of military intervention to secure oil imports.

32. TotalFinaElf

Mr Bauquis of this major French company has made a presentation on future energy supply.

2000	3.7 Gt	(74 Mb/d)
2020	5.0	(100 Mb/d)
2050	3.5	(70 Mb/d)

also pointing out that estimates of Ultimate recovery have changed little over the past thirty years. He forecasts that oil will provide about 40% of world energy until 2020 when it falls sharply, being replaced by growing nuclear energy. By 2050, coal, nuclear, oil and gas are expected to provide roughly equal proportions, with hydro and renewables contributing only slightly more than today. While the forecast appears to be much too high, it does at least imply peak and decline.

33. Debate in Denmark

In Denmark, a public debate has been initiated by an article from Klaus Illum in the newspaper, *Information*, of April 26. It summarises the perspectives on peak oil and the IEA oil production and demand forecasts, as published in Energy Outlook 2000, concluding that there is reason for political concern. On May 16, the Vice-Director of the Geological Survey for Denmark and Greenland (GEUS), Mr. Kai Sorensen, responded in an article saying that there is no reason to worry about the exhaustion of oil resources – missing the point that the debate is not about exhaustion but about the peak of production taking place before the peak of demand. Mr. Sorensen referred to the USGS as the "only institution which seriously seeks to form an opinion about the size of the global oil and gas resource".

In his article Mr. Sorensen refers to Campbell's analyses in a rather dismissive tone. Therefore, Campbell responded in a letter to the editor of *Information*. An abbreviated version of Campbell's response was published on May 22. This week, a summary of Campbell's response appeared on the News list on the home page of the Ministry of Environment & Energy, indicating that the Ministry is aware of the ongoing debate.

Also members of Parliament have become aware of the peak oil issue. On June 6 a seminar on oil revenues and the depletion of the Danish oil and gas fields in the North Sea will be held in the Parliament building. Dr. Roger W. Bentley has kindly agreed to speak at this seminar.

The exchange is reproduced below: INFORMATION, 16 May 2001 (translated by Klaus Illum).

If oil runs out
By Kai Srensen

Vice-director, Denmark's and Greenland's Geological Survey (GEUS)

The assessments of oil reserves vary from the pessimistic, as in Klaus Illum's article *When the Oil Crisis Comes*, April 26th, to the optimistic: *If* oil runs out. The pessimists often base their assessment on figures and arguments from Colin Campbell and King Hubbert's work in the 1950s: *"Predicting when oil production will stop rising is relatively straightforward once one has a good estimate of how much oil there is left to produce".* The prediction that the ultimate oil crisis is just around the corner is based on expectations as to a continuing growth in demand and a decrease in production soon to come. The problem is that this "good estimate" is not easily obtained.

Limits to Growth

Many of those who now worry about the energy future had their first AHA-thrill in 1972 when Meadows et.al. Published *Limits to Growth*. Considering the then known oil reserves, they predicted that the last oil would be produced already by 1992, assuming a continued (exponential) growth in consumption. The prediction did not come true. And, equally incomprehensible to the man in the street, the proven oil reserves are almost the double of what they were when *Limits to Growth* was published. Klaus Illum's assumption that oil consumption will increase by about 40 percent over the next 20 years should be seen in the light of an increase by 13 percent over the period 1974–94, 8 percent in the period 1979–99.

But is it not true that we are well on the way to the exhaustion of the World's oil reserves? By 1973 there were sufficient reserves for the production to continue unchanged for 48 years, by 1999 there was enough for 62 years! There is no simple explanation to this.

A very elastic concept

According to international consensus, the word reserve means the amount which can be profitably produced using presently known technology and under the present economic conditions. This is a very elastic concept. Most reserve assessment are based on the oil companies' estimates, which are, generally, conservative. Tracing the development in reserves in particular fields, it appears that over time oil fields generally 'grow' in size because of improved extraction techniques, which incorporate into the reserve sphere, resources hitherto assumed unprofitable to recover. Furthermore, over time additional resources are found in fields already known. This is the complex background for the growth in oil reserves.

One may say that the reserve assessments which for the last decades can be read in BP's annual energy statistics has the character of a kind of stock-taking. Many of those who are not familiar with this reserve concept, certainly believe, that reserves means something else. Something which is easily evaluated, as assumed by Campbell, for example.

Hence, there is substantially more oil in the underground than what appears from the reserve assessments made at any one time. In the world which occupies itself with the non-living raw materials one operates with yet another concept: the resources. These are the reserves plus the amounts which it is technically feasible to extract and which will probably be found. For oil and gas it goes as for coal: The amount of coal in the underground is some astronomical figure, which is of academic interest only. Therefore, it goes for resource assessments that they normally comprise only the amounts of raw materials which it will be economically feasible to exploit within a foreseeable future (say 30 years).

Today there is only one institution which seriously seeks to form its opinion on the size of the global oil-gas resource. That is the American geological survey (USGS), which currently has a *World Energy Project* running.. The latest report from this group arrives at the following resources (reserves plus the 30-year resource) for the

World as a whole: 410 billion cubic metres of oil. This should be seen in the light of already produced amounts of 110 billion cubic metres and an annual production of 4 billion cubic metres. As one can see, there is a very large resource to be added to the reserves.

The oil price moves up and down

In his article, Klaus Illum followed Campbell's foot tracks and reached the conclusion that the World faces a serious oil price crisis. The oil price moves up and down but this is primarily due to political events, not to circumstances which have to do with nature and size of the resource.. There are many good reasons to reduce the consumption of oil and natural gas, e.g. environmental, but the day when we shall really experience oil and natural gas a scarce resource lies far away in the future. So far away that scenario may never become a reality.

Long before that day we may produce energy in a way presently unknown and oil will remain in the underground as much coal presently does. If one needs to worry it is not the oil resource quantities one should pitch into. Because then the worry will turn into disappointment.

If Oil Runs Out – A Reply from C. J. Campbell

The very title of Mr Sørensen's article *If Oil Runs Out* demonstrates a failure to grasp the essence of this debate. Although oil, being a finite resource formed in the geological past, must one day run out, no serious analyst thinks that it will do so for very many years to come. The point is not when it will eventually run out, but when production will peak. That will clearly be an important turning point with far-reaching consequences, given the world's heavy dependence on cheap oil-based energy. It is not helpful to speak of optimists and pessimists, when what are needed are realists basing their assessments on valid data and the extrapolation of genuine discovery trends.

Mr Sørensen is right to comment on the elastic nature of reserve reporting. The industry has systematically under-reported the size of discoveries for good commercial and regulatory reasons; and several OPEC countries reported huge spurious increases in the late 1980s when they were vying with each other for quota based on reserves. It is precisely why serious analysts reject the published data, as given for example in the BP Statistical Review, referred to by Mr Sørensen, which simply reproduces unreliable information compiled by the Oil and Gas Journal and does not rely on the company's own considerable knowledge. World discovery, based on the industry's own data, has declined since the 1960s, despite all the advances in technology, a worldwide search and a deliberate effort to test the biggest and best remaining prospects. It is easy to understand this pattern of discovery. The larger fields, which hold most of the oil in any province, are found first because they are too large to miss. The peak of discovery has to be followed by a corresponding peak of production: that is realism, not pessimism.

Mr Sørenson falls into the common trap of comparing oil with coal. Oil is a liquid not an ore, being concentrated into a few accumulations by Nature. It is either there in profitable abundance, or it is not there at all. If Mr Sørensen studied the production statistics of the world's major fields, plotting annual against cumulative production, he would observe in most cases a straight-line decline, which conclusively shows that technology has had little impact on the reserves themselves.

It is understandable that Mr Sørensen should applaud the recent report of his sister organisation, the US Geological Survey. But if he studied the methods used carefully he could hardly fail but conclude that it is little more than pseudo-science. Take for example the case of East Greenland, which is not only treated as part of North America, but is considered to be one of the World's most prospective areas. The USGS made a number of alternative estimates of possible field-size distributions of the world's basins. The results were then given to statisticians who after many iterations concluded, in for example the case of East Greenland, that there was a 95% chance of finding more than zero and a 5% chance of finding more than 111.815 Gb. A Mean value of 49.684 Gb was computed from this wide range. It offends common sense to quote to three decimal places the slim chance that East Greenland might contain almost double the North Sea. Furthermore, the USGS admits in its text that it was exceedingly uncertain about the assessment of reserve growth.

Government institutions that simply accept published reports of this sort fail in their responsibility. What they should do is study the matter themselves, using valid industry data. The results will clearly speak for themselves without appeals to optimism or pessimism. The peak of discovery in the North Sea was in 1973 and peak production is now close. It should surprise no one but Mr Sorensen that the world discovery peak in the 1960s, which is fact not speculation, heralds the corresponding peak of production that cannot be long delayed. About half of what remains lies in just five Middle East countries.

It is significant that the International Energy Agency, which had also been infiltrated by the USGS, now delivers a clear message in its recent policy statement. Denmark has an important role to play, and its Geological Survey could contribute usefully. Thorough studies are needed more than expressions of blind faith.

34. Gas study by L-B Systemstechnik

Dr Zittel has produced an excellent study of Europe's gas supply and demand in a world context, from which the plot (Figure 5) is reproduced, showing cumulative discovery, which is markedly flattening against rising demand. Europe's will have little left by 2020.

35. Clausthal University

Professor Blendinger writes of his continued interest in the subject of depletion, of discussions with other interested parties in Germany, and of his quest for research funding. His website is attracting much world interest.

The university has put on its website a video of a lecture on oil depletion given in December. It is now attracting much interest and attention especially in the USA, being reproduced on several other websites. It can be seen on www.rz.tu-clausthal.de/realvideo/event/peak-oil.ram.

It has also been translated into Portuguese, and may be seen at www.amerlis.pt/oil/lecture

Newsletter No. 6, June 2000

36. American Policy evolves

In the last newsletter, we commented both on President Bush's admission that the country did now face an intractable energy crisis, and the veiled threats from Washington of military intervention to keep the "sea-lanes open" for US oil imports.

We have now been alerted to a curious link between the Balkan conflict and the construction of a pipeline to bring Caspian oil to the West. The pipeline, named AMBO, is planned to link the Bulgarian port of Burgas with Vlore on the Albanian coast, and will be operated by Anglo-American oil companies. The author of the study, Michel Chossudovsky, of Ottawa, Canada, sees a US strategic objective to secure control of this transit route. If true, it may have been based on earlier wildly exaggerated US hopes that the Caspian might hold reserves rivalling those of the Middle East. It may also speak perhaps of a certain military momentum that, once triggered, cannot easily be reversed when circumstances change or new knowledge appears.

There is a curious contrast in the United States that deserves comment. On the one hand is an admission of an energy crisis by the Government.

Recognition that oil and gas supply could not continue to meet demand led to a decision to step up coal and nuclear production, which in turn meant that it had to junk Kyoto, despite antagonising much of the rest of the World. On the other hand, we have the amazingly unrealistic forecasts of oil and gas abundance by the USGS and Department of Energy, which are evidently ignored by the Government itself.

It has been suggested that this contradiction may relate to US concerns about the future of the dollar. The country is heavily in debt, and needs to do everything possible to shore up confidence by which to encourage the inward flow of money, fearing perhaps that the Euro might prove a more attractive haven, despite its poor performance to date

This line of reasoning suggests that these outrageous forecasts are primarily for foreign consumption, while the US government itself follows a much more rational policy, which evidently ignores the forecasts.

One of the most remarkable of such forecasts is that by the Department of Energy in its International Energy Outlook 2001.

Its Reference Case forecast makes some utterly implausible production estimates including:

Mb/d	1999	2005	2010	2020
Persian Gulf	24	26	30	45
Indonesia	1.7	1.5	1.5	1.5
Venezuela	10.2	12.5	13.9	17.9
USA	9.3	9.0	8.7	9.3
Mexico	3.5	4.1	4.2	4.4
N.Sea	6.3	6.6	6.5	6.0
WORLD	79	88	97	122

Even Sir (now Lord) John Browne, the CEO of BP, who is known for the extreme care with which he chooses his words on this subject, admitted at the Davos Summit to a "maximum" production of 90 Mb/d by 2010. (For "maximum" read "peak".)

The accompanying plot (Figure 5) from the IEA report shows alternative implausible peaks, followed by unrealistic precipitate falls. Perhaps foreign investors will be impressed, but no serious analyst could give any credence to such material. Yet again, this stresses the importance of ASPO's mission to counter such dis-information, whatever its motive. Compare it with, for example, the scientific approach of Laherrère, who relates actual discovery with production, giving something substantive that can be verified.

It is noteworthy that the EIA calculates that the low USGS case gives an Ultimate of 2248 Gb, which is not too unreasonable for all liquid hydrocarbons, depicting it as having a 95% probability. In fact, the USGS itself refers to the case as F95, meaning that it had a frequency of 95% in the statistical iterations of its pseudo-scientific procedure. To be generous, one could say that the USGS was obliquely telling us that this low case had the highest probability of being right, but that would certainly be stretching a point. The study is in fact fundamentally flawed, so it probably serves no purpose to try to attribute any value to its findings. See Figure 6.

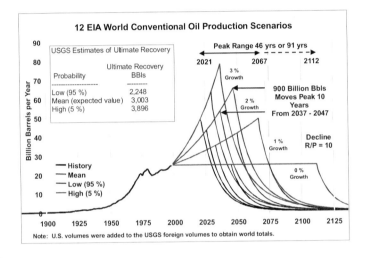

Figure 5

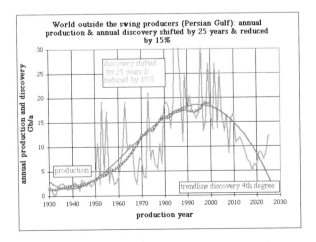

Figure 6

37. L-B Systemstechnik Studies of Oil and Gas

Dr Zittel continues his review of oil and gas production, looking particularly at US imports.(Figure 8.)

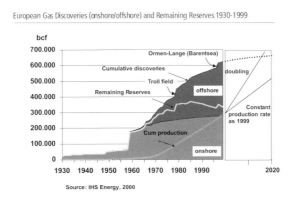

Figure 7

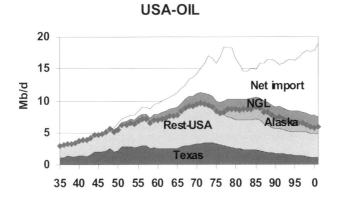

Figure 8

38. Ireland

C. J. Campbell gave a talk at the Energy Ireland Conference on June 11th in Dublin, which was opened by the Minister. The talk was entitled "An Imminent Energy Crisis – Ireland's Response". It looked at the rapid growth of energy demand in Ireland during the economic boom of the 1990s, which has done much to destroy the sustainable rural economy of the past. Ireland, with no oil and a nearly depleted gasfield, is heavily dependent on imported oil and gas. Electricity generating capacity, which now relies mainly on gas imported from Scotland, is barely able to meet peak demand. The talk tried to alert the country to the risks to supply as North Sea production peaks and as Europe is forced to rely increasingly on distant imports. It was stressed that the country is particularly vulnerable being very much at the end of the supply line. Ireland is an enthusiastic and loyal member of the European Union, but that is likely to make it a particular victim of flawed and out-of-date energy and general economic policies.

The conference concentrated on liberalisation of energy markets under conventional economic thinking and general EU guidelines, having the objective of reducing costs to the consumer, when a better policy should aim to reduce demand. Efforts to stimulate exploration on the Irish shelf, despite past failure and declining interest, were discussed. At the same time, a certain undercurrent of concern appeared in several of the presentations. Fears were expressed that the boom might be coming to an end; and that California might not be the last place to experience blackouts from faulty management and a failure to take due note of resource constraints.

39. A lesson from Brasil

It is noteworthy that Brasil, which relies heavily on hydropower, is now facing a very severe energy crisis, because exceptional drought has left the dams empty. The government has been forced to impose draconian rationing, which is severely affecting the economic life of this nation of 170 million. It incidentally raises doubts that the foreign companies recently invited to explore and produce oil and gas off Brasil will ever be allowed to export it.

It is a foretaste of what is to come unless governments pay more attention to energy issues and implement sustainable policies.

A realistic assessment and explanation of the situation and the policies that should be adopted are contained in an excellent review, recently published by *The Foundation for the Economics of Sustainability* (www feasta.org). It points out particularly that money supply comes from interest charged by commercial banks on lending money that they neither have nor own. In an epoch of growth, debt may be profitably employed, but at the same time depletes resources; increases the gulf between the rich and poor; encourages the managerial kleptocrats; and creates social tensions. The repayment of debt in a declining economy, as may be imposed by rising energy prices and shortages, may become an intolerable burden exacerbating the in-built instability of the system.

It is reassuring to read the words of this new breed of economists who have seemingly moved beyond the flat-earth views of their predecessors.

Newsletter No. 7, July 2000

40. European Fuel Cell Forum

C. J. Campbell gave the opening presentation of the Fuel-Cell Forum in Luzern, Switzerland on July 2nd. This three-day event attracted a wide international participation from those interested in promoting fuel-cells as one of the promising solutions to the impending world energy crisis. It seems that the motor industry has accepted the need for new more efficient vehicles, using new fuels, which will be introduced in the coming years. Already, several bus lines are running on fuel-cells, powered by hydrogen. Furthermore, there is scope for domestic and community combined heat and power generation with fuel-cells. So far, motivation for fuel-cell applications has come principally from environmental concerns, but it now appears that the growing realisation of the looming energy crisis will give development a new impetus.

41. Solar Today

Oil and gas depletion are covered in the July/August issue of Solar Today (a US journal) in articles by respectively C. J. Campbell and R. Udall & S. Andrews, which have received favourable comment on various websites.

42. Depletion Protocol

The Foundation for International Environmental Law and Development, which represents international lawyers concerned with climate change, globalism and such issues has taken an interest in the proposed Depletion Protocol. This protocol would be an international agreement whereby producers would limit production to the current depletion rate and importers would not accept infringements. In practice it would mean that the consuming countries would co-operate with the producers in managing depletion. More work is needed to define the mechanics of the proposal and evaluate its implications. If demand could be managed to match supply, it might be possible to prevent damaging excessive price rises.

A separate initiative by the Commodity Producers Association is also under way to counter various anomalies relating to the use of the US dollar as the prime currency for oil transactions.

43. Agriculture's fuel dependency

A paper by Tony Boys at http://www.net-ibaraki.ne.jp/aboys/pfe/dprkfc.htm gives a chilling account of what happened in North Korea when the inputs to modern industrial agriculture, including fuel, became unavailable. Twenty-two million people starved and three million died. The conclusions are reproduced below:

The experience of the DPRK [North Korea], and perhaps Cuba, points to several closely interlinked lessons that need to be learned by countries which currently operate a modern industrialized agricultural system based on commercial chemical and energy inputs. Agriculture has now become simply one adjunct of the overall economic-industrial matrix of the human global social-economic entity. This matrix is a highly complex web of financial and industrial relationships backed up by fairly precisely timed operations, such as transport of raw materials, fuel, components, and so on. Adjuncts to the matrix are therefore sensitive to disruptions and other irregularities. Thus the modern agricultural system can very quickly get into deep trouble if we do not have the ability to:

- fuel, maintain, repair, and replace agricultural and distribution-related machinery and infrastructure (trucks, tractors, transplanters, harvesters, irrigation pumps, fuel and chemical delivery systems, and so on)
- fuel, maintain, repair, and replace factories and factory equipment for the manufacture of agricultural machinery and inputs, e.g. regularly replaced items such as spark plugs and filters, spare parts, fertilizers, herbicides, pesticides, plastic sheeting and so on.
- ensure steady supplies of fuel, raw materials, and feedstocks for agricultural operations and inputs, such as petroleum, natural gas, coal, potassium and phosphorus minerals, and so on.

Again, the final answer is to convert to low-input, yet land and labour intensive, organic farming. Crucially, this would require perhaps a ten to twenty year transition period; something the DPRK has not had the luxury of.

As a final general statement, it can be said that once a country takes the decision to abandon traditional agriculture and switch to a modern agricultural system (a mechanized system making use of commercial chemicals and fuels), then in order to maintain food production levels, it is essential to ensure that levels of fuel and other inputs are maintained, and that machinery and equipment is kept in good working order. Shortages of fossil resources (oil, natural gas and coal) can result in productivity collapses when soils are mined, and eventually destroyed, due to crop production without replacement of essential nutrients, and where agricultural machinery and equipment can no longer be kept operational because of lack of fuel and maintenance.

A transition to organic and/or traditional and sustainable forms of agriculture is not easily carried out in a short period of time (for instance due to lack of livestock and lack of sufficient numbers of farmers with the requisite knowledge and skills). Meanwhile the population must be fed; a population that has ballooned on food produced by the modern industrial agricultural system that has been built up thanks to fossil resources (39). This is now the paradoxical complex of problems faced by almost all of the world, including the great food-producing areas of North America, Europe, South America and Oceania; how to maintain high agricultural productivity with decreasing amounts of the central element that has made that productivity possible, oil. The end of cheap and abundant oil and other fossil resources means the end of our current methods of food production and thus it possibly spells the end of advanced industrial society as we know it. The DPRK is an exceptional case only in that due to political miscalculation and mismanagement of its economy it has manifested these symptoms before fossil resource shortage becomes a serious concern for most of the world.

Newsletter No. 8, August 2001

44. The EIA changes its tune

The Energy Information Administration (an arm of the US government) has put out a new production forecast, termed "Accelerated Depletion", which reduces the forecast for US-48 oil and gas production, admitting to a peak for gas by 2015. US oil production is shown to rise from 4.3 to 4.7 Mb/d by 2020, (down from 5.0 previously), with an alternative scenario of the "slow-technology growth case" showing production declining to 4.0 Mb/d. It generally maintains a flawed "flat-earth" view that production is driven by technology to be extrapolated ever onward, but now as the reality of depletion begins to cast a darker shadow, it begins to contemplate "slow technology growth", which is at least a step in the right direction.

45. Further recognition of our efforts

Someone occupying a central position observing the oil business has offered the following exposition of the situation, but has to remain anonymous for obvious reasons.

A land without maps

Alice in the well-known children's story, Alice in Wonderland, was only required to do two impossible things before breakfast – ASPO members are required to do four:

- Verify beyond reasonable doubt that the world does have a looming and intractable oil crisis;
- Persuade at least some of the richest and most resource extravagant generation ever to walk the planet that the cheap oil party is over;
- Help persuade politicians and decision makers that actions have to be taken now, and;
- Avoid being branded as belonging to a 'Doomsday cult' by suggesting as many steps as possible that could ameliorate the adverse consequences of peak oil.

Now its only weapons are wit, humour and the truth. They are actually much more powerful than they appear to be because the reality is that large numbers of people actually know the size of the problem but have their own reasons for not acknowledging it. In short, ASPO is like the little boy in the story of the Emperor's new clothes. The little boy pointed out that the Emperor had no clothes. Everyone else knew the Emperor had no clothes but was terrified to say so. ASPO aims to point out the oil crisis. It is unlikely to be popular but the news will be less of a surprise than its members may suppose.

If so many know, why do they say nothing? Top politicians (with many dishonourable exceptions) actually have some idea of the problem. But what are they to do? Their constituents certainly don't want to know that the party is over. And they, equally certainly, don't want to tell the people because there are not many votes in that. They also have an acute sense that the messengers of bad news are rarely treated well.

Similarly, the top oilmen know only too well. But they can't tell because they're all hoping to become rich beneficiaries of all those share options they've awarded themselves. So, the great aim is to make sure that any crisis in sight will come after they retire or quit, which is commonly not far distant Only the retired have risked telling the truth (as Bernabe and Bowlin, the oil company chief executives, did on leaving office). No wonder top executives surround themselves with economists who declare resources to be infinite. (One is tempted to suggest that vendors of perpetual motion machines would also get a sympathetic hearing).

Any economist harbouring the slightest doubts that resources may not be infinite will drive back such doubts with two observations:

1. The oil companies have always misled the public about oil reserves.
2. They have always made a good living doing so. Any of them who have somehow recognised that depletion has now reached the point

where the myths are no longer sustainable may turn to ASPO as a source of new inspiration.

The other major group that has little interest in the likely outcome are the energy bureaucrats whether employed in national energy departments, the EIA, the IEA or other such organisations. Their problem is that they have families to support and pensions to look forward to. They have seen the career damage inflicted on those who extrapolated the energy crises of the 1970s. They don't want to risk being seen as doomsters crying 'wolf'. The idea that this time the wolf is actually at the door has not necessarily occurred to them, and in any case is a notion they'd be very reluctant to accept. Remember these people are highly political, however ineffectual. If Saudi Arabia says it has lots of reserves – who are they to question a government? If the oil companies say the Gulf of Mexico will revive US production, do they challenge, argue, and question? Yes, but very lightly.

Now the ASPO members may be feeling rather lonely as part of a small group who actually want to know. Don't worry you can be very influential for one very simple reason – someone has to say it.

In the medieval world the greatest challenge was always 'but who will tell the king?' whether it was lost battles, traitorous brothers, dead queens or empty treasuries. In those days, the bearer of bad news seriously reduced his life expectancy, so there was a quite natural reluctance to tell. That problem was solved by the development of the role of the court jester. He was a man of no wealth or power who used wit and humour to win the trust of Kings and Princes – *and so was able to tell them unpalatable truths and survive.*

This is precisely what ASPO can do and already does very well.

To give one simple and straightforward example – the Caspian. A persistent myth, generally attributed to the CIA, is that the Caspian contains 200 billion barrels of reserves – enough to make it a viable counterweight to the Middle East. Variant folk tales offer 50 to 400 billion barrels. The reality is that ten years of exploration and development effort have produced (being as generous as we can) 4–6 billion barrels in Azeri-Chirag-Gunesheli, 6–10 billion barrels in Kashagan, 1–3 billion barrels in the two Lukoil finds off Dagestan and 2–3 trillion cubic feet of gas and condensate in Shah Deniz. The last six wildcat wells have been dry; and several companies are said to be buying their way out of drilling commitments. So, ten years of intensive work on the best prospects has produced well under 20 billion barrels. In the very unlikely event that the area has the mythical 200 billion barrels of reserves, at current discovery rates they will take 100 years to find.

It means that the Caspian will not rival the Middle East. It will not even rival the North Sea. Probably a little bigger than the Gulf of Mexico or Angola, it is by all means nice to have and a lot better than that earlier great white hope – the Falklands.

Now everyone loves a fairy story and none is more popular that the magic porridge pot where no matter how much you eat, the pot is always full. In the minds of many, who should know better, the Middle East is the oil industry's equivalent. If I am allowed to contribute again I will explain why this is dangerous nonsense. But as a parting shot let me ask how many are aware that Opec's installed capacity was 5 million barrels/day larger in 1974 than it is today? Yes 5 million b/d *larger*.

46. North Sea Peak

United Kingdom production reached an irregular plateau at around 2.7 Mb/d in 1995, which lasted until mid 1999 when decline set in. It is now at 2.3 Mb/d, being 18% down in the twelve months ending in June. Most production still comes from the declining old fields, found long ago. However, there are reports of the discovery of a large stratigraphic trap, which may provide a minor late stage subsidiary peak on the decline curve. The midpoint of depletion, based on an Ultimate of 29 Gb, was passed in 1996, a few months after the actual peak in late 1995. Evidently, the desperate efforts to hold production high by bringing in small satellite fields and tapping subsidiary accumulations in producing fields have been successful, giving a plateau after the actual peak, but the natural consequence is that the eventual decline has become steeper.

The Norwegian Petroleum Directorate paints the same picture of an imminent peak and steep decline, commenting that some of the recent small developments have come in below expectation. This confirms general experience. The industry has systematically under-reported the size of large fields, being able to do so because they have the low economic threshold. But it no longer has that luxury in the case of small fields, being often forced to go with optimistic assumptions to make the projects fly at all, which lead to predictable disappointments.

We are too close to see the overall peak of the North Sea clearly, but when we come to look back at it from the other side of the hill, we will probably realise that we passed it around the end of the millennium. We will then be quite sure that the slope we are standing on is steep. Within ten years we will be half way down.

47. Industry pleads for yet more tax benefits

Dr Wolfgang Schollnberger, Chairman of the upstream oil lobby, OGP, pleads for even better tax terms, which he claims will encourage exploration, leading to sufficient new discoveries to provide for Europe's growing needs for twenty years. This implausible claim ignores the fact that exploration is already normally taken as an expense against high marginal tax rates, meaning that the oil companies are in fact commonly spending 10c dollars, or even less when the benefits of tax treaties and sophisticated accounting are taken into account. It is noteworthy that the drilling boom that accompanied the high oil prices of the early 1980s, which was substantially tax driven, failed to deliver corresponding discoveries. Ironically, the higher the tax burden, the easier it is to explore as there is more to offset the cost against. It is geology, not tax, that determines discovery; and discovery determines production. It is obvious that the sub-commercial deposits that require even more favourable tax treatment do not hold much oil.

Schollnberger, who was previously with Amoco, has evidently failed to learn from that company's experience of dwindling discovery.

The Director of the same lobby went so far as to claim that there were "ample reserves for the second half of the 21st Century", which is in blatant contradiction with the past discovery trend that peaked in 1964. It is no surprise that lobbyists should try to influence government to their advantage by making such claims: what is surprising is that responsible entities such as the European Parliament Energy Committee should give them a hearing.

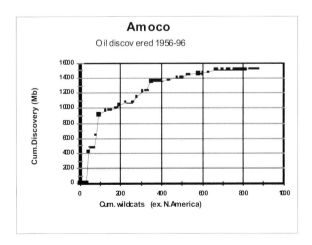

Figure 9

48. Perceptive presentation to Petroleum Exploration Society of Great Britain

The August-September Newsletter of this society, which represents the British exploration community, contains a first rate and surprisingly forthright article by M.R.Smith, explaining the resource limits and inevitable depletion. More telling perhaps than the article itself is the fact that it was published by this knowledgeable institution.

49. New Base-Case Scenario

The ASPO depletion model for conventional oil is based on several scenarios. The Base Case has assumed that demand increases at 1.5% a year, until Swing Share from the five Middle East major producers exceeds 35% of world supply. That is held to prompt a price shock curbing demand growth, which gives a plateau of production until Swing share passes 50%, when production is assumed to decline at the then depletion rate.

This is no longer seen as the most likely case, and in the forthcoming update more consideration will be given to the Low Case, which holds production flat until Swing Share reaches 50%. Even that may not be too realistic.

Oil price rose 300% during 1999 and 2000. It looked as if prices would continue to grow because there was clearly insufficient spare capacity to maintain the past growth of demand. But in fact they weakened at year-end, remaining down for the past six months. The explanation seems to be that the high prices prompted (in whole or in part) an economic recession, curbing demand, which is now about flat. This brought supply and demand into better balance, reducing pressure on prices. This is perhaps what Lord Browne, the BP Chief Executive, meant when he said that the market was working efficiently.

The logic is that if the economy picks up, oil demand would likely increase, and prices would again rise from capacity constraints that get ever more severe from the iron grip of depletion. Higher prices would in turn be likely to re-impose recession. However since demand is not infinitely elastic, continued recession may not be able to hold down prices indefinitely. The investment community seems to be grasping this new reality, with the CIBC bank of Canada predicting global peak production next year with oil at $40 a barrel see http://www.ottawacitizen.com/business/001006/4643011.html

50. Energy Saving

The primary mission of ASPO is to determine the date of peak oil and to raise awareness of this fundamental event as the world's primary energy source heads into decline. It is by all means an important message, but its impact will be strengthened if possible solutions were offered at the same time. Prime amongst them must be energy saving. A valuable small book has appeared on this subject, emphasising the gross wastage associated with the generation, transmission and usage of electricity, offering many practical simple solutions.

(*When the lights go on – Understanding energy* by Cari Spring: www.emeraldrs.org)

If every household had an energy audit with energy wastage or efficiency being reflected in rebates or surcharges on the utility bills, it could lead to new life patterns that would make a huge difference with little pain. It calls for government intervention as a profit-driven market can only encourage waste. The market could however play its part in giving people competitive options by which to improve their energy performance.

Newsletter No. 9, September 2000

51. Oil and the "Third World War"

On September 11th, the West learnt that it had lost an exclusive prerogative in inflicting *Collateral Damage*, which in plain language means killing innocent people. For the suffering victims, there is little difference between the consequences of the ten-year bombing of Iraq, the missile attacks on Belgrade or the suicidal assaults upon New York and Washington. The perpetrators acted with a comparable sense of mission and belief in the justice of their cause. It is evident that suicide bombers are at least not in it for money.

Figure 10. Archduke Ferdinand and his wife an hour before their assassination in Sarajevo, which led to the First World War

However, the latest assault was a departure in that it has led to a declaration of War of Retribution, although the enemy has yet to be precisely identified. It also came with a deeper meaning as a gesture against the concentration of World power and financial domination at a time of uncertainty and transition. It furthermore differed from other assaults in that it prompted an outpouring of grief and condolence throughout the World, with albeit muted regrets even in countries perceived to be hostile.

It may well be turn out to be one of those catalytic events in history, which change the World.

At the root of this transition lies oil, which has driven Western economic prosperity for more than a century, leading in recent years to an extreme concentration of wealth and globalism, as represented by the selected targets.

Approximately half of what is left of this precious commodity lies in just five Middle East countries, which have been forced into a swing role making up the difference between World demand and what the other countries can supply within their natural depletion profiles.

In 1997–8, the production of conventional crude oil reached a peak in the non-swing countries as an inevitable consequence of an earlier peak in discovery. The scene was then set for a gradual increase in oil price, resulting from the decline of spare capacity and the increasing call upon the swing producers. But instead, there was an anomalous fall due to various transitory factors, including an Asian economic recession, which depressed demand, and a devaluation of the rouble, encouraging Russian exports. It was a short-lived collapse before the underlying pressures of depletion and falling capacity re-exerted themselves in a 300% increase in price over 18 months ending in December 2000. This rise in oil price was accompanied by the onset of economic recession caused by the coincidence of the burst

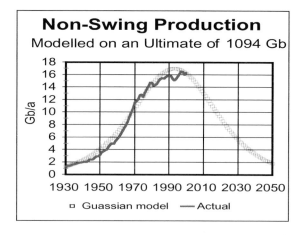

Figure 11

of the dot.com bubble and the high oil price itself. As Professor Oswald points out, all previous surges in oil price have triggered recessions. (www.Sunday-times.co.uk/news/pages/sti2001/0902/stibusecm3002)

Recession dampened oil demand, bringing supply into better balance, despite declining spare capacity. That in turn reduced pressure on prices, which have been weak for most of the year, until briefly surging above $40/b in the wake of the September 11th incident.

If recession continues to deepen, as many commentators now fear, the demand for oil and its corresponding price will likely remain flat or even decline. But if the economy should show signs of improvement, this stability would end, leading to higher oil demand accompanied by soaring prices as the market soon comes again to reach the sloping ceiling of spare capacity. That in turn would likely re-impose recession in a vicious circle.

However, oil demand is not infinitely elastic, as it becomes increasingly difficult to cut essential needs, especially for agriculture, so that recession may not be capable of holding down price for very long. Meanwhile, low oil price will presumably inhibit investments in exploration, non-conventional oil and gas, renewable energy and the all-important priority of energy-saving. In effect, delay will make the eventual crisis even worse.

That scenario sounds bad enough, but it has the edge over an alternative arising from an implementation of the declaration of what may be seen as the 'Third World War' that calls for military reprisals against not only those responsible for the incident but any country that may be deemed to be sympathetic to it. It is difficult to avoid the conclusion that Middle East supplies would be interrupted in such circumstances, which, in the absence of adequate spare capacity elsewhere, could not fail but trigger a mammoth surge in oil price with devastating and incalculable consequences for the World's economy and political stability.

As we view these scenarios and their variants, it begins to appear that the World may indeed have experienced a fundamental discontinuity when the economic growth of the past heads into long-term decline, as the principal fuel that made growth possible becomes, first, expensive and, later, in increasingly short supply.

It remains to be seen how myopic governments, who have so-far had the greatest reluctance to focus on the impact of oil depletion, will react to these unfolding events. In any case, they underline the extreme importance of our mission to raise awareness, as we are already beginning to do successfully. The World is not about to run out of oil. At peak, there is as much left as we have used so far, but we do need to use the high supply, while it lasts, to achieve an orderly transition. Energy-saving not only offers a solution but serves to reduce the gross disparities and power politics that led up to the declaration of the 'Third World War'.

52. Brasil

There is no standard agreement on the definition of the boundary between so-called *Conventional* and *Non-Conventional* oil. The US Geological Survey, for example, treats all oil as *Conventional* other than that lying in disseminated deposits lacking a clear oil-water contact. For the purposes of determining peak production, however, it is better to identify the different categories so as to determine their respective depletion profiles and calculate their possible contributions to global peak production. We therefore distinguish deep-water and polar oil because they occur in hostile environments, under special geological conditions, and because they are less well known. We use the 500m isobath as the cutoff, which although an arbitrary boundary, does generally serve well to distinguish in geological terms the normal continental shelves from the deepwater domain.

Brasil has long had a desperate need for oil to supply its population of 160 million, and with necessity being the mother of invention, its state company, Petrobras, led the industry into the deepwater domain when its onshore and shallow water territories failed to deliver significant discoveries. In the 1970s, it invited the major companies in to conduct exploration over its extensive shelf under service contracts. Seismic surveys and several unsuccessful boreholes, drilled during this campaign, provided Petrobras with a knowledge of the basic geology, alerting it to the possibilities of the deepwater Campos Basin. Exploration there began in earnest in the 1980s, leading in short order to the discovery of several giant fields. The oil is sourced in Cretaceous rifts that formed during the initial opening of the South Atlantic, and has migrated upwards to collect in Lower

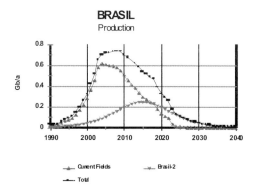

Deepwater production scenario

Figure 12

Tertiary sands. They originated as *turbidites*, which may be compared with submarine avalanches, but were locally subjected to re-working by long-shore currents that served to remove the clay components, leaving dune-like deposits with excellent reservoir properties. The discoveries depend on a unique combination of geological circumstances involving generation, migration, entrapment and the critical relative timing thereof.

Four giant fields were found during the 1980s followed by a fifth in 1996, which with other smaller finds have yielded a total of about 10 Gb (billion barrels). Approximately 100 deepwater wildcats have now been drilled. Operating conditions are however extreme, and small setbacks and accidents can have catastrophic results, as demonstrated by the sinking, with loss of life, of a production platform after an explosion. Furthermore, the oil itself was partly degraded in its tortuous migration; and the reservoirs may be deficient in natural drive. All of these circumstances have combined to hold down production to levels below what might have been expected. Annual plateau production averages about 6% of Ultimate, which is well below that achieved in the North Sea. However, the silver lining of slow depletion is that production will last longer. It is too early to know if other parts of the shelf will be as productive as the Campos Basin, but a prudent estimate would not anticipate more than about 5 Gb from new discovery. If so, production is likely to peak around 2004 at about 2 Mb/d, before falling sharply.

Looking farther ahead, Brasil faces a devastating energy crisis, even if oil demand can be held steady despite a growing population. As illustrated in the graph, non-deepwater production is in long term decline with little

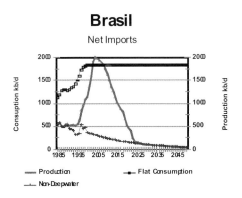

Brasil faces soaring imports after the brief respite from deepwater production

Figure 13

hope of reprieve, while the deepwater provides no more than a few years of respite. Imports would have the meet half the country's needs by 2015 and three-quarters by 2020. World prices will by then have soared to high levels under any realistic scenario, meaning that the country will have to severely curtail its use of oil. Although blessed with sunshine, offering scope for solar energy, and an already well-developed hydroelectric capacity, twenty years is not long to effect the transition. In fact, as mentioned in a previous newsletter, its hydroelectric generation failed to meet demand even this year as a result of low rainfall, giving a foretaste of energy rationing.

53. Report by the US News & Report
This popular US journal with a large circulation carries the following article

A new book argues that the world's oil will soon peter out, but recent numbers are reassuring.

By David Lagesse
You may have thought you were in the clear when the high gasoline prices of the summer eased in recent weeks. Think again, say a few well-known geologists. They believe the recent energy shortfall may have been a foretaste of a steadily worsening crisis that will set in as world oil production hits a peak this decade, then starts to fall, wreaking economic havoc.

Predictions of a permanent oil crisis are nothing new and are largely dismissed by government, industry, and most academics. But the debate is heating up with the approach of what the pessimists believe is an inexorable deadline, outlined in a book called *Hubbert's Peak – The Impending World Oil Shortage*, due next month from Princeton Press. At the same time, studies from the U.S. government and others have gotten rosier, saying we'll have plenty of oil for almost four decades – time to shift to alternative energy sources.

The debate is the legacy of a petroleum geologist named M.King Hubbert, who bucked industry dogma in the 1950s and correctly predicted that U.S. oil production would peak around 1970. Hubbert, who died in 1989, said oil discoveries and oil production follow similar trajectories, hitting a maximum and declining in bell-shaped curves. He noted that U.S. oil discoveries peaked in the 1930s; based on an estimate of total domestic reserves, he predicted that the production peak would follow about 40 years later. Sure enough, domestic oil production topped out in 1970 and has declined ever since, except for a couple of plateaus when prices spiked temporarily. U.S. wells now pump 40 percent less oil than they once did, and the country imports about two thirds of what it consumes.

World discoveries peaked in the 1960s. If world oil follows the same pattern as domestic oil did, production will max out about 2010, and maybe sooner. "The peak of production has to be a mirror image of the peak of discovery," says longtime petroleum geologist Colin Campbell, now living in Ireland, who is outspoken in his warnings. Since production growth barely keeps ahead of demand even today, supplies will quickly run short and prices will soar for everything from gasoline to plastics and synthetic fabrics. Fans concede that Hubbert's math was somewhat suspect, but because he was right about domestic oil, "it's scary to bet against him now," says Kenneth Deffeyes, a Princeton geologist and the author of Hubbert's Peak.

Different math is behind the buoyant view of the U.S. Geological Survey, which last year doubled its estimate of the world's remaining oil, to 2.3 trillion barrels. The world is consuming about 28 billion barrels a year, a figure that has grown an average of 2 percent annually over 30 years. But the USGS estimate leaves a comfortable cushion: The U.S. Department of Energy looked at the data and estimated that remaining reserves could meet demand until 2037. And the reserve estimate is conservative, says DOE petroleum analyst John Wood. "There are a lot of things USGS could do to get even higher numbers, but they didn't," he says.

The USGS started with the roughly 900 billion barrels in "proved" reserves. Then it tried to estimate how much oil has yet to be discovered. Committees of geologists laid odds of finding oil in areas now being explored. The longer the odds, the less weight the scientists gave an area in their estimate of total undiscovered reserves: about 700 billion barrels. Much of the new oil would come from the Middle East, already home to most of the world's reserves, and from new prospects off the Atlantic shores of Africa and South America.

Every drop

The USGS also included, for the first time, a phenomenon called reserve growth, in which industry finds it can pump more oil than expected from a field because of new technology and methods. The USGS looked at the U.S. experience, where technologies such as horizontal drilling and digital seismic imaging retrieve 50 percent or more of a field's oil. In the rest of the world the average is about 30 percent. Exporting U.S. technology and know-how should raise the yields in other countries by about 700 billion barrels, says Thomas Ahlbrandt, the head of the USGS study.

Campbell is skeptical of both these wellsprings of new oil. He criticizes the USGS for including fields with a 1-in-20 chance of producing oil in its estimate of undiscovered reserves, saying those odds are too long to be given any weight. Only about 150 billion barrels remain to be discovered, in his view. He also dismisses reserve growth, saying U.S. laws forced companies here to understate reserves, resulting in what appear to be higher recovery rates than in the rest of the world.

Still, the USGS's upbeat estimate spun around the International Energy Agency, whose members include the major industrialized nations. The Paris-based agency predicted alarmingly in 1998 that oil would peak before 2020 but changed its mind after the USGS report. The U.S. view is also echoed by the IHS Energy Group, an international consulting firm that owns the best database on oil exploration and that once held a more pessimistic view. "Recent data suggests there is plenty of oil out there for the near term," says IHS spokesman Pete Stark.

"Near term" – that's the consolation of his view. Sure, oil is finite. One day the world will need to shift from oil to other energy sources for transportation and heating, reserving what's left for uses where there is no substitute, such as in some chemical and plastics manufacturing. But four decades would give us a chance to prepare – providing it doesn't just deepen our petroleum addiction.

The quotation from the US Geological Survey is revealing. It evidently believes that the industry outside the United States is being operated by primitives with limited technology, so that as much as 700 Gb would be added if US "know-how" were brought to bear. It may come as a surprise to the many American oil companies and contractors, who have long been operating throughout the World, to learn that they are using less than cutting-edge methods. On the other hand, no one need dispute the observation of the IHS spokesman that there is "plenty of oil out there for the near term", which is consistent with an early peak and onset of gradual decline, as indicated by his data.

54. A further insight

Our anonymous contributor provides another insight from his central position in the oil industry, adopting the oil industry's practice of delivering uncomfortable messages in code and camouflage.

View from the Future

I recently met a time traveller who told me many things of great value (these things happen). He also showed me a document that I believe to be of great relevance. There is a problem, however, in that the translation of some words is ambiguous, and there is some confusion as to whether parts of the document refer to power in the military sense or the energy sense.

I will summarise the background before I deal with the specific report. It seems that in the world of the 1960s there were large and important private armies and that these in turn levied troops from their colonies. The mightiest armies were found in North America while the main colonial troops were levied in the Middle East. Considerable discontent amongst the colonial troops led to the formation of a radical group – Our People Expect Comforts (O.P.E.C) – to lobby for better pay and conditions. The generals of the great private armies took no notice of O.P.E.C's demands at all, and, as there was much fighting to do, levied more and more troops from the Middle East and South America. It was around this time that a brilliant but slightly eccentric military strategist and planner called Colonel Hubbert formulated his law. Hubbert's law states that once any fighting unit – army, battalion, regiment etc – has been reduced to half its original complement it enters into irreversible decline. And further, that the decline can be slowed but never reversed. He even suggested this fate would overtake the great American land armies around 1970. Needless to say, the generals were very angry declaring him to be mad and dangerous, and that such things could never happen to their armies as more troops could always be found. Luckily their attention was distracted by insurrection in the Middle East and the comforting discovery that they could recruit an Alaskan army, otherwise they might have noticed that he was right.

By the early 1970s, insurrection amongst the colonial armies had become so great that they were now effectively run by the local states, many of whom were very supportive of the aspirations of O.P.E.C. In consequence, the private armies had to pay much more for military services but as demand for fighting remained strong they had to pay while privately determining to minimise their dependence on colonial armies. This plan worked well for 30 years until it became clear that they really had run out of potential non-colonial recruits. They had made strenuous efforts to overcome the problem. First in North America and later elsewhere they recruited increasing numbers of General Auxiliary Services (G.A.S.). These were fairly lightly armed and lacked the punch of the Ordinary Infantry Levies (O.I.L.). Although they became increasingly important their primary weakness was that they could not be used for transportation or logistics. Repeated attempts at retraining and converting them to transportation duties proved expensive and ultimately unsuccessful. Another and rather more successful alternative came from relaxing the weight limit. As a result there were successful moves to raise heavy brigades the most notable of these being the Athabasca and the Orinoco. The generals also gave repeated assurances that they would be able to raise another Alaskan army if only the government would be a little less squeamish about their recruitment techniques. They also indicated that a shakedown of the mountains and parks would also produce more recruits.

As the situation deteriorated and the followers of Hubbert became more assertive in their predictions of problems ahead, the generals came to place more and more emphasis on advisors widely known as PYTIMFs (Place your trust in market forces). PYTIMFs also found great favour amongst governments even though they had no interest in military matters and could never quite explain why price would bring forth infinite recruits without raising the cost. (My informant tells me they fell from favour in 2005 and effectively disappeared, with many denying they had ever said such things)

I am led to believe that the following was a confidential analysis of military strength and prospects at end 2001/early 2002 for the Middle East. The report notes that in 1987 and 1989 the tally of Middle East reservists rose from 400 battalions (bn) to 550 bn and then on to 650 bn without any substantive reason for the changes.

The report specifically notes that while there was a well substantiated tendency for improvements in training and equipment to allow more fighting to be extracted from reservists than might otherwise have been expected, the 1987 and 1989 changes were well outside this range and essentially inexplicable. It also noted that, as reservist numbers were regarded as state secrets, there was a real possibility that the 1987 and 1989 increments were fictional and reflected political pressures amongst the group to be perceived as reservist rich. The report's conclusion was that if the reservist gains were politically motivated correcting them would be essentially impossible as any revised figure would require independent audit which was seen as politically unacceptable.

The report's stated aim was to address the issue of fighting strength without contesting the reservist issue. It aimed to achieve this with an audit of the number of operational units, noting how long ago they were formed and assessing their current fighting status. The report made the assumption that any unit formed more than 30 years earlier i.e. before 1971 was unlikely to offer anything more than static or declining fighting potential.

The key summary table is reproduced below:

Country	Fighting units	Post 1971 units	Potential new units
Abu Dhabi	13	1 Arzanah (disc '73)	None known
Algeria	74	10 (70s), 2 (80s), 2 (90s)	2 or 3
Iran	41	5 (70s), 2(80s)	Several
Iraq	20	5 (70s), 2(80s)	Several, large potential
Kuwait	8	none	limited potential
Libya	76	19 (70s), 3 (80s), 3 (90s)	Some
Neutral Zone	5	none	limited
Nigeria	over 130	30 (70s), 20 (80s), 5 (90s)	potential seatroops
Qatar	8	1 (70s), 2 (90s)	limited
Saudi Arabia	50	3-4 (70s),? (80s),? (90s)	unknown
Venezuela	over 200	26 post 1971	mainly heavy
Indonesia	several hundred	very limited	very limited

The report concluded that the expansion potential of the colonial troops was, with the limited exception of Orinoco style heavy brigades, very limited and almost certainly much less than most private army generals had been assuming. It noted that around 70% of the units were formed before 1971 and that most of the subsequent units were very much smaller than the earlier ones. It also drew attention to the fact that in 1974, the group could supply 37 million brigade days (b/d) but that by late 2001/early 2002 its capacity was no more than 32 million b/d. The report's conclusion was that for most of these countries, the chances of maintaining their fighting capability were 'heavily circumscribed' and 'unlikely to be maintained for more than a few years'. It did however, suggest that there may be significant potential for raising the more lightly armed G.A.S. troops, a source that, so far, had only been lightly exploited in the countries analysed.

55. Submission to UK Cabinet Office

ODAC's submission to the UK Cabinet Office may be found on the following website. The Cabinet Office is carrying out an Energy Review for the UK government, with a mandate to look at UK energy supplies over the next fifty years. Several oil companies have also made submissions. Reading between the lines, it seems that our anonymous contributor to the last newsletter was entirely correct in saying that the industry understands the situation only too well. They have no particular reason or obligation to discuss it openly.
http://www.cabinet-office.gov.uk/innovation/2001/energy/submissions/ODAC.pdf

Newsletter No. 10, October 2001

56. Oil and the "Third World War"

The war rumbles on with downright opposition in many countries and not more than muted support in the so-called coalition. It is now compounded by an anthrax scare in the United States of unknown origins. So far, there has been no particular direct link with oil, although to judge from the following quotation from a 1998 interview with bin-Laden, it indirectly may be one of the principal causes.

> On the U.S.-backed fight against the Soviet presence in Afghanistan (bin-Laden said): "Those who waged jihad in Afghanistan… knew they could, with a few RPGs (rocket-propelled grenades), a few anti-tank mines and a few Kalashnikovs, destroy the biggest military myth, humanity has ever known. The biggest military machine was smashed and with it vanished from our minds what's called the superpower."
>
> Bin Laden claimed the United States has carried out the "biggest theft in history" by buying oil from Persian Gulf countries at low prices. According to bin Laden, a barrel of oil today should cost $144. Based on that calculation, he said, the Americans have stolen $36 trillion from Muslims and they owe each member of the faith $30,000.
>
> "Do you want (Muslims) to remain silent in the face of such a huge theft?" bin-Laden said.

By Donna Abu-Nasr, Associated Press Writer

The following quotation from the *Spectator* of 20th October points to the wide popular sense of injustice, especially in Middle East countries, which lies behind the conflict.

> Throughout the cafes of the Muslim world, hundreds of thousands of young men are saying… "We have all this oil, yet what happens? It is sold cheaply to westerners, who despise us, to pay the night club bills of decadent pseudo-Islamic rulers. Given our control of oil, we could squeeze the world economy's windpipe. Yet we have not even been able to dislodge the Israelis from the lands they stole. Our current leaders are wasting our substance and our opportunity; let us rise up against them". Bin-Laden's aim is to compress all that café hot air until it explodes.

57. The Times of London sees the oil risks but doesn't question the reported reserves
The Last Oil Rush (October 25 2001)

Could the West survive without Saudi oil? The war on terrorism means that we may have to. The former Soviet Union could fill the gap, but this would bring its own set of pitfalls. It is mid-February, 2002. North America is in the depths of a bitter winter. Consumption of heating oil is at an all-time high and petrol use is back to prewar levels thanks to a long slump in world prices, but the war on terrorism drags on.

Contrary to most forecasts, Osama bin Laden has been captured alive and airlifted to the USS Carl Vinson by triumphant US Marines. In line with other forecasts, the terrorism has not stopped. The Strasbourg anthrax outbreak appears to be contained but a smallpox scare is unfolding in Los Angeles and well-sourced Pentagon leaks say that Saddam Hussein has assembled a "dirty" nuclear bomb with enriched uranium packed around a Scud warhead. Range: 1,300 miles.

The Bush-Blair coalition is intact but under intense pressure from Washington hawks who want to take the war to Baghdad. The nuclear leaks win the argument for them and, with Blair's regretful non-cooperation, B2 bombers of the 509th air wing resume their 22-hour raids from Whiteman Air Force Base in Missouri, this time on Saddam's revivified military infrastructure and key Iraqi oil assets.

Saudi Arabia erupts. The new offensive persuades millions in Riyadh and Jedda that the war on terror is in fact the war on Islam against which their imams have railed for months. Following the lead of a prominent dissident cleric, tens of thousands take to the streets to condemn the royal family's tacit support of the American attackers.

To restore calm, the Saudi Government suspends oil sales to the US in what it privately assures Washington is just a temporary move. But Iraqi exports under the UN-approved oil-for-food programme have already dried up and the damage is done. With a third of the world's known oil reserves in jeopardy, global prices zoom to $44 a barrel.

President Bush authorises an emergency withdrawal of 200 million barrels from the Strategic Petroleum Reserve held in underground caverns in Texas and Louisiana. It will make up the shortfall in US imports for barely a fortnight unless he can persuade voters to switch overnight from conspicuous consumption to manic conservation – a trick he is loath even to try. Instead, flanked by his energy secretary and an uneasy-looking clutch of oil executives gathered in the Roosevelt Room of the White House, he announces an historic ten-year plan to wean the US off Middle-Eastern oil and meet its energy needs elsewhere. "My proposals," he says, choosing words that would have been unimaginable six months earlier, "will end the Arab world's unhealthy dependence on the petrodollar. They will boost export-led growth for our friends elsewhere in the world. They will bolster our national security and transform how we define it. They may even transform the health of the planet we call home."

This scenario could be triggered in any number of ways besides the bombing of Iraq. Al-Qaeda terrorists could sink a supertanker in the Strait of Hormuz. Saudi Arabia could be overtaken by a full-blown revolution, or slapped with embargoes for failing fully to condemn future atrocities.

The result would be a seismic shift in patterns of oil procurement that would define the coming century. The losers, at least in the short term, would be the Gulf states of the Middle East. The winner, in the supreme irony of the post-Cold War period, would be Russia. In fact, it is already happening. Immediately after the September attacks, President Putin endeared himself mightily to President Bush by ordering his armed forces to stand down from the heightened alert they would otherwise have adopted in such circumstances. But he also offered to make up any shortfall in Middle East oil exports to the West that might result from the war on terror.

As if on cue, an Italian tanker left the Russian Black Sea port of Novorossiysk last week with the first load of oil to flow through a new 990-mile pipeline linking the Tengiz field in Kazakhstan to the open seas. To the west, a Russian oil terminal is to open before the end of the year at Primorsk on the Gulf of Finland to bring more crude from western Siberia, Russia's booming oil zone, to Europe via the Baltic. In the Far North, Lukoil, Russia's biggest oil producer, is building an Arctic Coast terminal from which to ship 250,000 barrels a day straight across the Arctic Ocean in a fleet of icebreaking tankers.

Plans for former Soviet Central Asia are even more ambitious. Starting in Azerbaijan, at least two pipelines will eventually carry oil and gas to the outside world via Georgia and Turkey, and in Turkmenistan, a land of scorching deserts and vast gas reserves bordering Afghanistan to the north, the current fighting has paradoxically revived hopes of long-term stability making possible the most Herculean undertaking of all: a gas pipeline over the Hindu Kush to Pakistan and India.

These are the outlines of the last great oil rush; a race to open the Caspian basin in the hope that it may replace the Middle East as filling station to the world – and the expectation that even if it doesn't, its oil will find a market somewhere.

The stakes could hardly be higher. With America alone spending £100 million a day on imported crude, oil remains the world's great wealth-creator. The rise of the personal computer notwithstanding, it still drives every industrial economy, provides profits for the world's largest corporations, pays for most of the Middle East's armies, and funds a sprawling culture of gilded vulgarity stretching from Dubai's seven-star Burj Al Arab Hotel to the subterranean swimming pools of Kensington Palace Row.

"Access to large sources of oil has long constituted a strategic prize," writes Daniel Yergin in The Prize, his seminal study of oil politics. "It enables nations to accumulate wealth, to fuel their economies, to produce and to sell goods and services, to build, to buy, to move, to acquire and manufacture weapons, to win wars."

It also forces importing nations to do business with regimes they would otherwise condemn, and the race to the Caspian could lead the West into an array of new strategic relationships every bit as problematic as those now under strain in the Persian Gulf.

Azerbaijan, key to the Caucasus and the oil-drenched Apsheron Peninsula, is one of the most corrupt nations on earth. At the start of the 1990s its capital, Baku, was hailed as the next Houston and enjoyed a brief boom, depicted with surprising accuracy by Robbie Coltrane and a host of dancing girls in 007's The World Is Not Enough. More recently, the multinationals have been pulling out in droves rather than adapt to Baku's rising violence and bribery. Kazakhstan is still run by its former communist chieftain, Nursultan Nazarbayev, ten years after the Soviet collapse, while his three daughters hold those levers of power that he does not. One is married to the son of the President of neighbouring Kyrgyzstan, another to the head of Kazakhstan's oil and gas monopoly. The third controls state TV. And Turkmenistan has degenerated from its previous incarnation as a Soviet Socialist Republic (something few thought possible in 1991) to a parody of a Third-World dictatorship under the deeply eccentric guidance of Saparmurad Niyazov, who likes to be known as "Father of all the Turkmens" and has anointed himself President for life.

Qualms over democracy and human rights have not impeded the hunt for oil in the past. A more important question, as Western leaders reassess their energy policies in the light of September 11, is whether the former Soviet Union has enough of it.

Broadly speaking, it does. According to figures from the US Energy Information Administration and the London-based Petroleum Argus, the Middle East produces about 16 million barrels of oil a day, of which Saudi Arabia pumps 7.5 million. The US relies on the region for 2.6 million, or about a third of its imports. The former Soviet Union pumps four million barrels a day, projected to rise to seven million over the next five years and much more within a decade as the Tengiz field and the even larger Kashagan reserves in the northern Caspian come on stream. Kazakhstan, by the most conservative estimates, is sitting on more than 20 billion barrels of recoverable oil. Russia has nearly 50 billion barrels, and exploration has barely begun in some of the remoter reaches of Siberia.

For Putin and Nazarbayev, that is the good news. The bad news is that Saudi Arabia's energy reserve remains the biggest and most accessible on the planet by such a margin that it would take a full-blown revolution there to end its dominance of Opec and the global oil business. "Stick a straw in the ground there, and oil gushes," says Ian Bourne, the editor of Petroleum Argus. "Then you put it in a tanker and ship it for $2 a barrel. It's almost as simple as that." At 262 billion barrels, Saudi Arabia's known reserves are still biblically huge. Its infrastructure is so extensive that if Iraq were to shut down production altogether, it could summon enough reserve capacity within 90 days to make up the shortfall and stabilise world prices. Over time, its shimmering sands have yielded so many new fields that successive predictions of a peak in production followed by decline have turned instead into a series of peaks – a plateau, as Bourne says, with no horizon in sight.

Iran, Iraq and Kuwait are similarly blessed. This is why, despite the region's record of war, sanctions, ecological devastation and grotesque abuse of human rights, most major Western oil companies were returning there before September 11 in the hope of winning new access to old but reliable reserves. Before the world changed irrevocably, Western companies were competing fiercely for new gas extraction contracts in Saudi Arabia that they still hope to use as toeholds in the Saudi oil business. In Iran, the prospects of an end to the national oil monopoly's supremacy were better than at any time since the 1979 revolution that toppled the Shah. Even Iraq looked a good long-term bet, as pressure from Russia and elsewhere mounted for a complete end to sanctions.

Now Big Oil has fallen silent, sometimes to the point of hostility. No company I phoned would comment publicly on what the war on terror might mean for its business. Bourne says: "They're holding their breath and crossing their fingers." One British spokesman insisted on anonymity before saying: "Nothing will change."

Analysts agree it is highly unlikely that Saudi Arabia will stop selling its oil to the West, or that the West will stop buying it. Yet if nothing changes within the world's only oil superpower, it could detonate a demographic time bomb. The Saudi Royal Family has cleaved to power since the 1930s thanks to an unwritten social contract by which its subjects remain politically submissive in return for free, oil-funded education and healthcare and an average annual income of $7,000. That contract is crumbling. Saudi Arabia's population is young, fast-growing, underemployed and increasingly resentful of the institutionalised corruption that is said to siphon the revenue from 600,000 barrels of oil a day to fund the louche lifestyles of the country's 15,000 princes.

The Saudi exchequer needs an oil price of $24 a barrel for the foreseeable future to put the economy back on a sound footing. The price is now $19 a barrel – barely enough to meet the country's immediate expenses and service its debts – and Opec is loath to raise it for fear of being seen to profit in a time of crisis.

Next to most Middle Eastern governments, Putin's Russia is a model of progressive development, even if the same cannot be said of his Central Asian neighbours. He has a vision of his country as a Eurasian commercial behemoth selling its oil to the highest bidder and earning transit fees on most of Kazakhstan's as it flows from Tengiz to the Black Sea. In this vision, Moscow's profits are limited only by the bore of its pipelines and the size of tanker that can squeeze through the Bosphorus.

There is a catch, of course. As James Bond learnt on his latest adventure, every pipeline is a potential terrorist target. And as Hitler showed with his murderous advance on Stalingrad – and, he hoped, the "oily rocks" of the Caspian shore – a thriving oilfield can drive the world to war even if it is embedded in the heart of Russia.

It is February 2002 again. The pundits are digesting President Bush's brave switch away from the Middle East in search of apolitical oil. They ask if he has found the answer to America's latest energy crisis and conclude that he has probably not, because oil, by its nature, will always be political. Instead they paint a picture of an America turning away from oil altogether in favour of liquefied natural gas, methanol, solar and wind power and hydrogen, the holy grail of alternative fuels. The Wall Street Journal says that America can lead the technological revolution that will lead the world into the post-oil era. Al Gore, with beard, emerges from obscurity to note that this might save the planet. This is a future that could work, the pundits say.

Whether Bush is the man to embrace it is another matter.

Time will tell what the outcome of the struggle will be. President Wilson at the end of the First World War adopted the slogan "Peace without Victory". It did not exactly enthral his allies, but perhaps offers a useful formula today before more collateral casualties inflame passions further. Some hoped for a Ramadan face-saver, but that has now been lost.

One positive outcome of the war is that the separatists in Northern Ireland appear to be laying down their arms, possibly having realised that their traditional sponsors will be less keen to fund their activities, which for many years have included bombings with collateral casualties.

Meanwhile, World recession, now widely recognised as such, deepens, reducing the demand for oil. It will be necessary to take this into account in updating our depletion models when the new data become available from the *Oil & Gas Journal* in December. There might be a case for adopting the following scenarios:

High Case (previously the Base Case): the World recovers rapidly from the present turmoil with oil demand resuming its previous upward growth at 1.5% a year until Swing Share hits 35%, which is taken to trigger another oil price shock, leading to a plateau of production until it reaches 50%, when production commences its long term decline at the then depletion rate.

Base Case (previously the Low Case): mild recession holds demand flat until Swing Share reaches 50%, as above.

Low Case: deep recession causes demand to fall at 1.5% a year until Swing Share reaches 50%, as above. (It means that peak oil occurred in 2000).

Prices are likely to remain weak under the Base and Low Cases as the low demand is in better balance with supply. The High Case Scenarios does not seem very likely because any economic improvement would lead to a parallel increase in oil demand, which would soon again reach the sloping ceiling of spare capacity, causing a new price shock, which would re-impose recession.

The low prices are however a mixed blessing, for they allow governments to continue to ignore depletion, delaying the introduction of effective energy saving policies. They also inhibit exploration, the development of non-conventional oil and gas, and renewable energy. Oil demand is not however infinitely elastic, meaning that prices will have to firm in due course, as essential needs, especially for agriculture, are faced.

58. Shell Scenarios

Shell is famous for its scenarios, but how plausible they are is another matter. Certainly, the latest pronouncement includes a bizarre case, termed *"Dynamics as Usual"*, with oil production reaching 105 Mb/d in 2050, implying an Ultimate of 7000 billion barrels, as depicted in the attached graph by Jean Laherrère. This is far out of the range of all published estimates over the past fifty years, including the 2000 Gb estimate by Shell itself in 1998 (published by Bookout). It is even far removed from the recent excessive USGS Mean and High estimates. It probably means that whoever compiled it simply extrapolated the past trend oblivious of the resource constraints. Oil company planning departments evidently have their fair share of "flat-earth" economists, and are often isolated from the exploration departments, whose members have little motive to draw attention to the limitations that their technical knowledge reveals. (For the Shell scenarios, see: http://www.shell.com/files/media-en/scenarios.pdf)

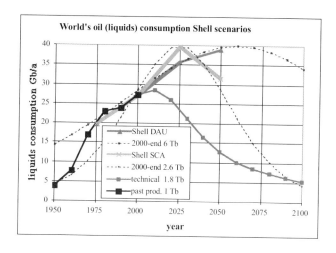

Figure 14

59. North Sea Peaks

Two studies have been received confirming our own assessment. One by Simmons underlines the bleak future for UK production, which is shown to decline at close to our estimated depletion rate of 7%. The other report documents the evolution of the official reported UK reserves, illustrating that several of the major fields were initially under-reported by as much as one-third. As illustrated by the representative example of the Thistle Field, these fields are now so far depleted that extrapolation of the decline curve leaves no doubt as to what the size of the field is. The first year of decline was in 1983, and its extrapolation would have given a clear indication of the ultimate size of the field, illustrating incidentally the minimal impact of technology on the reserves.

The explanation for the under-reporting is that whereas the explorers made confidential volumetric assessments of the size of the prospect, the initial published reports were based on the planned commercial development with a given number of wells. During the later life of the fields, every effort was made to tap subsidiary reservoirs and satellite traps, bringing the reported reserves closer to the original volumetric estimates.

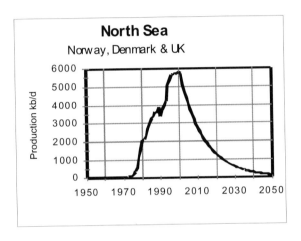

Figure 15

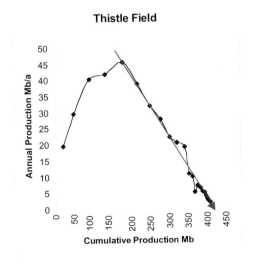

Figure 16

No particular new technology was involved, although knowledge of the details of the field naturally evolved as work proceeded. It is most unlikely that the skilled engineers would have systematically under-estimated the real size of the fields by such a large factor. It was all in the reporting, because they were reporting an initial phase of the project and not the full size of the field. There is however much less scope to under-report the later smaller fields, which accordingly run the risk of giving disappointing results, as Norwegian experience confirms.

"Reserve growth", as under-reporting is confusingly described, is one of the most misunderstood elements of depletion studies.

60. Hubbert's Peak
An excellent new book with the above title has been written by Professor K.S.Deffeyes, and published by Princeton University Press. The author started his career with Shell, working with the legendary King Hubbert, before moving on to teach at Princeton. He writes in a lucid, at times humorous, non-technical style, concluding that oil production will peak around 2005. He draws attention to the absurdity of the USGS estimates. His explanation of hydrocarbon source-rocks is particularly useful.

61. The Economist begins to grasp the hard truth
The Economist, as the principal organ of the flat-earth community, cannot be expected to admit to resource constraints, but in an interesting article on November 1st it did, after some rather derisory comments about our endeavour, nevertheless begin to accept the notion of depletion. It ends by referring to a statement attributed to the IEA that it would take an investment of a trillion dollars over ten years for the industry to make good the decline in non-OPEC production. Translated, that means peak and decline.

62. Yes, even the economists are rational
Our anonymous contributor from the heart of the oil industry writes:

Ever since my involvement with ASPO started, one feature has both puzzled and troubled me. Why are otherwise rational economists so violently opposed to even considering the compelling evidence that hydrocarbon resources are limited and that peak production is close? In the outer reaches of web where the geniuses and the madmen compete for space, there is much to find, and sometimes exciting things to learn. In this case, the relevant article at http:// dieoff.com/page173.htm cannot be accused of being particularly well written or even very coherent, but by the third reading, its devastating message comes through. It reveals that energy has now become so fundamental and so integral to any sort of developed economy that, without it, there can be no economy at all in any meaningful sense. Thus, energy can no longer be traded and valued within an economy. It is an external because a developed economy is impossible without it. In the absence of energy, no stock, share or bond can have value as it simply represents a share or a charge on something that has no value. Everything then becomes as valuable as Confederate dollars or Russian bonds.

Now, no one is talking of an absence of energy but we are talking of a shortfall and an increasing shortfall in the world's key energy source, namely oil, and that from some date between 2006 and 2012. We have tended to concentrate rather narrowly on the immediate implications for the oil industry and for supply. It seems that economists have subconsciously grasped that once energy supplies become severely restricted, economics, as we (and they) understand it, cease to have purpose or relevance. Therefore, for them, denial becomes not just important but vital.

To expand a little further, there is near universal acceptance that fossil fuels will provide the bulk of energy supplies for the next 20 years, with oil easily the most important because of its transportability, energy content and ease of use. Gas may be environmentally more acceptable but in terms of transportability and energy content, it comes second by a wide margin.

The economists tell us that a shortfall in supply of any commodity leads to an increase in price, which will re-balance supply and demand. This is fine and we can all agree that this is indeed the best way to allocate goods *within* the economy. If the price of chocolate bars go up, we buy and eat less of them, and the market re-balances. However, if the product is *external*, like energy, rather different things happen. Energy is fundamental and intrinsic to the whole of the modern economy. It is a necessary condition *for there to be an economy at all.*

If the price of energy goes up, the whole economy goes down. It shrinks, as every company becomes less profitable, less valuable. Share prices fall and keep falling. We had a glimpse of this in 1974 when oil supply was restricted and the price rose. In the 1979 crisis, it was purely a price effect as there was no actual supply shortfall, just the fear of one. Now we have the realistic prospect of increasing shortfalls and sustained price rises once oil supply passes peak. The rest I leave to your imagination. So my case is the belief in endless and bountiful energy supplies is a *sine qua non* if you want to be an economist for more than the next five years.

A funny way to run a railroad!

Our faith in market economics is very great. The oil companies are passionate believers, even though $10 oil nearly wiped them out, and so are most western governments. It absolves them from thought, action and guilt. But for how long? The market is currently driving oil prices down. If prices were to remain significantly below $20 for any length of time, two things would happen: firstly, high cost exploration and development in the non-Opec countries would rapidly fall. (If you don't believe me just observe how rapidly US gas drilling is falling now that the gas price is right down). Secondly, a number of oil producer governments would run out of money, leading to debt default, instability and quite possibly revolution. Sober and sympathetic analysts have calculated that for the Saudi State to cover all its costs, and service its debt, it needs production of 8 million barrels/day and a price of $25/barrel. Is our enthusiasm for unfettered market pricing about to become very expensive indeed?

A very good question.

The editorial writer in an oil magazine that recently came through my door clearly thought he was being whimsical and clever with the suggestion that there may not be enough carbon available for global warming to be significant. Initially, I dismissed the thought, but it nagged away so I jotted down what I thought were the key points.

1. There is a weak linkage between carbon dioxide concentrations and global temperatures.
2. There is wide range of known and unknown influences on global climate. Carbon dioxide concentrations are by far the easiest to measure.
3. Carbon dioxide concentrations were 285 part per million volume in 1850 and reached 360 parts per million volume in 2000.
4. Global temperature rose in an erratic pattern over the period averaging 0.5 degrees C higher in 2000 than in 1850.
5. All the climate change modelling is done on the basis of carbon dioxide levels doubling to 500 parts per million.
6. Since 1850, we have burnt 800 billion barrels of oil, 2200 trillion cubic feet of gas and many million tonnes of coal. The human population has increased sixfold, and cattle tenfold. Vast areas of forest have been burnt. But the carbon dioxide concentration has increased by only 75 parts per million.
7. The oil burnt is 35–50% of the most likely reserves, the gas burnt is

15–30% of the likely reserves. For coal we have probably burnt around 15–30% of reserves.

What then becomes clear is that the whimsical journalist, who visited me, is probably right. There just isn't enough available carbon to reach 500 parts per million even if we burn all the fuel, burn the forests and have lots of cattle with excessive flatulence. The IPCC squares this awkward circle by assuming future hydrocarbon supplies in such abundance as to make even the flattest of flat-earth economists blush. Everyone appears to have been too embarrassed to challenge this (all too convenient) nonsense.

The real challenge for ASPO-ODAC is this. Governments are cheerfully spending fortunes on good, bad and indifferent climate change research. Some are already making rules and regulations, and imposing costs on their industries and taxes on their enterprises *without even knowing if there are enough hydrocarbons for the threat to be realistic.*

Unless and until governments have a rather more realistic view of the global hydrocarbon resource:

- there is no chance of building plausible climate change models
- there is no way of knowing whether climate change is a threat that warrants expenditure

As explained above, insufficient energy is many orders of magnitude more important a threat to human welfare than anything else. Yet, governments are either unaware or have their heads in the sand unable to face so large a threat. The delicious irony is that the oil companies (at least those that understand) might like to point out the reality but that would mean telling shareholders, which wouldn't do a lot for the share price or the directors' share options.

It looks as though global warming provides a convenient diversion for governments and oil companies alike. However, for a tiny fraction of the cost of climate change research they could find out the reality of the global hydrocarbon resource and the true magnitude of the problem. Otherwise, they could all find themselves like the British in Singapore with the guns pointing out to sea and the 'problem' coming at them from behind.

63. House of Lords Select Committee on the European Union

The following submission has been made by ODAC to a House of Lords Committee in the United Kingdom. It triggered a series of intelligent questions, and was followed up by a presentation delivered in person by Dr Bentley and Prof. Meyer.

1. The world faces almost certain near-term hydrocarbon supply shortages. The reasoning is as follows:

1.1 Global oil supply is currently at *political* risk. This is because the sum of conventional oil production from all countries in the world, except the five main Middle-East suppliers, is more-or-less at the maximum set by physical resource limits.

1.2 World oil supply will soon be at *physical* risk. This is because the Middle- East countries have themselves little spare operational capacity, and this will be increasingly called upon as oil production declines elsewhere.

1.3 Large investments in Middle-East production, if they occur, could raise output, but only to a limited extent. The main exception is Iraq, but even here there would be significant delays before prospects are confirmed, and infrastructure put in place.

1.4 In any event, global output of conventional oil will soon decline. The date of the peak depends on the size of Middle East reserves, which are poorly known, and unreliably reported. Best estimates put the global conventional oil peak between five and ten years away.

1.5 The world contains large quantities of non-conventional oil and various oil substitutes, but the rapid decline in the output of conventional oil makes it unlikely that non-conventional sources could come on-stream fast enough to compensate.

1.6 For conventional gas, the world's original endowment is probably about the same, in energy terms, as its endowment of conventional oil. Since less gas has been used so far compared to oil, the world will turn increasingly to gas as oil declines. But the global peak in conventional gas production is already in sight, in perhaps 20 years, and hence the global peak of all-hydrocarbons (oil plus gas) is likely to be in about 10 or so years.

2. The above views have been amplified in a document submitted to the Cabinet Office's Energy Review Team. This is available at: http://www.cabinet-office.gov.uk/innovation/2001/energy/submissions/ODAC.pdf.

3. The background to these statements is that a group at the University of Reading has been studying, for some years, the question of global hydrocarbon supplies [Ref. 1]. The group has seven members (including the author of this submission), and contains petroleum geologists, engineers and physicists. This research has included extensive discussions with oil companies, the UK government, the IEA, the EU and the US Geological Survey. In addition, the group has had sight of the main oil industry resource data set.

4. The conclusion of the group is that the report by Campbell and Laherrère [Ref. 2] represents the best calculations to-date on the future of global hydrocarbon supplies. Their calculations lead to the

conclusions summarized above.

5. It is worth stressing that the Campbell/Laherrère calculations are based on:

- the authors' extensive geological knowledge;
- full access to the standard industry oil and gas resource database;
- detailed analyses of current hydrocarbon reserves (where, particularly for oil, neither FSU nor Middle East reserves can be taken at face value);
- a range of *statistical* approaches to assessing the yet-to-find;
- the use of models of future hydrocarbon production rates that the group at Reading has assessed as adequately accurate and robust.

6. The group at the University of Reading discussed these finding with the EU's DG-TREN, and in the course of these discussions had significant input into helping to draft portions of the Technical Background document published as Annex 1 of the EU's Green Paper: *Towards a European strategy for the security of energy supply*, (COM(2000) 769). The latter is currently the focus of Sub-Committee B's attention.

7. In the light of the foregoing, the specific responses of the Oil Depletion Analysis Centre to the questions raised by Sub-Committee B are as follows:

Q1. The EU faces severe near-term energy security issues.

Q2. Both supply side and demand side policies will be required.

Q3. There is a need for an EU Energy policy. A well-thought-out policy can help lessen the shocks from supply limits.

Q4. Producer/Consumer dialogue will be essential for a peaceful transition.

Q5. The risks of in-use and after-use of nuclear power need close examination.

Q6. The EU suggests that taxes on conventional fuel could support renewables. Many technologies for large-scale use of renewables are not far from cost-effective.

Q7. Detailed analysis is required, see ODAC's submission to the PIU.

Q8. The lessons of the 'fuel protests' is that tax harmonization is vital; the public will understand hydrocarbon limits, but will want to see fairness in the burdens.

References:

[1]. R.W. Bentley, R.H. Booth, J.D. Burton, M.L. Coleman, B.W. Sellwood, G.R. Whitfield. Perspectives on the Future of Oil. Energy Exploration and Exploitation, Vol. 18, Nos. 2 & 3, pp 147–206, Multi-Science Ltd., 2000.

[2]. C. J. Campbell and J. H. Laherrère. The World's Supply of Oil, 1930 – 2050. report from Petroconsultants S.A., Geneva, 1995.

18 September 2001

R.W. Bentley, Co-ordinator, The Oil Depletion Analysis Centre, Suite 12, 305 Gt. Portland St., London W1W 5DA. (Tel: 020 7436 6544; e-mail: odac@btconnect.com)

The Oil Depletion Analysis Centre aims to supply independent information on global hydrocarbon resources.

Patron: Mr. David Astor. Trustees: Mr. & Mrs. Richard Astor, Dr. Colin Campbell, Mr. Richard Hardman, Mr. Roger Harrison, Mr. Stuart Kemp.

Advisory Board: Mr. A.M.S. Bahktiari, Senior Analyst, National Iranian Oil Company; Mr. R.H. Booth, Visiting Professor of Sustainable Engineering, Oxford University; Mr. B.J. Fleay MIEAust MAWA, Murdoch University, Western Australia; Mr. R.F.P. Hardman, Global Exploration Advisor, Amerada Hess; Mr A Ianiello, former Managing Director, Agip UK; Dr Klaus Illum, Energy Consultant, Denmark; Mr. L.F. Ivanhoe, M.K. Hubbert Center for Petroleum Supply Studies, Colorado School of Mines, USA; Mr. J.H. Laherrère, Consultant and formerly Deputy Head of Exploration, Total, France; Mr. R.C. Leonard, Director of Exploration, Yukos Oil, Moscow; Dr. S. Peters, Professor of Political Science, Geissen University, Germany; Mr. M. Simmons, Investment Banker and Member of the President's Energy Committee, Houston, USA; Mr. C. Skrebowski, Editor of 'Petroleum Review', Institute of Petroleum, London; Mr. D. Strahan, Producer, BBC, London; Mr. W. Youngquist, former Exxon geologist, and academic; USA; Dr. W.H. Ziegler, former Senior Exploration Geologist Exxon, and Petrofina; Switzerland.

64. Deepwater Potential

A new data-set suggests that about 30 Gb of oil have been found in water depths of more than 500 m. It lies mainly in Brasil (36%), the Gulf of Mexico (28%) and Angola (26%), confirming that a very special geo-tectonic setting is responsible. The sum of the individual modelled production profiles suggests a peak of about 8 Mb/d by 2012.

Newsletter No. 11, November 2001

65. Oil and the Afghan War

This newsletter is intended to report on progress in the study of the natural peak and decline of oil production, and the consequences, but since oil is such a central factor in the world's economy, it is difficult to avoid touching on wider issues.

In the last issue, we commented on some of the oil-related factors behind the Afghan War. It prompts sympathy for the editors of the western media in their predicament. On the one hand, they have to depict themselves to be loyal to the cause, while on the other they can hardly avoid letting slip concerns about fomenting civil war in a sovereign country and the mindless slaughter of innocent people by bombing and in associated atrocities, which now reach proportions far in excess of the outrage that prompted the action. It would have been so much easier for them to know how to react had the Russians mounted the attack.

It is also difficult for them to cover retaliatory actions as may have been represented by, for example, the New York air crash and the explosion at the Toulouse fertiliser plant. These incidents may have to be depicted as accidents to sustain popular morale. Fertiliser does not explode without being primed, and it is curious that one of the victims was found to be wearing five pairs of underpants, seemingly the hallmark of the Islamic suicide bomber as he prepares for his idea of Heaven.

Several analysts, seeking some plausible explanation for the war, see a hidden motive in US control of Caspian and Central Asian oil, drawing attention to a proposed Unocal pipeline project through Afghanistan. While indeed it is possible that planners and military strategists were misled by exaggerated claims for Caspian oil, and modelled their policies accordingly, it seems doubtful if there is much substance to such interpretations. Certainly, the chimera of Caspian oil would be a poor justification for the bloodshed.

66. Attitudes begin to change even in America

A series of articles in Business Week of Oct 29th give hints of changing attitudes even in America. There is a growing realisation that US oil production is set to continue to fall, and that opening up the environmentally sensitive areas wont make much difference. It means that the growing dependency on imports will continue. There is a new realisation that Middle East imports are politically vulnerable, leading to desperate hopes of finding other suppliers whether in deepwater off Africa, in the partially illusionary reserves of the Caspian or from mother Russia. It seems to be a time of changing friends: the once vilified Russians now sit at high table, while the Sauds fall from grace. If sums about depletion were to be done, the commentators would realise just how wise they are to end with a call on Detroit to improve the efficiency of vehicles and cut national demand.

There is no point in being a superpower without an enemy: Islam seems to be taking the place of Communism as justification for military prowess – good news no doubt for the arms industry.

67. The Nemesis Report

Under a new pseudonym, our contributor for the heart of the oil industry writes:

Market madness?

International oil markets are both volatile and unstable. One barrel more than is immediately wanted – and the price sinks. One barrel less – and it soars. We've all got used to it. It's the way it is. Traders wouldn't have it any other way. Actually, it's a rather strange way to manage such an important commodity. Imagine what it would be like shopping if the prices went up every time there were more people in the store and fell every time there were less.

All oil pricing is now intimately related to spot prices. And this has arisen because no one has the courage to buy oil at other than a spot related price. You can bankrupt your company without endangering your pension by buying or selling spot but if you contract at non-spot prices the moment spot prices go below the level of your contract, you'll be fired. I parody a little but not that much. The so-called triumph of markets has spread far and wide. Few dare oppose them. But, perhaps, we should remember that one of the key features of the California electricity disaster was that state regulations effectively meant all the electricity had to be bought at spot prices – it was great when prices were low but *a disaster when supply was limited.* Californians will have many, many years of paying for their leaders' belief in market forces.

Now, ASPO's basic message is that the time is approaching when oil supplies will be limited. I am beginning to believe that the current method of pricing oil on a spot-related basis is both dangerous and destabilising and likely to become more so as supply becomes restricted. In fact, we have inverted proof of this idea as the market for crude is now very weak. Demand, since 11 September, has fallen dramatically. The November issue of the IEA's Oil Market Report estimated demand as being down by 750,000 b/d or roughly 1%. (If this doesn't sound dramatic, remember it is the largest fall recorded since the 1973/4 crisis). Crude prices have already fallen by $6–8/barrel, and commentators are now talking of a slide to $15/ barrel or even $10/barrel once the winter demand peak is over. The big slide will probably start in late January or early February as any crude loaded after that date can't turn up as products in the main markets until the late Spring, when demand falls away naturally.

According to the IEA, the main OPEC producers already have spare or unused capacity of over 5mn barrels/day, and that's before any further cutback they might finally agree. Historically, OPEC has never managed to maintain quota discipline once unused capacity reached this sort of level.

How has it come about and quite so rapidly? On the demand side, we've had the big post 11 September hit, initially impacting obvious sectors such as aviation (the major airlines have grounded up to 30% of their fleets while keeping the most fuel-efficient flying). Recently, the decline has become more broadly based as the world's three largest economies – the USA, Japan and Germany – are all officially in recession. On the supply side, significant increases have been seen in Russian and North American production while a number of OPEC countries have, according to the IEA, managed to expand their capacities. Demand for 2002 is being steadily revised down, while capacity, both OPEC and non-OPEC, is being revised up.

Barring economic miracles, oil prices in 2002 are going to be grim. It poses a challenge for the members of ASPO-ODAC, because it's going to be very difficult to get people to worry about future energy supplies when their immediate experience is low prices, plentiful supplies and OPEC in crisis. The obvious recourse is to make the analysis as tight and keen as is humanly possible: to point out the essentially transient and reversible nature of the emerging glut; and to underline the immense damage low prices will do in terms of restricting investment in oil production, alternative energies and the more efficient usage of oil and gas.

Poor dears

Many of you will have noticed the oil companies' grim enthusiasm for using that complete 'apples and pears unit' the barrel of oil equivalent (boe) when measuring their reserves. It is hard to be certain, but the practice seems to have started in the Gulf of Mexico, possibly by Shell. It is a nonsense unit, quite apart from the fact that

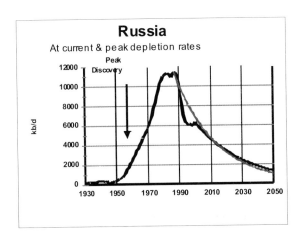

Figure 17

there at least two different but widely used conversions to get from gas units to boe. It's only possible purpose is to obfuscate. A field that is 90% gas and 10% oil is radically different from one that is 10% gas and 90% oil but their boe's will be the same.

Some time ago I was talking to a member of the Shell main board and asked him about this strange enthusiasm for boe. He replied, quite straight-faced, that it wasn't Shell but pressure from the analysts for a single reserves unit. Now, I freely admit to having a fairly low opinion of stockbroker's analysts, a view clearly held in common with the board member, but surely, the poor dears really aren't that stupid are they?

So the choice is yours – unbelievably stupid analysts or oil companies seeking to confound and confuse over their reserves, with the analysts as willing accomplices.

Myth of the month

Even cursory readers of the newspapers in Europe and North America cannot fail to have noticed that the Russians are now our greatest chums, the people who are going to supply us with such rivers of Russian oil that the dread OPEC will be kept firmly in their place. This happy and convenient myth is based on the fact that Russian production has recently been rising at around 7%/year, and that virtually all of it has been available for export. The implication is that this state of affairs can continue forever or at least for a long time.

Alas, like the fabulous oil wealth of the Caspian, the Russian oil flood is a gross exaggeration. After the collapse of the USSR in 1991 a total lack of maintenance and investment led to a precipitate fall in oil output. It was far faster and steeper than the decline curve that had already started on when production peaked in 1989.

By 1999, large-scale investment had recommenced. Naturally, the largest and richest company – Lukoil – was first. Soon, it was recording output gains of 7%. Already these have faded with the Company planning for less than half that level in 2002. Next was Yukos, which has recently been recording 7% output gains. Surgutneftegas, TNK and maybe even Tatneft and Bashneft will follow the same pattern. Meanwhile, Russian economic growth, which has been running at a sizzling 6–8%, will translate into increased oil products demand (most of those inefficient tank factories have now been closed). So, barring large-scale discovery or development of a secret cache of finds, this output surge will last for two or maybe three years before being overwhelmed by expanding local demand and the return to the decline curve.

68. Evora Conference

This two-day conference in Portugal covered a wide range of issues related to climate change, energy supply and energy policy. It is evident that research into climate change has a great momentum, yielding fascinating scientific results about the astronomical cycles that have caused climate changes in the past and the balance between the carbon emissions and sinks. The scientists are working with remarkable data on many aspects of the matter, ranging from polar ice thickness, tree rings, the size of rain drops to the isotopic composition of fossils, but face immense difficulties of interpretation in determining the interaction of all these many factors. As one commented *"All models are wrong, but all are useful"*. It was also revealing to hear about the political and commercial reactions to the risks that the observed changes may relate to man-made influences. If the Dutch should plant a forest in Portugal to sequester carbon, should the credits go to Portugal or Holland? And where do the commercial interests of using the forest to make pulp for the paper industry fit into the picture? The monumental scale of the negotiations needed to settle the issues in an equitable fashion is self-evident.

C. J. Campbell offered a talk *"Peak Oil – a turning point for Mankind"* on behalf of ASPO, and Professor Rosa closed the meeting with a wide-ranging review of the energy options, recognising that the production of oil and gas is set to decline due to depletion.

It was encouraging to hear that two ministers in the Portuguese government are bent on designing a national energy policy aimed at the more efficient usage and bringing in more alternative sources, such as wind and solar. The message of the conference was also not lost on one of the delegates, representing a Portuguese electricity generating utility, whose plans have to have a 30-year time frame. He admitted to having been misled by the flawed and misleading information provided by the responsible international agency, and expressed appreciation for the new insights coming from the conference. He returns to his drawing-board, better equipped. The climate scientists too appreciated the need to use more realistic information on oil and gas supply than had been available to them.

69. Two New Books

Two important new books have appeared. *Eco-economy* by Lester R. Brown of the Earth Policy Institute provides a lucid blueprint for a new world order. The pernicious role of the flat-earth economist is exposed with a call that presently hidden environmental costs should be integrated into the market economy. He provides nice examples of how previous civilisations, from the Sumerians to the inhabitants of Easter Island, collapsed when they over-stressed their natural environment. Although the book is written in an optimistic style, it is chilling to learn, for example, that 480 million people are being fed by grain produced at low cost by an unsustainable over-pumping of the aquifer, or that two-thirds of the world's fisheries are collapsing from by over-fishing. It leads to the conclusion that the decline of oil, which is responsible for so much of the excessive behaviour, may carry a certain silver-lining.

The following quotation sums up the message: "*In a world where the demands of the economy are pressing against the limits of the natural systems, relying on distorted market signals to guide investment decisions is a recipe for disaster*"

The other book s by D.Riley and. McLaughlin of the Alternative Energy Institute gives an excellent summary of the peak oil issue, drawing heavily on material from analysts associated with ASPO-ODAC, before going on to explain the scope for alternative energy. It is aimed primarily at US readers and explains the urgent need for a new energy policy in that country, which has become so vulnerable, having severely depleted its own oil and gas resources.

70. A pilgrimage to Jean Laherrère

A visit was made to Jean Laherrère at his home near Preuilly-sur-Claise in France, where he works single-handedly, producing monumental studies. His desk-top computer is a gold mine of information, gathered from diverse sources, all duly compared, analysed and recorded in spreadsheets and graphs. His particular achievements have been the identification of the parabolic fractal to describe the natural size distribution of oil fields, and in demonstrating the close link between discovery and production after a time shift. References to, and some of results of, this remarkable work are to be found on www.oilcrisis.com and other sites. His single-handed effort dwarfs what many major institutions would be proud to have achieved.

71. Deepwater Potential

The study of deepwater oil, which is here treated as *non-conventional*, continues as a prelude to the year-end update of the depletion model. It is burdened by the lack of a consistent definition of the boundary, here taken at 500 m, and by conflicting information in the various data sources. While definitive conclusions prove elusive, present evidence suggests that the deepwater domain is severely restricted in geological terms. Two factors dominate the evaluation. The first is the requirement for an effective underlying source-rock, as provided in early rifts in divergent plate-tectonic settings. The second relates to the development of turbidite reservoirs far from land where there are two special requirements: the improvement of reservoir quality by the winnowing action of long-shore currents; and the

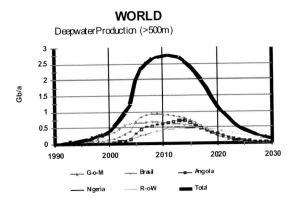

WORLD
Deepwater Production (>500m)

Figure 18

ponding of sediment behind contemporaneous sea-floor relief to provide what amount to stratigraphic traps. For these reasons, it seems likely that the deepwater potential is largely confined to the Gulf of Mexico and the margins of the South Atlantic. Elsewhere, deepwater prospects may have to rely on source-rocks within the delta-fronts themselves that are likely to be gas-prone. Apart from geology, the technological challenges remain daunting, even in such superficially simple matters as heating the flowlines in the freezing ocean depths to allow the oil to flow. The smallest accident or setback can have devastating consequences.

The study concentrates on the Gulf of Mexico, including a possible Mexican extension, Brasil, Angola and Nigeria. It appears that about 28 Gb have been discovered so far in the world as a whole, out of a tentative ultimate of 60 Gb. Modelling depletion at this early stage is difficult but a provisional peak around 2010 at about 7.5 Mb/d is now indicated, falling to about 3 Mb/d by 2020. More study is called for.

72. Oil Supply, Money Supply and Interest Rate

It is difficult for anyone, least of all a humble geologist, to grasp the workings of national and international money supply. It seems however that money is not simply a medium of exchange but carries its own economic dynamic. Commercial banks charge interest on lending money that they neither have nor own. This is money created out of the clear blue sky and fed into the system. The debtors have to work hard not only to repay the loan but meet the interest charges. In some mystical fashion these are all book transactions built on a confidence factor that economic growth will somehow justify it all. Professor Watt in California finds remarkable links between interest rate, the cost of oil imports and the size of consumer debt in the United States. The rising cost of oil imports has apparently closely matched the growth of domestic debt. If King Fahd keeps his oil revenue with Chase Manhattan in New York, perhaps it forms, without his knowledge, the collateral for US domestic debt.

It looks as if recession was triggered, at least in part, by high oil prices last year, brought about by falling spare capacity due to depletion. The central bankers quickly reacted by cutting interest rates, ostensibly to shore up the flagging economy, but indirectly to reduce money supply to match the falling consumption of oil from declining economic activity. Low interest rates penalise the geriatrics on pensions and in fact inhibit commercial initiative, as lenders become very risk averse unless rewarded by high returns. If any attempted economic recovery proves to be frustrated when the consequential increase in oil demand hits the falling ceiling of production capacity leading to soaring prices that re-impose recession, it may well be that the entire foundations of this virtual economy, built on debt and interest charge, disintegrate. That truly would be a turning-point for Mankind, but one flounders in the dark in trying to understand these arcane processes.

The role of the dollar in oil transactions is a related phenomenon, which is now being addressed by the Commodity Currency Exchange through a proposed alliance of core OPEC and non-OPEC countries to mitigate the adverse currency and volatility effects.

Meanwhile, the 62 billion dollar collapse of the energy company, Enron, confirms that real business is not all conjuring with smoke and mirrors. And the merger between the oil companies Phillips and Conoco lets us know that it is an industry in contraction despite the brave words to the contrary. The line between courage and folly is a narrow one.

73. Chevron Confesses

In a paper on the possibilities for increasing heavy oil production, Chevron confesses to an Ultimate of 1800 Gb for Conventional Oil, which is only slightly below our current estimate of 1850 Gb. Our resident detective has duly noted this piece of evidence. http://www.synergytechnologies.com/docs/heavyoil.pdf

74. Oil Depletion Protocol

An earlier proposal for an Oil Depletion Protocol has been taken up by an organisation of international lawyers, who were responsible for drafting the Kyoto Protocol, and a meeting is proposed. The idea is that the producing countries would limit their production to their current depletion rate and that the importers would refrain from accepting infringements. There could be exemptions for gas, non-conventional oil and gas, and possibly for new or small producers. The inclusion of the importers in the Protocol would have the effect of balancing international trade and reducing market-induced wild fluctuations in price, which are widely perceived to be damaging. The importers are left with the task of allocating their available imports under the protocol as they see fit through fiscal measures, auction or in other more socially-conscious methods. If fiscal methods were used, they would lessen the call for revenue from other sources. The protocol would indirectly stimulate energy saving and the increasing use of other energy sources. Countries could appeal the assessment of their depletion rate by opening their data to objective technical audit, which would have the added advantage of improving the database, which is currently grossly unreliable. The idea was first mooted at a conference a few years ago, where it attracted the keen interest of the then Secretary-General of OPEC, Lukman.

Newsletter No. 12, December 2001

75. The New Year
Dr Bakhtiari, an Advisor to the National Iranian Oil Company, continues to provide invaluable insight into the Middle East situation, as seen from Tehran. The Afghan War is apparently viewed as but one step in a ladder, with a likely escalation during 2002. Dr Bakhtiari fears that Iran maybe an epicentre of conflict, sandwiched, as it is, between the oilfields of the Persian Gulf and the Caspian.

76. Are the principles of classical economics discredited?
The following article thinks so:

What's wrong with economics?
Five fundamental errors
The Short Version, by Jay Hanson
(The Long Version is archived at http://dieoff.com/page241.htm)

Any ONE fundamental error in neoclassical theory should be sufficient reason to reject conclusions based upon that theory. Here are five fundamental errors in the theory:

#1. A fundamentally incorrect "method": the economist uses "correlation" and "post hoc, ergo propter hoc" (after-the-fact) reasoning, rather than the "scientific method".

#2. A fundamentally inverted worldview: the economist sees the environment as a subsystem of the economy, rather than the other way around. In other words, economists are trained to believe that natural resources come from "markets" rather than the "environment". The corollary is that "man-made capital" can substitute for "natural capital". But the First Law of thermodynamics tells us there is no "creation" – there is no such thing as "man-made capital". Thus, *all* capital is "natural capital", and the economy is 100% dependent on the "environment" for everything.

#3. A fundamentally incorrect view of "money": the economist sees "money" as nothing more than a medium of exchange, rather than as social power – or "political power". But even the casual observer can see that money is social power because it "empowers" people to buy and do the things they want – including buying and doing other people: politics. If employers have the freedom to pay workers less "political power", then they will retain more political power for themselves. Money is, in a word, "coercion", and "economic efficiency" is correctly seen as a political concept designed to conserve social power for those who have it – to make the politically powerful, even more powerful, and the politically weak, even weaker.

#4. A fundamentally incorrect view of his raison d'etre: the economist sees "Homo economicus" as a "Bayesian utility maximizer", rather than "Homo sapiens" as a "primate". In other words, contemporary economics and econometrics is *wrong* from the bottom up – and economists know it. The entire discipline of economics is based on a lie – and economists know it. Moreover, if human behaviour is not the result of mathematical calculation – and it isn't – then in principle, economists will *never* get it right.

#5. A fundamentally incorrect view of economic élan vital: the economist sees economic activity as a function of infinite "money creation", rather than a function of finite "energy stocks" and finite "energy flows". In fact, the economy is 100% dependent on available energy – it always has been, and it always will be. See a synopsis of the current energy situation at http://dieoff.com/synopsis.htm.

What possible role could a discipline as totally screwed-up as neoclassical economics (essentially nothing more than ancient Catholic religious dogma. See: http://dieoff.com/page243.htm) play in our society? It's obvious when one thinks about it... The role of the economist is simply to "prevent the American public from achieving a correct understanding of the actual workings of the American economic system." See Robert H. Nelson in http://dieoff.com/page235.htm: "The members of the American economics profession, as Arnold contended, performed a vital practical role in maintaining this unique system of corporate socialism – American style. It was their role to prevent the American public from achieving a correct understanding of the actual workings of the American economic system. Economists instead were assigned the task to dispense priestly blessings that would allow business to operate independent of damaging political manipulation. They accomplished this task by means of their message of 'laissez faire religion, based on a conception of a society composed of competing individuals.' However false as a description of the actual U.S. economy, this vision in the mind of the American public was in practice 'transferred automatically to industrial organizations with nation-wide power and dictatorial forms of government.' Even though the arguments of economists were misleading and largely fictional, the practical – and beneficial – result of their deception was to throw a 'mantle of protection… over corporate government' from various forms of outside interference. Admittedly, as the economic 'symbolism got farther and farther from reality, it required more and more ceremony to keep it up.' But as long as this arrangement worked and there could be maintained 'the little pictures in the back of the head of the ordinary man,' the effect was salutary – 'the great [corporate] organization was secure in its freedom and independence.' It was this very freedom and independence of business professionals to pursue the correct scientific answer – the efficient answer – on which the economic progress of the United States depended."

77. Not all economics discredited

We have already speculated that the current world recession was triggered last year when oil demand was hitting the ceiling of supply capacity due to depletion. Demand fell, putting it into better balance with supply. That in turn reduced pressure on prices, which have since collapsed. It demonstrates yet again what an unsatisfactory volatile market it is, over-reacting to small surpluses and shortages. Confirmation for the relationship between recession and oil price shocks comes from the newsletter of the prestigious Argonne National Laboratory, reporting on a study.

It points out that although money supply is normally blamed for recession, the transport sector represents 10% of the economy. High oil prices prompt a fall in the sale of automobiles and products, as well as increasing the cost of transport affecting the entire economic system. The model has apparently demonstrated a close relationship, after a time lag, between all previous recessions and oil price. No other relationship as close has been identified. (see www.anl.gov/OPA/logos19-1/econ01.htm)

If this is true, the logic is that any attempt at economic recovery will be

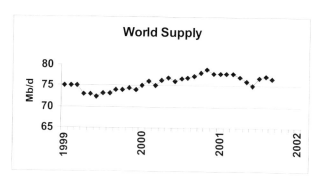

Figure 19. World Oil Supply reported by *World Oil.* Was global peak reached in November 2000?

frustrated in a vicious circle, because it would lead to an increase in oil demand, which would soon again reach the limit of supply capacity, causing prices to soar and re-impose recession. We have already speculated that we may live to discover that global peak production actually occurred in November last year as a combination of falling demand from recession and progressive depletion makes it increasingly difficult for production to ever again stagger up to last year's level. If so, it would have come only a few years before the midpoint of depletion, which is a normal pattern demonstrated in several mature countries.

78. The Economist is grasping the truth

The Economist, once the organ of the flat-earth heresy, now sees the oil situation clearly. In another important article *A Dangerous Addiction*, published on December 15th, it concludes that the United States is indeed dependent on Middle East oil as neither technology nor investment can materially increase production from other sources. It tacitly admits to the failure of the open market when it recognises that only higher taxes on consumption can reduce this dependence. It points out that US military expenditure to safeguard Middle East supply is running at $30–60 billion a year, a charge that can only rise as a consequence of the Afghan War.

79. Dutch Government Sued over Depletion

Although we do not have details, it is understood that the Dutch Government has been sued by a youth group claiming that its Government has been irresponsible in allowing the depletion of the country's gas without making adequate provision for energy-saving and substitution to replace it when it is gone. Evidently, this group accepts the strength of the adage: *"we don't inherit from our parents, but steal from our children"*

80. Report from Australia

Dr Brian Fleay reports on developments in Australia. Immigration is always a sensitive issue, and particularly so in Australia which adjoins a very populous part of the world. The government has commissioned a new study on the level of population that is sustainable in a country where indigenous oil supply is now at peak and set to decline. The issue of oil depletion is also being covered in a new two-year policy review being undertaken by the Premier's office.

Although large, much of the Australian continent is a stressed and sensitive natural environment with limited soil fertility, adverse climate and water shortage. It is both natural and encouraging that the country seems to be taking a leading role in planning for a sustainable, post-materialistic future.

It is also noteworthy that Les Magoon, the maverick member of the US Geological Survey, who warns of an imminent peak of oil production, has made an extensive lecture tour of Australia, briefing Ministers, who in turn have alerted the Premier. Given that another branch of the USGS has put out such a thoroughly flawed study, claiming an abundance of oil far into the future, it is perhaps surprising that he has not been subject to the attentions of the CIA. It is especially curious because the USGS seems to go to exceptional lengths to publicise its flawed report, sending emissaries to conferences all over the world and infiltrating the International Energy Agency.

It is encouraging that the CSIRO, the government agency responsible for scientific and industrial research in Australia, puts out a website drawing attention to the impact of oil depletion: http://www.bml.csiro.au.bigrol.htm

81. Excellent report in Le Monde of France

On December 6th, the prestigious Le Monde Newspaper of France carried an excellent article explaining the approach of the post-petroleum age. The article contrasts the bland words of the oil companies and the IEA with the warnings given by other experts. On balance, the article leans towards a looming crisis.

82. The Nemesis Report

The study of oil depletion is such a sensitive issue that knowledgeable experts in industry and government find themselves at risk if they should discuss it openly. The Nemesis Report is therefore to be widened to include non-attributed commentary from these muzzled sources.

Winners and Losers

The Third World War, which began on September 11 with the tragic terrorist attacks on New York and Washington, will be a protracted, bloody and cruel war. It will engulf the Middle East in a shattering cyclone of violence and destruction. But, there may be a positive side insofar as war can accelerate the resolution of deeper long-term tensions.

The victors and losers can, already, be clearly seen. On the victorious side will be the so-called Allies, with the US taking the lion's share of the spoils of war in the Middle East. Britain will play second fiddle in the Middle East despite having had 400 years of experience in the region; and Russia will have to be content with a back-seat role. She will get more than she deserves, but much less than she had hoped for – having been less useful to her allies than she had been in World War II. All seven 'Stans' – from northern Kazakhstan to southern Pakistan – stand to be substantial winners, albeit relative to their starting positions in 2001. Afghanistan will embrace peace after a quarter of a century of warfare, and Pakistan will reap the fruits of its timely co-operation. The central Asian 'Stans' will finally come in from the cold of the old Soviet Empire. Also on the winning' side will be the major oil and gas companies, standing to benefit proportionally to their size, when called in to sign lucrative Production Sharing Agreements. Finally, the very poorest among Middle East groups and individuals will emerge as winners, getting the bare minimum as the Allies launch a 'Marshall Plan', partly financed by the region's crude oil exports.

Both the Israelis and the Palestinians could end up as winners or losers. Their intricate problem is the most convoluted Gordian Knot in the world: and any solution, however limited, could bring a degree of stability to the whole of the Middle East. But the prospects are not good, given that half a century of warfare and negotiations has so far utterly failed to reach even square one.

On the losers' side will be all Middle Eastern regimes supporting (openly or covertly) terrorist activities – justly reaping the seeds they have sown. Another casualty might be the 'Euro' whose launch has come at the worst possible time; further adding to the woes of the European Union apart from Great Britain who will, however, have to sheath her bid for superpower status.

Also among the losers will be the main oil importers in the Asia/ Pacific region, once again denied control of petroleum reserves – with the behemoths India and China topping the list. But neither of these powers could have afforded not to climb aboard the anti-terrorism train whatever their apprehensions over the consequences – and however unpalatable these might prove to be, especially for

China,. It will now have to bide its time and wait for the end of the present decade to make its gambit for control of oil reserves (at a time when its oil imports have eventually surpassed its domestic output and its muscles are more ready to be flexed). But this will fall outside the scope of the present war; maybe to become the stuff for a Fourth World War – bearing in mind Walter Lippmann's prophetic words "China on the loose…". Did anyone say there was an "End to History"?

Of Math and Seesaw

Most, if not all, of you will have played on seesaws as children – a rather overrated experience as there is little fun to be had unless the two children are of very similar weights Quite small imbalances produce skyward hikes and spine-jangling crashes. It is a bit like crude prices in the oil industry. However, it is not prices I wish to discuss but production decline.

Of the world's 800–900,000 producing wells, all are experiencing flows that are either rising, roughly constant, or declining. In fact the overwhelming majority are in the static or declining categories. For the present, the declining flows are offset by the increasing flows. However, just like the seesaw this can change very rapidly indeed. Overall decline occurs when the loss of volume from declining wells exceeds the gains from the wells with increasing flows.

To labour the point a little, there can be significant new discovery and development but total output can still be falling. If your boat is shipping water faster than you can bail it out you will sink.

So now to some simple math. At the moment the world is producing around 75 million barrels/day of liquids (oil plus condensates). Typical onshore decline rates are 3–5% while offshore they are double that level at 6–10%. The state of each element in total production is difficult to assess but an intelligent guess would be that around one-third of production (25 million barrels/day) is experiencing decline. Onshore the US (lower 48 +Alaska), much of western Canada, most production in continental Europe plus a lot of older Russian and Chinese production, Syria, Colombia and Egypt fall into this category. The next 25 million barrels/day are in areas struggling to maintain production expanding some years and contracting in others. Examples here would be Argentina, most of the remaining onshore production in Asia, Europe and America with the North Sea now in this category. The final 25 million barrels/day come from areas of expanding production essentially the new offshore areas plus some but not all Middle East production plus a few new fields scattered round the world.

Now to hold production steady the expanding sector has to match the declining sector. As virtually all the decline is currently onshore getting it to add up is quite easy. The 5% decline is offset by a 5% gain in the expanding sector. If we have a little growth of say 1–2% this can easily be met with growth of 5–6% from the expanding sector.

Now suppose we have no growth at all but all the production that had been holding steady has now slipped into the declining sector. On Colin Campbell's analysis this could be the case as early as 2004/5. Now we have 50 million barrels/day declining at 5% and just 25 million barrels day of capacity with expansion potential. Just to hold output steady this now has to expand at 10%. This means long before we run out of new fields or even expansion capacity the problem is becoming harder and harder as the remaining expandable fields are called on to produce faster and faster. This of course means more and more of them become depleted and drop into the declining output category, which in turn means that those that are left have to increase output even faster still.

Now if demand growth returns to the historic 2%/year then 2000's 75 million barrels/day becomes 90 million barrels/day by 2010. Suppose 60 million barrels/day was declining by 7%/year (to take account of the faster offshore decline rates) then the expanding 30 million barrels/day would have to expand by a demanding 14%/year. If, as is more likely, only 15 million barrels/day had expansion potential then it would have expand by a fairly ludicrous 30.33%/year (65 divided by 15 times 7%).

The conclusion is simple. At some point between now and 2010 the required expansion of production from fields with expansion potential will be so great as to be effectively impossible. Decline becomes inevitable and like the boat that can't be bailed out rapidly enough nothing can stop it sinking.

83. Discovery in 2001

As the year comes to an end, it is difficult to think of any particularly significant discoveries during 2001. Some large deepwater finds were made on the West African margin in well-established provinces, but otherwise there has been nothing particularly noteworthy. No important new provinces have been identified. As a tentative guess, perhaps total discovery will amount to around 10 Gb/a with about half coming from deepwater areas, here defined as *Non-Conventional*. If so, the long-term downward trend will have resumed after the spikes of last year. Generally modest gas finds continue to be made in many areas. There are some indications that Siberia may have more potential for oil and gas than so far appreciated, but in terms of timing, its main impact is likely to be in ameliorating the post-peak decline rather than delaying peak itself.

Newsletter No. 13, January 2002

84. Some Comments on Reserve Definition

The Norwegians have a love of honesty and openness, and so it is not surprising that the Norwegian Petroleum Directorate has one of the world's best reserve reporting procedures, which is open to the public on its website. It recognises as many as eight classes of reserves, based on different stages of development, giving detailed data by field. It rightly accepts a degree of uncertainty, providing Low, High and Best estimates over a comparatively narrow range.

85. The Practice and Consequences of Reserve Reporting

Explorers map the size of a prospect with the help of sophisticated seismic surveys but they have to estimate the reservoir characteristics, including the net thickness, porosity, saturation, fill and recovery, based on data from neighbouring wells and regional knowledge. They normally make a range of estimates for each parameter and then compute the alternative combinations with a technique known as Monte-Carlo simulation. The result is a plot showing estimated reserves against subjective probability rankings, normally recognising a Low Case, having 95% chance of being greater than the indicated value, known as P_{95}; a High Case, termed P_5; with *Mode*, *Mean* and *Median* values in between. The explorers aim to make a "best estimate", which equates in probability terms with a *Mean* estimate, although in the real world they are often under pressure to exaggerate. The trick of the trade is to raise the extremely uncertain P_5 value, which then lifts the *Mean* value.

If drilling on the prospect is successful, attention turns to designing an optimal initial development plan to drain the main part of the field with a given number of wells. The expected production from such wells is reported as *Proved Reserves*, it being normal to quote conservative estimates. As production experience is gained, the reserves can be reported with more confidence, leading normally to an upward revision. Later, when draining the prime part of the field is far advanced, attention turns to tapping subsidiary reservoirs and additional compartments that were not originally produced, leading to further upward revisions. These late-stage secondary operations are much influenced by economic considerations, more being done in more profitable situations.

Proved Reserves are commonly defined as "those quantities which geological and engineering information indicates with reasonable certainty can be recovered from known reservoirs under existing economic and operating conditions". The key words are *operating conditions*: meaning in practice the currently producing (or planned) wells at any given time, not the field as a whole. They are commonly equated with a certain Probability ranking, (P_{90-95}), but it does not seem to be a sound practice because the term *Proved Reserves* refers to a particular stage of development whereas the probability ranking refers to the field as a whole. The probability ranking of any particular phase is not the same as that for the field as a whole. A large field will have several phases of development, with the consequential growth of *Proved Reserves*, whereas a small field with a short life may have only one. A cardinal mistake is to apply the reserve growth of the large fields of the past to the smaller more recent ones.

Proved Reserves are reported for financial purposes and enter the public domain, whereas *Proved & Probable*, which do relate to the field as a whole, are commonly not reported. This simple matter is at the heart of much of the confusion surrounding the subject.

86. The Treatment of Condensate

There is no standard agreement on the definition of *Conventional Oil*, meaning that each study must state how it applies the term. The writer has so far excluded *Condensate* on the grounds that its depletion is driven primarily by gas production and not oil production, but this may have been a mistake. Thinking further about the issue, we may note that an oilfield commonly contains hydrocarbons in both gaseous and liquid phases in proportions partly depending on ambient pressure and temperature, which can change over its life. The liquid, known as *Condensate*, condenses from the gas on being brought to the surface, where it is separated in order to meet the sales specifications of the gas, being commonly then fed into the oil stream, with which it is often metered.

During the early phases of development, efforts are made to maximise the production of oil, in part using the expanding gas cap as a drive mechanism to expel the oil. The production of *Condensate* is accordingly low. But later, when most of the oil has been extracted, more and more gas is produced as the pressure falls, eventually converting the field effectively into a gasfield, with an increasing yield of *Condensate*. In some cases, *Condensate* may even have developed in the reservoir itself in cases where it has been uplifted by earth movements into a lower pressure environment. The *Condensate* yield of gas also ranges widely, depending on geological circumstances related to source-rock and the maturity of generation. Norwegian practice is to meter *Condensate* with oil where it is fed into the oil pipeline system, which seems a sensible procedure. It is distinguished from Natural Gas Liquid (NGL) which is extracted by processing and metered separately. In future, we will follow the Norwegian practice in principle, although in global studies the available data are normally far too unreliable to attempt such fine distinctions. In short, Condensate from oilfields will, in principle, be included with oil; whereas NGL from gasfields will not.

87. New Study of Giant Fields

Matt. Simmons has made an excellent new study of the world's giant fields. Whereas internationally, it is normal to define a giant field as one having more than 500 Mb of ultimate recovery, Simmons prefers to use production alone, defining a giant field as one producing more than 100 000 b/d. He was shocked to find how difficult it was to obtain accurate information on such fields, especially in view of their critical importance to world supply. He identified 120 such fields, which provide 47% of the world's production, with the largest fourteen contributing as much as 20%. He further notes that 36 of these fields, which were found more than forty years ago, contribute as much as 21% whereas the 12 found over the past decade provide only 2%.

The world's dependence on these ageing giants at an advanced state of depletion is self-evident, and should be a cause of great concern.

88. Good Scenario of Supply by TotalFinaElf

P.R. Bauquis of this French oil group has published an excellent pragmatic review entitled "A Reappraisal of Supply and Demand in 2050" (Oil & Gas Science & Technology, Rev. IFP 56-4 389-402). It takes note of falling fertility rates, suggesting that the world population may not exceed 8 G by 2050, which will have an impact on energy demand. It pragmatically assumes that the good intent of Kyoto will in practice have little impact on climate change, so that we will have to live with the consequences whatever they may turn out to be. It notes that estimates of ultimate oil recovery have not changed much over thirty years being in the range of 2 to 3 trillion barrels, depending partly on whether *Non-conventional* liquid hydrocarbons, variously defined, are included. It suggests that oil production is likely to fall from around 2010 due to depletion, with a plateau of gas production following not long afterwards. It draws the obvious conclusion that there will have to be a radical change in the world's energy mix, speculating in an increase in nuclear power, possibly as a source of hydrogen fuel.

It is remarkable to find an oil company providing such a sound analysis, when so many are bent on denial and obfuscation, either claiming not to forecast the future or obscuring the situation with elegant multiple scenarios of minimal real meaning.

Certainly, it is in stark contrast with a recent paper by ExxonMobil, which is the latest to lumber in with the full spectrum of public relations platitude about market economics, technological impacts, and the need for government to let the industry run its own affairs. Any idea of government to government supply arrangements is roundly condemned. One searches in vain for valid data about the past trend of discovery, which could give an indication of future discovery, although careful reading does reveal that UK imports are set to rise. Could that possibly be an admission that indigenous production has peaked, as a very oblique reference to depletion? It is a sad commentary on the calibre of advice being given to the UK government by the oil industry. A similar document has been provided to the EU.

http://www.cabinet-office.gov.uk/innovation/2001/energy/submissions/ExxonMobil.pdf

89. Review by Jean Laherrère of K. S. Deffeyes's book *Hubbert's Peak*

Jean Laherrère has produced an excellent review of this important book, with a full suite of very telling graphs and a discussion amplifying the analysis. A few examples are reproduced to illustrate his work

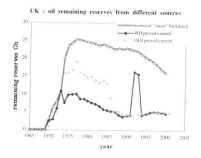

Figure 20

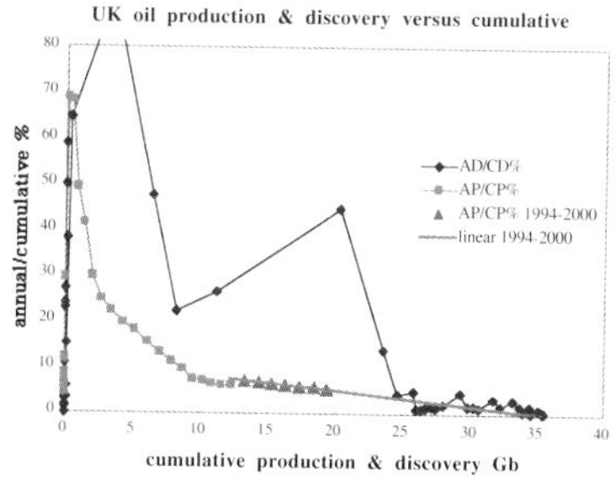

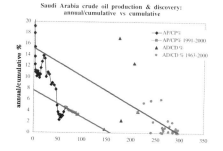

Figure 21

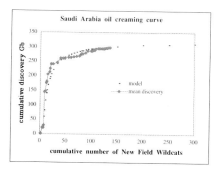

Figure 22

Figure 23

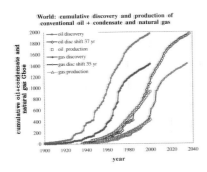

Figure 24

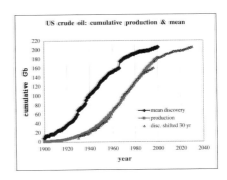

Figure 25

90. Depletion Statement by the BGR

The Biennial Report of the German Federal Institute for Geosciences and Natural Resources for 1999–2000 carries a telling statement on the risks of oil depletion

The End of the Petroleum Age and the Search for New Energy Sources

Energy drives our lives and societies. On the way to work, sitting in front of the TV, in the shower, we are consuming energy. Energy is such a natural part of our lives that we usually take it for granted. We seldom think about what it takes for us to have 10 litres of petrol to drive 100 km. That is 10 L. of a valuable mixture of hydrocarbons formed at several kilometres depth from the remains of living organisms deposited together with inorganic material millions of years ago. The formation of crude oil, from which gasoline is made, was an exceptional occurrence, and the accumulations of oil and gas in the Earth's crust are limited.

The rapid developments of the 20th Century were made possible by the apparently limitless availability of energy, mainly in the form of crude oil. Crude oil guaranteed economic growth, mobility and prosperity. Oil has also led to political conflicts. About 40% of the global energy demand is supplied by crude oil. The remaining oil reserves are not evenly distributed among the countries of the world. Whereas most industrial countries, including Germany and the USA, have passed the apex of their crude oil production, the OPEC countries can increase production. OPEC has more than two-thirds of the known global oil reserves.

Viewed globally we are approaching the time when half of the conventional – i.e. the easily obtainable oil reserves – have been used up. The estimates for this time range from a few years to 30 years. From this time on, oil production will decline and the price of oil will increase. The decline will coincide with a continually increasing global demand resulting form a two-fold cause: global population growth and increased demand from the developing countries. Thus, it can be foreseen that there will be insufficient supplies in the near and medium term future that could shake the foundations of our global economy.

If a situation in which supplies cannot meet demand is to be prevented, unconventional oil occurrences must be developed and other energy sources must be found and developed. Unconventional oil occurrences include occurrences that would be difficult to develop with today's technologies. Advancing technological developments may be expected to make it possible to produce from such deposits. For example it is becoming technologically possible to produce from continually increased water depths in the oceans. Whereas oil production was previously concentrated on the continents and shallow marine water depths, new target areas are being developed on the continental margins. With this in mind, BGR has been conducting studies of the hydrocarbon potential in deep water areas – eg, in the South Atlantic off Argentina, Namibia and South Africa. Such studies are conducted in the long term "fore-field" of industrial development. A major aspect of efforts to find a way to avoid the predicted scarcity of oil must always be environmental compatibility.

Newsletter No. 14, February 2002

91. The Coming Decline of Oil by Gerald Leach
Mr Leach, a senior research fellow at the Stockholm Environment Institute, has published an important article in Tiempo (Issue 42 of December 2001) in which he succinctly explains the peak and decline of oil, drawing on the data provided by the oilcrisis.com website. It opens with:

> *"Amongst the billions of words brought forth by the climate debate over the past years, remarkably few have touched on an issue that ticks behind it like an unexploded time bomb. This is the probability that world oil production will reach a peak sometime during this decade and then start to fall, never to rise again".*

It goes on to explain the colossal impacts of such a discontinuity in economic, political and environmental terms. Readers of this newsletter will find that many of the words, thoughts and expressions have an extraordinarily familiar ring to them.

92. Shell Results

The London Times of February 8th commented on the financial results of the Shell group of companies, which recorded a 71% fall in fourth quarter profits and a 74% fall in *Reserve Replacement* during 2001. The analysts were concerned that Shell was not spending enough on exploration, fearing that output could not be maintained.

When companies report *Reserve Replacement*, which impresses the analysts, they naturally refer to *Proved Reserves*, which represent what the wells in the current phase of development are estimated to eventually deliver, and does not necessarily reflect what the field as a whole is expected to provide. *Reserve Replacement* therefore has comprised both what has been found in new discoveries during the financial year and what is taken from the inventory of under-reported past discovery. It looks as if Shell has depleted its inventory from the past, and now has to rely on new discovery alone. This makes sense because its old fields are now so old that the reported *Proved Reserves* probably do reflect the total as there is no scope left for further development, and the new fields are too small and short-lived to support more than a single initial development. The Company probably is spending as much as it can on the viable exploration opportunities available to it, but sees no reason to drill dry holes to please flat-earth analysts, who are oblivious of the natural resource constraints and fail to grasp that the company is running out of opportunities.

Instead of inventing scenarios of abundance and enduring plenty, the corporate image-makers may soon find it expedient to explain their actions in terms of the truth. The analysts too may come to understand the position, as seemingly Goldman Sachs already did in their famous statement of 1999, which is worth repeating in this context.

> *"The rig count over the last 12 years has reached bottom. This is not because of low oil price. The oil companies are not going to keep rigs employed to drill dry holes.* They know it but are unable and unwilling to admit it. *The great merger mania is nothing more than a scaling down of a dying industry in recognition of the fact that 90% of global conventional oil has already been found."*

Goldman Sachs, August 1999

93. Call on Middle East Oil

Under the 2001 Update scenario of flat global demand and weak prices, Middle East Gulf share drops slightly for a year or two, before rising to a maximum of 24 Mb/d by 2010, when it supplies 40% of the world's oil. The model assumes that each country (except Iraq) produces its current share of

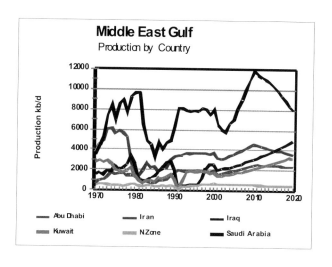

Figure 26

the demand on the region, until it reaches its midpoint, with the balance being provided by Saudi Arabia. It is also assumed that no country can achieve an annual increase in excess of 5%. It is thought that Iraq will gradually increase its production. It is uncertain that the heavy demands of the model on Saudi Arabia to increase from 2005 onwards can in fact be met. Kuwait may also be pressed although its northern fields may contribute enough. Looking at the figures, trends and relationships, especially the low depletion rates, gives the impression that the estimated reserves may still be too high.

This model is naturally tempered by uncertainty about what Russia can and will export, which needs further study. Higher Russian exports would depress the call on the Middle East.

94. Report by LBST on European Union Green Paper

Dr Zittel and his colleagues at LB Systemtechnik of Munchen have issued an excellent report, commenting on the EU Green Paper "Towards an EU Strategy on the Security of Energy Supply". It divides the world into three groups: countries close to peak; countries after peak, which are set to continue to decline; and countries not yet at peak, analysing the contributions that each can make. It supports the argument with some telling graphs of individual field profiles for Norway and the United Kingdom, also addressing the US situation. In fact, it seems to suggest that the EU's quest for security of energy supply is something of a vain hope, at least in terms of fossil fuels.

95. Mbendi Website

A statement of the current state of oil depletion, and the factors affecting it, has been published on a prominent African website: www.mbendi.co.za/indy/oilg/p0070.

96. The UK Cabinet Office remains in the dark

Despite having received representations from ODAC, the final energy report of the UK Policy Initiative Unit (PIU) makes depressing flat-earth reading. http://www.piu.gov.uk/2002/energy/summary.shtml

The authors evidently succumbed to Establishment pressures to provide a misleading impression of an abundant supply of oil and gas without a cloud on the horizon. The closest the report could come to recognising the peak and decline of UK oil and gas production was an oblique admission that imports are set to rise, and that supply lines will lengthen, carrying geopolitical risks. Missing the essence of the matter, the best it could do was to offer the pathetic suggestion that we should at least try to be friendly to these distant suppliers.

97. Exploration Highlights

We await with interest the results of discovery for 2001. There was a flurry of comment about some large alleged discoveries in Tibet, of which little has since been heard. Otherwise, the picture seems to have been dominated by two rather unexpected late stage discoveries in mature areas in the UK and Trinidad, which together probably add not more than 500 Mb. Shell also found an oil accumulation testing 8000 b/d beneath the Malampaya gasfield in the Philippines. It sounds as if we return to the underlying falling trend of finding about one barrel for every four that we consume, after the spikes

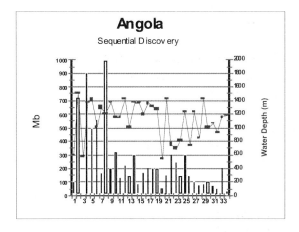

Figure 27

of last year and the spate of deepwater finds as the prime prospects were tested during the early phases of exploration. The plot of declining sequential discovery in deepwater Angola certainly confirms this picture.

98. When Recession become Depression

Peter Bernstein, a premier Wall Street analyst, has written a book "*Against the Gods – the remarkable story of risk*". It contains a chapter about discontinuity. It seems that the general advice to investors has been to try to track the Mean over a long period. This is seemingly confirmed by statistics showing that, although individual sectors gave differing results, the average performance of all mutual funds has been constant over long periods.

Business activity had fallen in only seven years between 1869 and 1929, and so it was understandable that President Hoover dismissed the downturn that heralded the Great Depression as a temporary event with the words "Prosperity is just around the corner". But GDP fell by 55% between the 1929 peak and the low of June 1932. A second more subtle discontinuity came in 1959 in the relative performance of stocks and bonds. The ratio of bond interest to bond price soared whereas the ratio of dividend to share price collapsed. It took the economic stimulus of the Second World War to end the Great Depression, and it has been suggested that US plans to invade Iraq have the same objective, (see Saddle Up – 20.02.02.-ljmweir@hotmail.com)

Bernstein points out that "when discontinuities threaten, it is perilous to base decisions on established trends that have always seemed to make perfect sense but suddenly do not". He mentions climate change as a possible looming discontinuity. But the present recession, combined with the amazing changes in US foreign policy, which stem directly and indirectly from the supply of oil, may indeed prove to have marked the onset of an unparalleled discontinuity.

Newsletter No. 15, March 2002

99. Exploration Highlights

United Kingdom

J.Munns of the UK Department of Trade and Industry has given a lecture in Aberdeen explaining that the Government is desperate to attract new small companies following the mergers and general loss of interest by the major companies. He admitted that oil and gas production peaked in 2000 and is set to decline. He commented that 4000 exploration and appraisal wells had found 285 fields, adding that the recent Buzzard Field with 300–400 Mb was the largest find since 1993. He quoted the official Brown Book publication, which claims that between 4 and 27 Gboe of oil and gas combined are yet-to-find. This no doubt is another example of the misleading Probability Approach, as used by the USGS, based on the subjective ranking of prospects, whereby the upper end of the range is assessed to have no more than a 5% probability. But at the same time he revealed in a slide that the range for oil is only 1.5 – 9.8 Gb (excluding the unknown and dubious Atlantic margin). With no more than 2.4 Gb having been found over the past ten years, falling to 0.85 Gb over the past five years, it sounds that our own estimate of 2 Gb is not wide of the mark.

It is evident that the United Kingdom is having the greatest difficulty in accepting that its oil production is in terminal decline, leading to a loss of revenue and soaring imports. The hyperactivity induced by the policies of Mrs Thatcher squandered the national resource in two short decades at times of low price, which it helped induce. Now, the government is forced to trawl the back streets of Houston and Calgary looking for anyone willing gamble on a late stage surprise. There may be some but not many.

Greenland

Phillips Petroleum and Statoil (the Norwegian State Company) have withdrawn from West Greenland after abortive drilling. We don't know why they preferred West to East Greenland, where the USGS claims there is 47 Gb to be found. Since the government has now decided to reduce tax to 30% as an incentive, it does not appear that the industry exactly shares the USGS view.

Australia

A 300 Mb discovery, named Enfield, has been made off Western Australia in what is claimed to be a new province.

Caspian

In April, the countries bordering the Caspian are to meet to try to decide how to divide the offshore. With the small matter of ownership unresolved it is hardly surprising that BP, Statoil and Exxon-Mobil have withdrawn from ventures there. China meanwhile is buying into Azeri fields, with the help of funding from the European Bank of Reconstruction and Development, speaking of its desperate need to find foreign oil to offset decline at home.

Iran

Iran is inviting tenders to develop as much as 1.7 Gb in a shallow reservoir above the giant South Pars gasfield.

Iraq

Both Russian and Chinese companies have signed recent deals to drill as many as 50 wells in Iraq apparently with the agreement of the UN embargo committee, adding a further element to the consequences of the threatened US invasion.

100. Three Very Interesting Books

Three very interesting and important books have appeared. The first is entitled *Not by Money Alone – Economics as Nature Intended* by M. Slessor & J. King (ISBN 1-897766 72 6), which is a marvellous counter to flat-earth economics, explaining the critical role of energy and the corresponding impact of its depletion. It proposes a radical shift in taxation from labour (income and sales taxes) to energy. The second is the *Death of the West* by Patrick Buchanan, the former American presidential candidate, who discusses the decline in western population and the consequences of massive immigration in times of economic decline. It is a remarkably erudite, courageous and outspoken account, far from the bland platitude so beloved of politicians. It is clearly relevant when considering scenarios of oil demand and the political and social reactions to the decline in supply. The third book is a racy novel *The Adjustment* by C.E.MacArthur set in aftermath of the ultimate power failure in New York. Manuscripts of two other excellent books on oil depletion and its impact have been received, and will be covered when published. It is encouraging that the subject is now receiving new attention.

101. US Fears of Rising Oil Price

The headline says it all.

"US says economic recovery at risk without more OPEC oil"

WASHINGTON (Reuters) – OPEC must boost oil production during the second half of this year or risk slowing an economic recovery in the United States and the rest of the world, a U.S. government energy agency warned Wednesday. The Organization of Petroleum Exporting Countries' current production quota – 21.7 million barrels per day – stands at its lowest level since March 1991, and OPEC Secretary General Ali Rodriguez has said he does not see the cartel raising output this year. "If this quota is observed, OPEC efforts to boost (oil) prices could result in prices well exceeding their target, just as the U.S. and world economies recover," the Energy Information Administration said in its monthly update on the cartel. Because OPEC quotas are very low, world oil markets "will tighten rapidly" unless OPEC members, excluding Iraq, agree to boost daily production in the latter half of the year by one million barrels, the Energy Department's analytical agency said...

"OPEC could be faced with a repeat of 2000, when OPEC misjudged the market by over-correcting with tight quotas for too long, except that this time world economies are much weaker," EIA said. "If OPEC production does not increase as expected, prices could rise further and dampen the expected U.S. and world economic recoveries," the agency said.

102. Australia admits to depletion

John Akehurst, the Managing Director of Woodside, a major Australian oil company, points out the dire position in which Australia finds itself. Indigenous oil production is set to fall from 724 kb/d to 200–300 kb/d by 2010, by which time it will meet only 40% of demand. The present surplus of $1.2 billion is set to swing into a deficit of $7.6 billion. He comments that this should set the alarm bells ringing in the Treasury and in other arms of government concerned with managing the economy. He notes that Australia has been consuming three times more than it has been finding over the past seven years.

It is truly remarkable to find the Chief Executive of an oil company make such a frank and honest statement in the national interest. Would that his counterparts in other parts of the World were as honest, for what he says is the unadulterated truth, and the national predicament of Australia is by no means unique. Britain too has passed peak, not that anyone cares to admit it.

103. The Unconstrained World

The plot in figure 28 is taken from another beautiful set of graphs by Jean Laherrère, illustrating the robust pattern of discovery and production in the

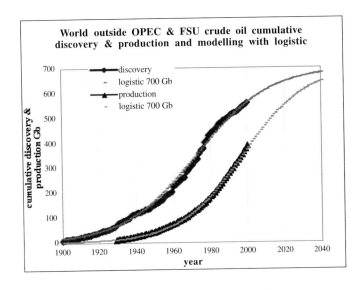

Figure 28

World outside OPEC and the FSU, which were artificially constrained by government policy. All the plots point to about 700 Gb for all liquids. In the current database of *Conventional*, circulated to ASPO members, the total comes to 612 Gb, but adding the *Non-conventional* brings it very close, which is welcome confirmation. All will have been produced when the production curve hits the discovery curve, and that will come around 2050 unless consumption can be reined in radically.

104. Controversy with the Flat-Earth fraternity

The energyresources website has contained a vigorous debate in which a Mr Lynch, who is a prominent member of the flat-earth heresy, has attacked the depletion models proposed by C. J. Campbell. It attracted the following response, not to try to convert Mr Lynch, but to provide agnostics with explanations.

Misunderstandings

1. Defining what to measure

It is absolutely true that the definitions of what to measure have evolved as this study has progressed. In the early days when working with public data I took a very generalist view of the matter and, as Mr Lynch sometimes likes to point out, got quite a few things wrong. No doubt I will continue to do so although hopefully also making progress. The first step in the analysis I have tried to make was to determine the endowment of oil and gas in Nature. It soon became evident that different types of oil depleted in different ways – obviously a tar-sand is different from a Middle East flowing well. Furthermore, knowledge of some types is better than for others. The terms *Conventional* and *Non-Conventional* have no fixed definitions, but nevertheless are widely used to roughly distinguish the easy from the difficult. Perhaps mistakenly, I used these terms in a particular sense which I defined very clearly so that everyone could understand the meaning attached to them. I would add a few words of explanation about each of the categories that I treat as *Non-Conventional*

1. Oil from coal and shales (actually immature source rocks) – no particular comment

2. Bitumen – principally the tar-sands of Canada, defined by viscosity, from which synthetic oil is made

3. Extra-Heavy Oil – defined by density <10 API and mainly in Venezuela and Canada

4. Heavy Oil. This is tiresome. Canada has a cutoff at 25 API whereas Venezuela uses 22 API. I started out with a rounded 20 API. but later moved it down to an arbitrary 17.5 because there are quite a lot of fields producing happily with gravities just below 20. Any cutoff is arbitrary,

but it does seem useful to distinguish heavy oils for two reasons a) production generally lasts a long time but reaches only a low peak and b) the recovery factor is low, meaning that there is particular scope for the application of enhanced recovery methods

5. Deepwater Oil – Again the cutoff is arbitrary but I use 500m water depth. I think it is useful to distinguish this category because a) the geology is very different for most of it, relying on special plate-tectonic settings and special reservoir circumstances; b) it is a hostile environment stretching technology and management to the limit, which in turn has an impact on what prospects can be handled; and c) it is less known, so there is more scope for surprises, even good ones.

6. Polar Oil – Yet again the cutoff is arbitrary at the polar circles, but it is useful to distinguish these provinces because a) Antarctica has poor geological prospects and is closed by agreement b) the Arctic has special geology making much of it gas prone, with the exception of Alaska and parts of Siberia d) it is a hostile environment and e) it is less well known.

7. Gas Liquids are another tiresome element. There are basically two types a) that which condenses naturally called *Condensate* and b) that which is extracted by processing. Previously I excluded both from *Conventional* oil on the grounds that they would deplete in relation to gas not oil. But I have changed my mind on that, recognising that an oilfield contains hydrocarbons in both liquid and gaseous phases in proportions that can change with depletion. So I now include condensate from the gas caps of oilfields with *Conventional* oil, recognising too that it is commonly metered with oil.

8. Others – there are a few other categories such as HTHP, high sulphur, unusual reservoirs etc that could also be distinguished but I haven't done so.

I assume that all feasible enhanced recovery practices will be applied, but more could be done to smoke out the details.

I find it useful to at least define what I try to measure, even if the data at my disposal does not always allow it to be done as thoroughly as it could and should be done. In practice the boundaries are fuzzy, but the total should be about right.

2. Forecasting Production

Clearly, each of the above categories can contribute differently to peak, but the contribution of all must be taken into account. It would be possible to do this more thoroughly with full access to the industry database, but I do attempt to show all production. I have spreadsheets for every country for *Conventional*, summed into Regional and World totals but add a separate global assessment for the other categories, sometimes showing all in a graph. I can easily supply anyone who is interested with the current breakdown. I won't go into all the reasons and details here, but can summarise the present assessment. I am still agonising over what Russia can supply, so there may well be revisions as new information or insight comes in.

1. Conventional oil production is flat at 60 Mb/d to 2010 when it begins its terminal decline at about 2%
2. Deepwater oil peaks in 2010 at 8 Mb/d
3. Polar oil declines to 0.6 by 2010
4 . Heavy oil etc rises to 3.6 Mb/d by 2010 and 4.6 Mb/d by 2020
5. NGL rises to about 9 Mb/d by 2010

    I also have gas rising to a plateau starting around 2015 at about 33 Tcf/a and provision for Non-Conventional gas, mainly Arctic gas and coalbed methane.

    In short, all liquids peak around 2010, meaning that total production need not fall below present levels for about 20 years.

3. Other Points

I do not diminish in any way the impact of technology and better geological knowledge and mapping techniques, but the study of most large fields shows a straight line decline once it sets in, lasting over many years, which implies that technology has had a negligible impact on the reserves. It evidently serves mainly to extract what is there at a higher rate. I also note that most of the known *Conventional* oil is in old giant fields, which are already efficiently exploited, so I don't think technology will have much impact on total reserves. If anything, it is likely to advance peak, by extracting the oil faster. It may however contribute more to the extraction of *Non-conventional* oil. No one can exclude the possibility of some miracle new technology but I don't make provision for it.

4. Reserve Reporting

I have abandoned the Probability ranking system, having seen the mess it got the USGS into. I conclude that *Proved Reserves* as reported for financial purposes refer mainly to what the wells at the current stage of development are expected to deliver, without necessarily saying much about what the field as a whole may provide over its life. In plain language, they are *Proved So Far*. I observe that the initially reported *Proved Reserves* of most large old fields understated what the field would ultimately yield by about one-third, simply because such fields were subject to successive phases of development, each of which added reserves. But in the case of the smaller more recent fields, reported *Proved* may indeed reflect the entire field, because there is not scope for more than the initial development scheme. It is therefore a mistake to apply the "Reserve Growth" of the past to the future. I failed to grasp the extent of this initial under-reporting in earlier studies, which explains why they underestimated both reserves and the potential for new discovery derived from the extrapolation of past discovery trends. This explains the valid criticism of Mr Lynch.

Proved Reserves, as reported in the public domain, have to be adjusted to remove any identifiable *Non-Conventional*, as herein defined, as well as any "political" component or simply the consequences of a failure to update (64 countries implausibly announced no change on 2001). The adjusted value has then to be multiplied by a factor to give a best estimate of what the fields, when fully developed, will eventually deliver.

5. Modelling Depletion

I assume that production in counties that have passed their midpoint of depletion will continue to decline at their current depletion rate, whereas it will rise to midpoint in those that have not yet got there. I also assume that the five major producers of the Middle East exercise a certain swing role around global peak making up the difference between world demand and what the other countries can supply under their depletion profiles, so modelled.

6. New discovery

It is evident that discovery of *Conventional* oil peaked in 1964, once the corrected and properly backdated reserves are applied. A smoothed trend has declined to approximately 6 Gb/y today although there have been occasional spikes, as occurred in 1999 and 2000 from two major discoveries in hitherto closed areas. Given that this has been the result of an industry diligently searching the world with the best technology and always deliberately testing the biggest and best prospects, it is hard to advance any evidence to suggest that it will improve in the future. The larger fields are normally found first for obvious reasons. There is scope for more discovery in the deepwater but it too is beginning to show the same eternal pattern of diminishing results. Likewise, few would dispute that the major fields of the Middle East have already been found, so stepping up exploration there would deliver results smaller by orders of magnitude.

7. Economic Consequences

I am not qualified in this domain but can appeal to common sense, suggesting

a) oil price rises when demand exceeds capacity or is perceived to be about to do so

b) high prices bring on recession, reducing demand, and thereby price

c) most cheap oil lies in the Middle East and Russia (the latter due to an adverse exchange rate)

d) *Conventional* oil costs less to produce than *Non-conventional* (as here defined)

e) tax distorts the picture in many ways, giving amongst other things a subsidy to Western companies for exploration, (many spending 10c dollars)

f) the high social costs of Middle East governments with growing populations bring serious political strains if prices are low, and also limit their ability to make major investments needed to expand capacity

8. Conclusion

Today, oil provides 40% of traded energy, and energy not money drives the economy. Production is set to start declining within about ten years. Since Hydrocarbon Man will be virtually extinct by the end of the Century, it might be a good idea to start planning how to use less and bring in such substitutes as can be found. Given the importance of the subject, it is surprising that more serious work is not done to resolve the matter. The obstacles are primarily political, tolerating ambiguous definitions and lax reporting practices, as there are no particular technical challenges in estimating the size of an oilfield or in assessing the potential for new discovery.

105. The Nemesis Report

Another contributor from the oil industry discusses the imminent peak of production in relation to falling discovery. He is obliged to remain anonymous because expression of these well-founded concerns would be an embarrassment to the corporate image-makers. He is at the same time somewhat anxious, lest revelations of the truth may cause panic in investment circles.

The Impact of Falling Discovery on Peak Production

It may seem strange to make such a request, but I earnestly beg you all to find errors in the following because, if you can, then we can all sleep a little easier in our beds.

As everyone is aware, politicians when confronted with awkward or difficult questions have a tried and tested response. They answer the question that hasn't been asked or they answer the question quite literally, deliberately evading the substance of what their interrogators are trying to ask. Now, in these tricks the world's politicians are mere amateurs when compared with the professional bureaucrats. This, I believe, is the explanation for all those reassuring reports and analyses from the IEA, EIA, USGS etc. But, before tackling them, let us look at a perfect contemporary example of the 'answering the question that wasn't being asked'.

There is growing concern in the UK about future North Sea oil and gas production, and questions are starting to be asked. However, because most people view oil supplies as analogous to a pot of honey, the question they always ask is 'When is it going to run out' while what they think they're asking is 'Can production levels be maintained'. Politicians, oil companies and bureaucrats cannot believe their luck and reply 'There's as much to produce, or more, than we've produced so far', which is a perfectly accurate but wholly misleading statement. The clear implication is that there's nothing to worry about. Even some of the analysts, who should know better, keep claiming that the decline to date (6% in 2000 and 12% in 2001) is a temporary phenomenon, an investment glitch, and a soon-to-be-reversed setback. They are plain wrong. We are past 50% depletion of the UKCS and sustained decline is now inevitable. Interestingly, a recent BP presentation in Aberdeen confirmed that the UK North Sea was past its peak, though rapidly adding that they (BP) would be able to maintain production at only a little under their peak for many years.

Turning to the global scene, careful reading of both the EIA and IEA reports establishes that they are *not saying* that particular production levels are *possible*, only that these are the *production levels necessary* to sustain a 'business as usual scenario'. It is actually quite easy to see how they generate their reports. They appear to put all the production numbers in a spreadsheet and extrapolate forward at the historic growth rate, save where they know an area is in decline, when they minimise any decline, or where they know of discovery, in which case they rev up the growth. If this isn't enough they prorate everything up to fit the required production level. Sometimes these methods produce quite amazing results. A couple of years ago, the IEA could only get to the required number (business as usual) by inventing large volumes of the quaintly named 'unidentified unconventional'. They clearly got severely reprimanded for hinting 'business as usual' might not be possible and have since reverted to earlier tried and tested number juggling techniques.

The news that the North Sea is in decline has not yet reached the EIA, who are still confidently extrapolating their spreadsheets forward to give a reassuringly impressive North Sea production of 7mn b/d in 2020. You can understand their problem. All that expanding North Sea production kept Opec in check and prices low. It is evidently not good politics to say the North Sea won't produce lots more.

Now, it is easy enough to make fun of bureaucrats but they do have a real problem. The politicians don't understand and don't want to face up to the reality. I'm not entirely convinced that even those who do understand the problem want to face up to the reality. I, myself, don't really want to face up to the devastating reality either. This is why I'm hoping someone will find great errors in what I'm saying.

Maths and Physics are clear and logical disciplines, with little room for fuzziness. I have tried to be as clear and logical as possible. I have drawn on three primary sources:

1. Colin Campbell's latest assessments as published in Newsletter 14
2. Jean Laherrère's 'Forecasting future production from past discovery'
3. Wolfgang Schollnberger's presentation to the IP Week conference (Feb 18, 2002) and available on the Institute of Petroleum's website at www.petroleum.co.uk This can be taken as the 'industry view' from a senior BP executive who is also the Chairman of the International Association of Oil and Gas Producers (OGP).

I have tried to stand back and look at the overall picture. Wonderful detailed work has been done in many areas but sometimes the big picture gets lost in the process. However, the conclusions I reach are so startling and, in a sense, so novel that the instinctive reaction is to deny that they are possible. It is difficult to believe. I have searched in vain for an error in the logic. Hopefully I have overlooked something.

In the first part, I am only considering liquids – oil and condensates. Later on, I will show that a key alternative, namely the oils that do not flow at ambient temperatures – heavy oil and tar sands – are not a viable 'solution' as the volume likely to be produced by the critical dates is far too small.

Returning to conventional liquids, the number we have most confidence in is Cumulative Production to date. Campbell gives 873 Gb for oil only (end 2001), whereas Schollnberger gives 900 Gb for all liquids (end 2000). *So we can say with confidence that the world had used a rounded 900 Gb of liquids by end 2000.*

We also have reasonable confidence in global demand, *which has been holding steady at around 75 million barrels/day or 27 Giga barrels/yr.*

Units are an endless source of confusion and may be one of the reasons why there is so little appreciation of the imminent crisis.

Measures of flow-rates in barrels/day are unhappily mixed up with measures of stock quoted in billion barrels (Gb). The easiest and most illuminating way to avoid the confusion is to work with stocks in Gb and production in Gb/year. Suddenly things become much clearer. For those trying to mislead, this makes things far too clear, which may explain the industry's enthusiasm for mixed units, mixing oil and gas by quoting reserves in barrels of oil equivalent, and other wilful obfuscations.

Perhaps most surprisingly of all, *there is an emerging consensus that the world's remaining reserves are around 1000 Gb*. Different sources arrive at this conclusion in different ways. The highly suspect public databases, as published by API, BP Statistical Review, World Oil, Oil and Gas Journal and Opec, all have reserves reaching a plateau at around the 1000 Gb mark at the end of a long rising trend. According to Jean Laherrère's graphs, backdated reserves from the industry database have also reached the same level but at the end of a 20-year decline from a 1200 Gb peak in 1980. Campbell's latest assessment gives remaining liquids reserves as 1077 Gb (1950–873). (We should probably add some condensates into the 873 Gb produced raising it to over 900 Gb, which would then give remaining reserves of around 1000 Gb).

Jean Laherrère's paper makes use of creaming curves (arguably the most accurate predictor of ultimate reserves) giving cumulative discovery of 1200 Gb outside the Middle East, 700 Gb in the Middle East and 50 Gb in deepwater. This totals 1950 Gb, which gives remaining liquids reserves of 1050 Gb after the removal of past production. We now have a truly remarkably narrow range. We can now say with some confidence that *the world's remaining liquid reserves are between 1000 and 1100 Gb.*

There remains only one area of significant disagreement – future discovery. Surprisingly, and quite counter-intuitively, this matters much less. The peaking of global oil production is likely to occur in the next 3 to 10 years. What is discovered in 2015 or 2040 may slow the rate of decline but it won't affect the peak. To illustrate with a homely example: if there is no bread and no flour available you will starve even though the farmer plans to grow a big crop of corn next year.

So what is important is the rate of discovery.

According to Petroconsultants/IHS Energy over the last decade discovery has averaged around 9 Gb/year with a couple of good years in 1999 and 2000 when levels reached 14–16 Gb/year. Figures in the Schollnberger presentation from Cambridge Energy Research

Associates (CERA) show the same pattern but the decade average from CERA is, if anything, lower than the IHS numbers at around 5–7 Gb/year (excluding Russia). It also shows the 1999/2000 upturn. It is worth noting that discovery in the 1960s averaged 43.8 Gb/year. (No I have not slipped a decimal point, it averaged 43.8 Gb/year). Demand rose rapidly over that decade from just under 5 Gb/year in 1960 to just over 11Gb/year in 1969.

Returning from the industry's 'Golden decade' to the current harsh realities. The public database may be highly suspect but, as it shows essentially unchanged reserves, what it is actually saying (rather improbably) is that liquids are being found at the rate they are being consumed. But this is after 150 years of always exceeding the consumption rate. These are massaged figures but even they are hinting at the problem.

Now, in fact, most serious commentators accept that net reserve growth is negative. But how negative? Returning to Jean Laherrère's graphs, we find that over the last two decades remaining reserves (backdated technical data) have declined by over 200 Gb or 10 Gb/year. However, the decline is accelerating with 170 Gb of the decline in the last 10 years, namely a decline rate of 17 Gb/year. Remember this is net of any discovery made!

Colin Campbell assesses undiscovered oil at 146 Gb, which with 10% added to give all liquids (oil +condensates) gives an undiscovered liquids of 160 Gb. Taking a 50-year view that means a little over 3 Gb/year. However Campbell sees discovery declining steadily from the current 6 Gb/year to zero. For our purposes we will take a rounded 6 Gb/year.

In sharp contrast, Schollnberger, as representative of the 'oil industry' view, anticipates discovery of a further 500 Gb (Suspend disbelief!) with improved recovery adding an additional 280Gb (Believe in miracles!). However once we take a 50-year view, this translates to 10 Gb/year of discovery and 5.6 Gb/year of enhanced recovery, namely 15.6 Gb/year in total. A highly optimistic view but not wholly absurd.

The current annual decline in the technical database of 17 Gb/year implies a discovery rate as 10 Gb/year (27–10=17). So, with some confidence, *we now know discovery is running at between 6 and 16 Gb/year with 10 Gb/year as a good working number.*

Currently liquids consumption is running at 27 Gb/year. So the pessimist Campbell predicts that the world eats into its store of discovered reserves at 27–6=21Gb/year while the optimist Schollnberger thinks the decline is only 27–15.6=11Gb/year (rounded numbers). The technical database indicates a decline of 17 Gb/year and rising. So we can confidently say the world is eating into its store of discovered liquids at between 11 Gb/year and 21 Gb/year with a 17 Gb/year decline as a good working number.

Now, the Central Limit Theorem tells us that the sum of any number of production profiles (any shape) is always a single hump curve with the peak at 50% of the total (Area under the graph). This is hard to grasp and always appears slightly mystical. However, half an hour with an Excel spreadsheet or careful viewing of those summed production profile graphs will confirm it to be true. Anyone still with doubts could look at Jean's graphs. The creaming curve for liquids discovery in the US lower 48 trends to 220 Gb. Half this is 110 Gb. US cumulative production reached 110 Gb in 1971, exactly when production peaked. UK offshore cumulative liquids discovery is trending to 36 Gb. Cumulative production reached 18 Gb in 1999. Production peaked in November 1999. In both these examples, discovery has fallen to minimal levels and so can be effectively ignored. The creaming curve has become flat. These figures could be calculated for all countries and all areas, which would be well worth doing. Only in those areas where there is significant levels of discovery will there be any real uncertainty.

The only other area of real unpredictability is the prospect of wholly new and undiscovered provinces. Jean Laherrère believes there could be an additional 30–50Gb in deepwater. Does anyone know of anywhere else where there might be significant undiscovered reserves? If they do, rest assured that every oil company on earth would pay large sums for the knowledge.

However, in a very important sense, to get bogged down in analysing countries and regions is to miss the big picture. What applies to the countries and the regions applies to the whole. *Once the world has consumed 50% of the reserves, the production level cannot be maintained.* It seems crazy, and is wholly counterintuitive, but it is true.

Now, to determine the fateful day when liquids production can no longer be expanded is quite easy. We have used 900 Gb and we are using 27 Gb/year. There are between 1000 Gb and 1100 Gb of reserves. By deflating consumption by the *net discovery* rate, we get the rate at which we are using up the reserve store.

Campbell's net reserves decline at 21 Gb/year (27–6) while Schollnberger's do so at 11Gb/year (27–16), and the technical database indicates 17 Gb/year. Now, 50% of global discovery figure is in the range 950 (900+1000 divided by 2) and 1000 (900+1100 divided by 2).

So, we only have 50 Gb to 100 Gb left to consume before production levels become unsustainable. *At best*, it could be 10 years (100 divided by 11), namely *2011*. *At worst*, it could be under two and a half years (50 divided by 21), namely *mid 2004*. *The most probable date on 1000 Gb reserves and 15 Gb/year net decline is 2005.*

A major offshore field development takes around five years from discovery to production, whereas a major onshore field takes around 3 years. So there isn't much time to push out the decline date. There aren't even that many major projects or discoveries in progress.

Now, you see why I really am hoping I have made some silly mistake.

We can easily examine the sensitivity of these predictions: The decline date would be pushed out if reserves were above 1000 Gb, if there was a rapid and sustained upturn in the discovery rate or if there was some evidence that production levels could be sustained beyond the 50% of reserves depletion point. We would also get more time if demand fell steadily or, better still, collapsed. On the other hand, the date would be advanced if reserves were below 1000 Gb, or if discovery fell back to early 1990s levels, or even if demand returned to its 25-year average of 2%/year growth.

Note the working assumptions are both that any discovery goes into the reserve development chain and that the chain is long enough to allow a steady flow of new developments. There is presumably some point at which there are so few undeveloped reserves that the time between discovery and production actually impacts the ability to even partially replace production. This seems unlikely to be other than a local problem in the time frame we are discussing.

Already throughout the non-Opec world, any significant discovery is being fast-tracked to production. It is obvious that past undeveloped discoveries must be so marginal that only very high prices will get them a place in the development queue. The only significant known fields I know of that are not actively being developed are: a number of fields in Iraq (embargo); some in northern Kuwait on the Iraqi border (security fears); some very high sulphur fields in Saudi Arabia (unsaleable); some fields in Eastern Siberia (no roads, pipelines or infrastructure) and accumulations in countries where the government operates a development queue (Angola). In short, any significant discovery currently known about is in the process of development with only minimal exceptions.

A 'quick and dirty' way of looking at the problem is to look at the age of oilfields. We know that 70% of the world's producing oilfields were discovered over thirty years ago. Few fields are able to maintain production levels after 25–30 years. The early Middle East giant fields are wholly exceptional in this respect, but they too decline.

We also know that the larger accumulations are discovered first, so that 70% of known discoveries accounts for at least 80–85% of known reserves. If there are 1000 Gb of known reserves, no more than 150–200 Gb are in fields where production could be expanded. So at best, there are 800 Gb in fields where production is static or declining, with just 200 Gb in fields where production could be developed or expanded to offset the decline and meet incremental demand. It is an area worthy of study. Reality could easily be rather worse than I am suggesting.

The fact that overall or global production levels cannot be sustained does not mean there will not be regions or areas where production is actually expanding rapidly. The most likely reason why there is so little appreciation of the imminent peaking is that people tend to focus on the areas of expansion and fail recognise the speed at which other areas will be declining. *Overall decline comes once the decline of production in fields in decline (post-peak) exceeds the production gains of fields with expansion potential (pre-peak). It is really that simple.*

So what about non-flowing oil supplies – tar sands and heavy oil? Production of these types of oil is more like mining than conventional oil production. It is the one area of the business where flat-earth economics applies. If the price is high enough more will be produced. The investment costs are huge, compounded by severe environmental problems and long lead times. Campbell's latest assessment is that Canada and Venezuela combined could provide 2.8 million b/d in 2005, 3.6 million b/d in 2010 and 4.6 million b/d in 2020. According to my calculator that's 1 Gb/year in 2005, 1.3 Gb/year in 2010 and 1.7 Gb/year in 2020. Nice to have, but its not going to save the day even if we were to double the estimates. At the moment, global gas production is being expanded very rapidly. It is not at all clear that it could be expanded more rapidly although at higher prices there might be even greater investment.

Perhaps we should start to consider what happens when liquids production can no longer be sustained.

To date, when an individual country, province or region reached this point (production decline), it simply bought from elsewhere. The economic harm was usually minimal, although it could be argued that Russia's inability to maintain production growth after 1988 denied the economy cheap fuel for its inefficient economy and led to the collapse of Communism. But what will happen when decline is a global phenomena?

Initially it will be denied. There will be much lying and obfuscation. Then, prices will rise and demand will fall. The rich will outbid the poor for available supplies. The system will initially appear to rebalance.

The dash for gas will become more frenzied. People will realise nuclear power stations take up to ten years to build. People will also realise wind, waves, solar and other renewables are all pretty marginal and take a lot of energy to construct. There will be a dash for more fuel-efficient vehicles and equipment. The poor will not be able to afford the investment or the fuel.

Exploration and exploitation of oil and gas will become completely frenzied. More and more countries will decide to reserve oil and later gas supplies for their own people. Air quality will be ignored as coal production and consumption expand once more. Once the decline really gets under way, liquids production will fall relentlessly by 5%/year. Energy prices will rise remorselessly. Inflation will become endemic. Resource conflicts will break out.

In 1975, the US Lower-48 produced 7 million b/d of crude. By 2000, production had fallen to just under 3.5 million b/d. Despite maximum financial incentives, the finest technology in the world and a complete openness to innovation, the US has been unable to slow, never mind reverse, this 2%/year production decline. Is there any reason to think the world will fare any better once peak is passed?

106. An Excellent Review

Dr R.W. Bentley has published an excellent review of the issues surrounding oil depletion entitled Global Oil & Gas Depletion – an Overview in *Energy Policy 30* (2002).

Newsletter No. 16, April 2002

107. Estimates of Ultimate Recovery

Many estimates of Ultimate Recovery have been published over the years, but we need to research the details to know where they drew the boundary between *Conventional* and *Non-conventional* oil, what they measured, and what their methodology was. However, the following figure shows 65 estimates at face value, as published by oil companies, government institutions and others, including the outlandish estimates by Odell in 1973 and the USGS in 2000.

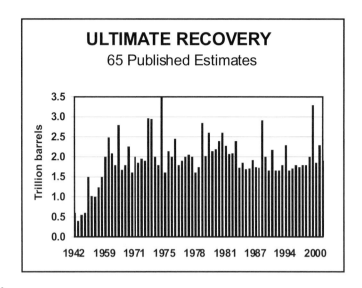

Figure 29

The average is 1.93 trillion, which by coincidence comes very close to the present ASPO value. Whether or not this estimate proves to be correct, we can at least say that we have the support of the consensus of qualified opinion. Accordingly, the onus rests on those with counter-views to justify their position with more than expressions of blind faith.

108. Halfway through the Industrial Revolution

Jean Laherrère has produced another useful graph plotting energy

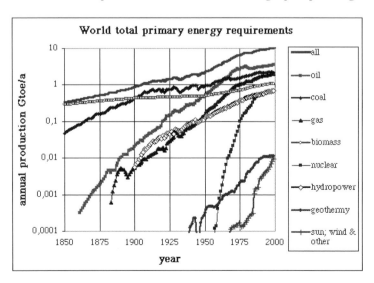

Figure 30

consumption over the past 150 years. The total over the first hundred years followed a straight upward trend followed by a surge from 1950 to 1975 after which it flattened, largely reflecting the levelling of oil. The most spectacular increases have been nuclear and renewables (sun, wind etc). We may speculate if the next 150 years will be the mirror image.

We may ask too what impact this will have on the "Death of the West" (lifting the term from the title of Patrick Buchanan's book). Will the energy-dependent industrial societies age and wither away in parallel with their energy supply to be replaced by immigrants and their descendants, willing to till the fields and work the coal mines from their new homes in equitable temperate climes? Revolutions by definition do not last for long, and there is no reason to think that the Industrial Revolution will be an exception.

109. Growing Gap between Oil Discovery and Consumption
It appears that total discovery in 2001 was about 8 Gb including deepwater oil, and NGL. Although there remain the eternal uncertainties about the reliability of the data, it appears that the world's oil account has been running a deficit since 1981, as it continues to eat into its inheritance from past discovery. This downward discovery trend was evidently not affected by the tax-driven surge of exploration drilling in the early 1980s, underlining that discovery depends on geology not economic incentive. Apart from two sizeable finds, the 8 Gb of 2001 came from some 300 discoveries whose average size must accordingly be approaching the lower limit of viability. It explains the continued fall in exploration drilling, as modern technology allows the companies to accurately map the size of

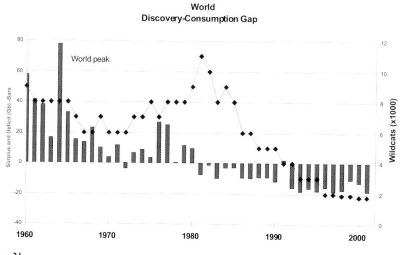

**World
Discovery-Consumption Gap**

Figure 31

prospects, enabling them to concentrate on the shortening list of viable opportunities. These results from the real world are compatible with the consensus estimate of the size of the endowment, discussed above, and confirm the thoroughly flawed nature of the USGS report of 2000 that has misled so many governments and international agencies. The USGS claimed (as a *Mean* probability estimate) that 732 Gb would be found between 1995 and 2025, which means an average of 24 Gb a year. In fact, the average has been only 10 Gb during the first seven years of the study period, when discoveries should be above average as the larger fields are normally found first.

In the face of such easily documented evidence, it is difficult to avoid the conclusion that those who still give credence to this report must have some ulterior motive for doing so. It is surprising that the World Energy Council should provide a platform for it at its Ankara meeting.

It is noteworthy too that Shell, which in its scenarios claims to support the USGS excessive estimates, prefers to purchase Enterprise Oil for its reserves – it having virtually no other asset of conceivable interest to Shell – rather than explore for the largesse it pretends is there to be found. Its actions speak louder than its words and scenarios. Shell's offer values Enterprise at $6.5 billion for reserves of 1.12 Gb of oil and 2.1 Tcf. of gas, which amounts to about $6/b for the oil alone. The addition dilutes the value of Shell's existing reserve portfolio, yet is still preferable to exploration.

110. Explaining the failure of the US Geological Survey

National geological surveys are normally staffed by serious scientists, and are not the most obvious places in which to expect to find conspiracies by sinister forces with ulterior motives to confuse and mislead. So how do we explain the flawed study of world oil discovery by the USGS that has so mislead the world's governments on this most critical of subjects?

The answer perhaps lies in mind-set and experience. The world has approximately 700 sedimentary basins having notional oil prospects, of which at a guess perhaps 200 have been found to be productive. Assessing the potential of the productive basins is a fairly straightforward task that can be achieved by extrapolating the experience to-date with the help of a series of robust statistical techniques, relating declining discovery to drilling activity, and field size distributions. Assessing the presently non-productive basins is a different challenge altogether, and can be based only on abstract geological reasoning. The USGS failed to distinguish the two categories, applying the latter technique to both, which was understandable, given that they are geologists and not practical oilmen.

The USGS proposes that, as a Mean probability estimate, as much as

732 Gb (billion barrels) will be found between 1995 and 2025, tacitly assuming that an infinite number of wildcats would be drilled to find it. It was very hard for them to categorically say that their expectation for any basin was zero, especially if it was a large one, although in Nature there are many very large and very barren basins. So, they found themselves having to hedge their bets with probability rankings, which in no case accepted the proposition of zero discovery. Their zero ranking with 95% probability was defined to mean that there was a 95% probability of there being more than zero, namely at least one barrel. Then they went to the other end of the range and asked what the 5% percent probability might be. Although the very concept of a 5% subjective probability is almost beyond the resolution of the human mind, the question demanded an answer, and in the case of a large basin it could well be a very high one. From this wide range, they computed a *Mean* value, which they summed to give the world total of 732 Gb.

An oilman, doing the same job, would approach it very differently, because he would know that he had only a finite number of wildcats to find it. He would no doubt dedicate most of his effort to the established provinces, which he would rightly conclude represented the best investment. He might dedicate a small part of his budget to confirming his negative assessment of the un- or under-tested basins, recognising that they are under-tested because they are perceived to have poor chances. If his first few wells confirmed the negative assessment, he would not waste more of his precious budget with abortive drilling. So instead of asking about subjective probability, it might be better to ask how many consecutive dry holes would be drilled before abandoning the search.

If we took this more practical approach, we would not assume an infinite number of wildcats but a decidedly finite number, because as the odds lengthen, fewer and fewer investors are likely to be tempted by the lure of black gold, strong as it might be. Some idea of the scale of things may be gained by looking back to the record over the past two decades to see what has actually been achieved, although, we face the eternal problem of unreliable data to know even how many such wells were drilled. For what it is worth, the information we have suggests that about 110 000 wildcats have been drilled since 1980, costing, say, 500 billion dollars, and they found approximately 300 billion barrels. This was less than the 500 billion that the world consumed, as it ate into its inheritance from past discovery. There was evidently plenty of incentive to find more oil, but the rate of wildcat drilling declined from an average of 77,000 over the first decade to 31,000 over the second.

If we now look ahead to the next twenty years, we may expect the amount of drilling to continue to halve each decade as the list of perceived remaining viable prospects continues to shorten. With an optimistic assumption that the same rate of discovery could be held, that means about 150 Gb of in the next decade and about 75 in the one after that. A more realistic assumption would expect discovery rate to continue to decline, yielding much less.

As the Nemesis Report pointed out last month, the answers we get often do not match the questions posed. We thought we were asking the USGS how much oil would be found between 1995 and 2025 in the real world, but the answer we got was that there was a Mean probability of finding 732 Gb assuming an infinite number of wildcats. It might indeed be a sound logical answer to the degree that anyone can truly grasp such a concept, but it was not the answer to the question we thought we had asked. As pointed out above, seven years into the study period, we find that the real world is indeed different to the hypothetical one that occupies the minds of the USGS. Their great mistake was to put a time-span on the study. Their estimates could have survived well enough if they had left the time-frame open ended, as no one can know what someone with a mop and a bucket might find a hundred years from now in some distant seepage.

111. New Study of Global Oil Supply to 2050

A new study of global oil supply has been produced by Dr M.R. Smith, giving comparable results to those proposed here. Contact address glow@lineone.net,

112. Newsweek contemplates a world beyond oil

Newsweek of April 15th 2002 carries an extensive article entitled *Beyond Oil*. It says much between the lines hinting that it perceives, but cannot quite state, the reality of the situation. It falls into the common trap of using Reserve to Production ratio, quoted in years, to deny depletion, as if it were remotely plausible that production could be held constant for a given number of years and then stop overnight when all oilfields are observed to decline gradually towards exhaustion.

It emphasises that US military adventures from Colombia to the Caspian have oil as their target, whatever the pretext. It treats the sound work of Deffeyes (author of *Hubbert's Peak – the impending world oil shortage*) with a degree of scepticism, verging on derision, yet gives prominence to it, almost as if to draw attention to it, without quite having the courage to endorse its unassailable logic.

113. The attempt to oust Venezuela's strong man

The Times of 16th April implicates the United States in the failed attempt to oust the Venezuelan President, who had taken a strong position on oil. It gives a further hint of the new US oil policy, even if the CIA does seem to have lost its touch since the days when it successfully did in Allende in Chile.

114. Denmark's change of direction

The journal *Science* (v.295 of 8 March www.sciencemag.org) comments on how a prominent flat-earth economist, Lomborg, has been put in charge of a new institute for Environmental Evaluation. The appointment is receiving strong opposition from Denmark's scientific community. Lomborg, who rejects the evidence of oil depletion, may find support from Denmark's Geological Survey, which has fallen under the influence of the USGS. It remains to be seen if the Danish government, which now holds the EU Presidency, will allow itself to be misled by the flawed advice it will be receiving from these sources. It is ironic that Denmark, which has given such a lead in renewable energy, should now find itself suffering this reversal.

Newsletter No. 17, May 2002

115. ASPO Workshop in Uppsala

The ASPO workshop on May 22–25th in the ancient town of Uppsala proved to be a great success, thanks to the hospitality of the university and the fine organisation of Professor Aleklett and his secretary. Approximately 60 delegates attended from Australia, Byelorussia, Denmark, Finland, France, Germany, Iran, Ireland, Norway, Portugal, Russia, Sweden, United Kingdom and the United States. Two days of meetings covered the resource base, the date of oil peak, the economic, geopolitical and environmental implications and the scope for substitution by gas, renewable and nuclear energy. Keynote speakers from the United States, Russia and Iran addressed the evolving energy situation and challenges in their countries and regions. Abstracts and details may be seen on www.isv.uu.se/iwood2002

On the final day, a meeting was held to plan ASPO's future policy. An ad hoc steering committee was appointed to implement a programme, which includes the expansion of membership, the establishment of Associate Membership, funding, the production of information packs and educational presentations, the development of a common database and depletion model, and preparations for the next annual meeting. It was decided to continue the present newsletter in broadly the same style.

The meeting attracted wide media interest. It appeared as a lead news item on Swedish television, and was covered by Associated Press with an article that was widely syndicated around the world's press (see www.oilnews.com for references).

116. Energy Mirror

Referring to an item the last issue, Paul Metz points out that the growth of renewable energy production is likely to continue.

> Thank you for another interesting issue. Some articles are not well spent on me:-), but I would comment the graph presented by Laherrere on 150 years primary energy sources (page 2). This is an interesting one, however, the remark on "mirroring for the next 150 years?" would fit much better with statistics for fossil sources only. The reader will certainly understand that for renewable sources no such depletion mechanism is in place and therefore no stabilisation or decline – but strong growth – for that part needs to be expected, but and let us be happy with the promotion of renewables, even without them telling exactly why... the security of supply. Such a graph will be more convincing for our critics – and for ourselves!

Indeed, renewables will undoubtedly grow, but energy savings, along with declining population and reduced economic activity consequent upon the depletion of oil, means that overall energy demand may decline, mirroring the past growth.

117. Monthly Country Assessments

It is proposed to cover a particular country's depletion situation each month, starting for no particular reason with Syria. A standard format will be adopted as follows. ASPO members are invited to audit the findings. Comments and better information from any source will be greatly appreciated.

Syria

> Syria, which covers an area of 186,000 km2 and supports a population of 14 M, lies on the northern margin of the Arabian Peninsula bordering Turkey and Iraq. In geological terms, most of the country falls on a shallow platform where prospects are confined to possible Palaeozoic plays, likely to be, if anything, gas prone, but its northeast extremity extends into the prolific Mesozoic Middle East oil belt. The first significant oil find, made in 1940, was followed after the War by several others, including Suwaidiyah in 1959, which is ranked as a giant field with an ultimate recovery of about 1.5 Gb. Shell has had a long-standing presence in the country, but exploration and production have been inhibited by an uncertain political climate. The geology is well known. As many as 270 wildcats have been drilled, suggesting that most of the large promising structures have been tested. Wildcat drilling peaked in 1992, falling to an average of no more than five a year over the past five years. On this basis,

exploration is set to end around 2020, with a total of about 380 wildcats. There remains some potential for the deep Palaeozoic play, sourced by the Silurian, but it is likely to be gas-, rather than oil-bearing. The country has no obvious *non-conventional* potential.

SYRIA		
Rates Mb/d		
Consumption	2001	?
Production	2001	0.515
	Forecast 2010	0.261
	Forecast 2020	0.123
Discovery 5-year average (Gb)		0.006
Amounts Gb		
Past Production		3.60
Reported *Proved Reserves*		2.50
Estimated Future Production to 2060		
From Known Fields		2.01
From New Fields		0.39
Future Total		2.40
Total Production to 2060		6.00
Current Depletion Rate		7.3%
Depletion Midpoint Date		1998
Peak Discovery Date		1966
Peak Production Date		1995

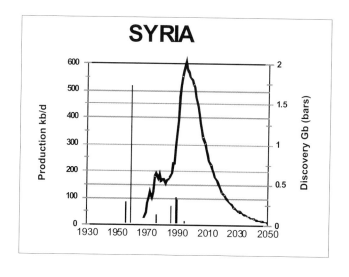

Figure 32

Production started in 1968, peaked in 1995 and is now declining at about 7% a year. Reported reserves of 2.5 Gb have been unchanged for nine years and are clearly unreliable. Important, currently closed, pipelines run through the country from Kirkuk in Iraq to the Mediterranean. They are likely to resume importance both for Syria and the West if and when the embargo is lifted or the country invaded, giving Syria the role of a transit country, which carries political implications for the region. No information on Syria's consumption has been located, but it is probably a net importer, mainly from Iraq. One may also imagine that Iraqi oil is smuggled out through Syria. The country will be ever more dependent on Iraqi oil in the future, as its own production declines.

118. Colombian Lecture

Professor Alvarez made a presentation entitled *La Politica Petrolera Colombiana frente a la proxima Crisis Petrolera Mundial* on behalf of C. J. Campbell. It pointed out that Colombia's production had peaked, and that even with flat demand it would become a net importer within twenty years. It was suggested that the national interest would be served by strengthening the State Company, and conserving the resource for domestic usage. It was proposed that a new currency for oil transactions be introduced, based on the new economic principles of Slessor and King (see Newsletter 15).

Newsletter No. 18, June 2002

119. Oil and Gas Equivalence

Some readers have questioned to use of the 10:1 factor for the conversion of gas into oil equivalent, when the calorific equivalence is about 6:1. The whole issue of units and equivalents is a complex one. Oil is sometimes measured in volumes and sometimes in weight, despite varying density. Gas has to be measured at given temperature and pressure, and it is not always clear whether the reported values relate to raw or marketable gas after the removal of non-flammable gases such as carbon dioxide and nitrogen, which are often also present. The treatment of flared gas and re-injected gas is also confusing. Furthermore, there is the issue of monetary value and net energy value after removing the energy used in transport etc.

There is a certain pragmatic merit in the 10:1 factor, but there are other valid approaches too. There would be great merit in standardising the units and practices of measurement, but for the present we live in a world of great confusion concerning these important matters.

120. The ASPO Workshop at Uppsala

The papers and presentations of the workshop are being posted on the Uppsala website www.isv.uu.se/iwood2002, and are attracting keen interest from around the world. A start has been made in assembling the ASPO Statistical Review of Oil & Gas to try to correct the misleading impression of another widely booklet with a similar title. Much more work on it is however required. It too has now been posted on the above site.

121. Middle East production constraints

Matt Simmons reports.

> Today's Oil Daily had a very interesting story on the oil production woes of Kuwait which has been forced to rely overly on the output from Burgan, the world's number two producing oilfield. According to the story, Kuwait has been forced to close three of its four crude centres over the past few weeks in the western part of the country, cutting production there by 243,000 b/d. The shut-ins resulted from a variety of bad problems, including toxic fumes, a collapsed line, a gas plant needing a redesign and routine maintenance, etc. This all highlights how old the major oil producing centres of the Middle East really are. We all know this, but most of the world still thinks Middle East oil is almost free.

> According to the story, Burgan is now producing 1.35 Mb/d of its assumed 1.8 Mb/d output capacity. Everyone should remember that Burgan was discovered in 1938. As young kids in America often say, "it is no spring chicken!!" The cost of merely maintaining flat production in the Middle East is now on the rise. The days of cheap Middle East oil are a distant memory. Most people have not a clue about all this.

Based on discussion with a key Middle East expert at the Uppsala Workshop, it transpires that some of the national Middle East oil companies are managed by ageing executives who are reluctant to retire. They have little understanding of their countries' reservoirs or the status of depletion, having been brought up in an age of plenty. It may be not so much that valid information on Middle East reserves is difficult to access, but rather that does not actually exist. Possibly, the OPEC secretariat is planning its strategy on information no more reliable than that in the BP Statistical Review of World Energy, which draws on the Oil & Gas Journal. It in turn relies on a questionnaire sent to foreign governments to which many do not reply, leaving the reserve reports implausibly unchanged, sometimes for years on end.

We have long suspected that Middle East reserves are greatly exaggerated. It begins to look as if we were right. Even if the reserves in the ground are there, the management and political environment suggests that these countries will have great difficulty in offsetting the natural decline of their ageing fields. It is not just a case of opening the valve, but calls for a great deal of skilled and careful reservoir management, nowhere more so than in Iran with its sensitive fields prone to gas coning.

122. The Peak of the Industrial Revolution

Chapter 10 of *Europe – A History* by Norman Davies is entitled *Dynamo – Powerhouse of the World 1815–1914*. It makes interesting reading as the following extracts from the opening paragraphs confirm:

There is a dynamism about nineteenth-century Europe that far exceeds anything previously known. Europe vibrated with power as never before: with technical power, economic power, cultural power, intercontinental power. Its prime symbols were engines – the locomotives, the gasworks, the electric dynamos. Raw power appeared to made a virtue in itself, whether in popular views of evolution, which preached the 'survival of the fittest', in the philosophy of historical materialism, which preached the triumph of the strongest class, in the cult of the Superman, or in the theory and practice of imperialism…

Seen in detail the process of modernisation can be broken down into an apparently endless chain of sub-processes and new developments, each interacting with each other…

Agricultural production benefited from the gradual introduction of machines, from McCormick's horse-drawn reapers (1832) to steam-driven threshers and eventually petrol driven tractors (1905)…

New sources of power were brought in to supplement 'King Coal', first with gas, then with oil, and later with the commercial use of electricity… Oilfields were opened up in Europe at Boryslaw (Galicia), at Ploesti (Romania) and at Baku on the Caspian. With time, the internal combustion engine (1889) was to prove as revolutionary as the steam engine. Electricity became widely available only in the 1880s…

Capital investments multiplied in proportion to growing returns. Private firms reinvested growing profits; governments invested a growing proportion of rising taxation. A bottomless demand for capital exhausted the possibilities of private borrowing, and revived the potential for joint-stock companies…

Domestic markets were boosted by population growth, by the greater accessibility of population centres, by expanding affluence, and by the creation of entirely new sorts of demand…

In the purely economic sphere, the growth of the money economy turned self-sufficient peasants into wage-earners, consumers and taxpayers, each with new demands and aspirations.

The last sentence is the most telling. What is so striking about this account is that it reminds us that the beginning of the modern world was so recent – my father, who was born in 1874, only one generation ago, witnessed it. Is it possible that the first tractor ploughed its furrow less than 100 years ago? The Industrial Revolution was built on power from fossil fuels, and it exploded during the 20th Century as more and more were tapped. Now as the 21st Century dawns, we face the peak and decline of these essential drivers. Logic proclaims that the entire economic, social and political fabric of the modern world is at risk, especially since the pace of change seems to have been accelerating.

It seems, further, that we are generally moving from the nation-state to a form of global kleptocracy, in which corporate and political leaders co-operate to their mutual advantage, although a few anachronistic ultra-nationalistic states remain under increasing pressure and violence. A remarkable website touches on the kleptocracy. The message will be too blunt for most tastes, but there is no smoke without fire.

http:/www.scoop.co.nz/mason/stories/HL0206/S00071.htm

123. Country Assessment Series

Continuing the series started last month, we take a look at Indonesia. It is intended that the series will stimulate an interest in tracking down more data and insight so that the assessments can be improved and assembled into a useful compendium.

Indonesia

Indonesia is an archipelago, stretching for about 3000 km from Asia to Australasia and including the large islands of Java and Sumatra as well as much of Borneo. It has a diverse ethnic population of 200 million, of which about 3% are Chinese, who have traded and settled in the area for centuries. It is a predominantly Muslim country.

It has had a long history being settled by peoples from Malaya and Oceania, but was also influenced by Arab traders in the Middle Ages. From 1602 until 1798, most of the territory was controlled by the Dutch East India Company, before it passed into Dutch colonial rule. It was occupied in the Second World War by the Japanese, whose motive for going to war was partly access to oil.

A move to independence followed under the leadership of Sukarno, being finally granted in 1949 under less than amicable terms. The western end of New Guinea, with its very different ethnic people, was added to the new republic in 1963, later being renamed

Irian Jaya. The former Portuguese territory of East Timor, with its predominantly Catholic population, was annexed in 1976, but has recently successfully seceded.

Sukarno, who had Communist leanings, ruled in an authoritarian style until 1965 when he was ousted by General

INDONESIA		
Rates Mb/d		
Consumption	2001	1.065
Production	2001	1.200
	Forecast 2010	0.83
	Forecast 2020	0.56
Discovery 5-year average (Gb)		0.29
Amounts Gb		
Past Production		19.41
Reported *Proved Reserves*		5.00
Estimated Future Production to 2075		
From Known Fields		9.00
From New Fields		1.59
Future Total		10.59
Total Production to 2075		30.00
Current Depletion Rate		4.0%
Depletion Midpoint Date		1992
Peak Discovery Date		1955
Peak Production Date		1977

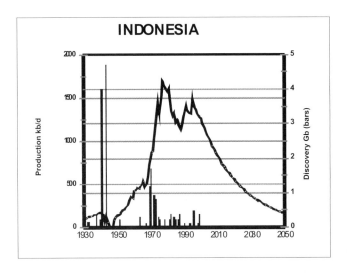

Figure 33

Suharto in a bloody conflict costing 500,000 lives. His rule was endorsed by popular elections in 1968, having adopted more Western-oriented policies, seeking overseas investment. Since his departure, the country has lurched from one political crisis to another under a somewhat uncertain administration. Further difficulties are likely to be experienced as economic conditions deteriorate, with the possibility of successful separatist movements in various islands.

Indonesia has had a long oil history, being the birthplace of Royal Dutch/Shell, with its early fields in Borneo. Sumatra, however, has the largest fields, Duri and Minas, which were found in the 1940s but not developed until after the Second World War. Duri contains heavy oil (20° API), being produced with low net energy yield by steam injection, putting it on the borderline of *Non-conventional*.

The country joined OPEC in 1962 and effectively nationalised the oil industry in 1965 with the creation of a state company Pertamina. It, in turn, entered into production-sharing contracts with foreign companies, bringing about a very successful and active co-operation.

It is bordered by a large continental shelf, but in geological terms, much of the country is strongly deformed and volcanic, so that its petroleum prospects are confined to a few well-known Tertiary sedimentary basins in Sumatra, the Java Sea, S.E. Borneo and locally in Irian Jaya.

Exploration is at a mature stage having commenced in the Nineteenth century. Some 3400 wildcats have been drilled, but drilling rate has been falling for some years, pointing to an end to exploration by about 2040. Even so, it is estimated that more than a billion barrels await discovery, coming mainly from ever smaller fields in the established producing areas. The country has some *Non-conventional* deepwater oil potential, as already confirmed by Unocal's work off Borneo, but generally the source-rock conditions for such are adverse. Whereas the prolific deepwater tracts of West Africa and the Gulf of Mexico are underlain rifts containing rich source-rocks, the possibilities in Indonesia are confined to the delta-fronts themselves that are likely to be lean and gas prone.

The production profile exhibits the typical OPEC saddle due to quota restrictions. Production accordingly peaked in 1977, fifteen years before the midpoint of depletion in 1992. A secondary peak was passed in 1995, partly making up what would have been produced naturally save for the OPEC saddle and partly coming from a second smaller cycle of discovery in the 1970s. Production has now commenced its terminal decline at a Depletion Rate of about 4 percent a year.

Consumption is 89% of production, meaning that the country is set to become an importer by 2004, assuming flat demand. Its continued OPEC membership must accordingly be increasingly in doubt.

124. US Production Forecast

The following plot (Figure 34), taken from the ASPO Statistical Review of Oil and Gas, shows how US imports are set to rise unless the government can somehow rein in demand. It has every reason to be concerned about access to foreign oil, including whatever it can get from the Caspian.

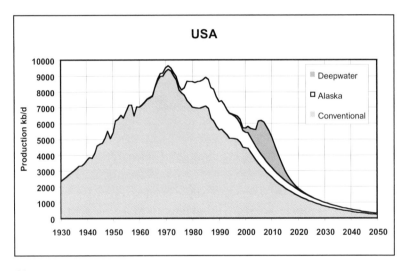

Figure 34

125. Gas Hydrates

Peter Gerling from the BGR draws attention to the following article on Gas Hydrates. There are good grounds for being very sceptical about the economic production of gas hydrates. The principal reason is that they mainly occur as thin laminae and disseminated granules such that the methane cannot migrate to accumulate in commercial deposits. Some of the reported thicker occurrences may be nothing more than seepages of normal gas on the sea floor. It is well said that they are "the fuel of the future _ and likely to remain so". Meanwhile they attract an enormous amount of unjustified research funding that continues to give equivocal results.

First qualitative assessment of the economic potential of offshore gas hydrate accumulations

Previously, the volume of the offshore gas hydrate resource was estimated – mainly based on data from Kvenvolden – in the range of 21×10^{15} m^3. In 2000, a Russian scientist came out with the number of 0.2×10^{15} m^3, approximately half of the world's conventional gas endowment. All these numbers are pure estimates without any economic consideration.

A.M. Milkov and R. Sassen from Texas A&M University have published this year (Marine and Petroleum Geology, Vol. 19, 1–11) an article about offshore gas hydrates considering geological, technological and economic factors. The authors distinguish between structural accumulations, stratigraphic accumulations, and a combination of both. Only structural accumulations occurring in areas with rapid gas transport from great depth, e.g. in the northwestern Gulf of Mexico, may contain economic gas hydrate resources. However, the authors clearly state that their preliminary qualitative economic analysis is based on many assumptions. They assume that gas hydrate may be profitably recovered from some accumulations.

126. False Scenario?

The base case scenario behind the depletion forecast of conventional oil depicted in the frontispiece of the Newsletter is that the world about ran out of spare capacity in late 1999, causing oil prices to soar, which helped trigger recession. The recession in turn cut oil demand, lessening pressure on price. Since depletion continues to reduce capacity, the scenario imagines that future price shocks are inevitable and that they will cause recurring recessions. Accordingly, the average production of conventional oil is assumed to be flat until 2010, when in practice the Middle East producers are no longer able to exercise their swing role, offsetting the natural decline elsewhere.

Mr McKillop, an economist, however questions that high oil prices trigger recessions, seeing them rather as a boost to liquidity, insofar as the high prices reflect revenue increase rather than rising production costs as such. Such a boost to liquidity may thus be a global economic stimulus rather than the converse, although there may be domestic set-backs in certain countries and time-lags in the effects. If this interpretation is correct, then perhaps a stimulated economy will after all lead to a rising demand for oil, which can only advance the peak and steepen the subsequent decline, making a bad situation much worse. (See McKillop A, 1989, On Decoupling: Int. Journ of Energy Research 14 83-105.)

127. The Fuel that Fires Political Hotspots
An article by C. J. Campbell with the above title appeared in *The Times Higher Education Supplement* on May 17th 2002, referring to the ASPO Uppsala Workshop.

128. More Evidence of Major Oil Company Downsizing
Two recent articles in *World Oil* touch on downsizing. The April issue reports that BP has decided against making political contributions in the USA, $834,000 having been spent in this way in 2001. The directors of companies, who have a fiduciary duty to make money for their shareholders, have no reason or justification for making donations unless commercial advantage results. Political corruption, however elegantly packaged and euphemistically described, is universal, and indeed essential to business success, so to end the practice speaks less of a moral awakening than of reduced activity.

The May issue reports on the same company's replacement of staff by consultants, the total for various drilling functions being 31%. The declared aim is 30–40%: above that the Company loses essential expertise and continuity; below the level, its own staff have to be shed. It is not exactly the image of soaring future drilling activity.

Newsletter No. 19, July 2002

129. Country Assessment Series – Colombia
Continuing the series started last month, we take a look at Colombia. It is intended that the series will stimulate an interest in tracking down more data and insight so that the assessments can be improved and assembled into a useful compendium. Depletion is modelled on the basis of the current depletion rate (annual production as a percentage of estimated future production).

Colombia
Colombia, covering more than a million square kilometres on the northwest corner of South America, is cut by three ranges of the Andes, which are flanked to the southeast by extensive plains in the headwaters of the Amazon and Orinoco rivers. Its coasts are washed by the Caribbean and the Pacific, separated by the narrow isthmus of Panama.

One of the chieftains of the ancient Colombian Chibcha civilisation had the habit of covering his body in gold-dust before bathing in Lake Guatevita in which emeralds and other precious stones were thrown to placate the gods. *El Dorado*, as he was known, stimulated the interest of the Spanish Conquistadores in the wake of Columbus, who reached the northern coast on his last voyage in 1502.

In a remarkable short span of fifty years, the Spaniards had established themselves throughout the country, building towns and monasteries high in the Andes. By 1739, Bogota had been established as the Vice-Royalty of Nueva Granada, holding dominion over what is now Venezuela, Ecuador and much of Central America, south of Mexico. But in 1819, Simon Bolivar, the great "Liberator" of South America, defeated the Spanish royalists, bringing independence to the region, which later fragmented into separate republics. Cornelius O'Leary, a mercenary Irish soldier from Cork, wrote the national anthem for the new republic. The last territorial adjustment came in 1903, when the United States engineered the secession of Panama after Colombia had refused consent for the construction of the Panama Canal. The United States tried to make amends in 1914 by paying an indemnity of 25 million dollars.

Independence brought the eternal conflict between federalism and centralism, exacerbated by physical mountain barriers and the fact that the several regions had been settled by immigrants from different parts of Spain. It sowed the seeds of the violence, often degenerating into banditry that has been endemic for two centuries before the narcotics drug trade brought it to its current extreme level. Large tracts of the country are now under the control of war lords, importing arms by air to support private armies, while surprisingly also sponsoring certain social programmes in their regions. The coca leaf has been grown since pre-Conquest days, without posing any particular problem, but that changed when cocaine became a commodity of global trade.

Colombia's population has more than doubled over the past thirty years to exceed 50 million. It is of mixed European, Indian and Negro origins, with little sign of racial discord. A few indigenous Indian communities survive in remote areas but probably face extinction or integration, The violent political situation has driven people to the cities, especially Bogotá, the capital, where more than 10 million live, many in desperate conditions. Their untreated effluent now flows over the once beautiful Tequendama Falls, putting up a bacteriological fog. But despite every adversity, the Colombians retain their vitality, courage and good humour.

Colombia has had a long oil history, starting in 1905 when General Virgilio Barco secured rights to the Colombian extension of the Maracaibo Basin, and Roberto de Mares took a concession in the Middle Magdalena valley. These two areas dominated Colombian oil production for many years, with the rights eventually passing to subsidiaries of Esso, Mobil and Texaco, before the creation of the

Colombia		
Rates Mb/d		
Consumption	2001	0.23
Production	2001	0.86
	Forecast 2010	0.39
	Forecast 2020	0.24
Discovery 5-year average (Gb)		0.07
Amounts Gb		
Past Production		5.54
Reported *Proved Reserves*		1.75
Estimated Future Production to 2075		
From Known Fields		3.94
From New Fields		0.53
Future Total		4.46
Total Production to 2075		10.00
Current Depletion Rate		4.9%
Depletion Midpoint Date		1999
Peak Discovery Date		1992
Peak Production Date		1999

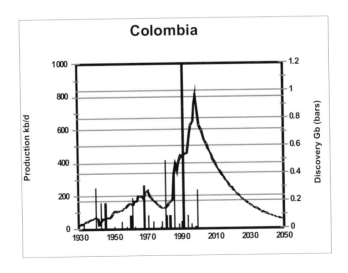

Figure 35

State oil company, Ecopetrol in 1951. Further discoveries were made in the vicinity by other foreign companies.

The prospects of the foothills of the Eastern Andes had been evident from the earliest surveys, being in part confirmed by oil seepages, but development was delayed by the exceedingly remote location and the fact that mountain ranges to over 2000m would have to be traversed by long export pipelines. However, Texaco and Gulf mounted a heroic campaign that was rewarded in 1962 by the discovery of the Orito field in the Amazon headwaters, close to the Ecuadorian border. An export pipeline across the Andes to the Pacific was eventually constructed. Next, Occidental brought in Cañon-Limon in 1983 at the northern end of the belt near the Venezuelan border, followed by BP and Total with Cusiana in 1988. These giant fields highlighted a second and last cycle of major oil discovery in Colombia. The geology of the country is well known, with exploration at a mature stage after the drilling of almost 1500 wildcats. Peak drilling was in 1988 with about 80 exploration wells, but activity has now fallen to about 15 a year. It is reasonable to extrapolate the trend, expecting exploration to end around 2025 after another 150 boreholes.

The source of Colombia's oil are a few hundred metres of Middle Cretaceous organic claystone. It was laid down about 90 million years ago in one of the World's prime epochs of oil generation, being also responsible for the finds in Mexico and the US Gulf Coast as well as the vast, partly degraded, deposits of Venezuela. In Colombia, it is confined to the eastern part of the country, meaning that other large sedimentary basins, especially along the Pacific Coast, which lack this essential source, are likely to remain barren, save perhaps for gas generated in the Tertiary. It follows that future discovery is likely to be confined to ever smaller finds in the established regions, with some upside possibility concealed by the complexity of the thrust belts and the difficult terrain. There is also perhaps a slim chance of some non-conventional deepwater discoveries on the delta front of the Magdalena River

Production, reflecting the two main discovery cycles, reached a peak of 816 kb/d in 1999 at the midpoint of depletion. A marked decline to 625 kb/d by 2001 gives a current depletion rate of just under 5% a year. Consumption stands at about 230 kb/d, less than 40% of production, meaning that the country is presently an important exporter, largely to the United States. This perhaps partly explains the latter's military intervention with the provision of helicopter gunships to patrol the pipelines from frequent dissident attacks. The widespread terrorism and drug trade provide an ample pretext for any further interventions to maintain the flow, although the success of such operations is far from assured in view of the mountainous terrain. In any event, falling production from natural depletion means that Colombia's export potential is set to decline, so that production will fall below domestic demand by 2020, or sooner, if consumption should increase above present levels.

Colombia may be among the first countries to enter post-globalism, with the reduction of central government and the emergence of small sustainable communities and local markets under a federal political structure. In its case, the transition is due not so much to the decline of essential fuel supply as a result of oil depletion, but to the impact of the international narcotics trade which gives it a particularly violent character. In energy terms, Colombia is in fact well blessed, with many years of, albeit declining, oil supply, which could be conserved for national needs, and a good renewable energy potential from hydroelectric and solar power. It is also endowed with substantial coal deposits. Left to itself, it would be well placed to find a way forward, but foreign demands for its cocaine and oil, as well as near term risks of US military intervention, place obstacles in the path to an otherwise sustainable future.

130. Another BBC Broadcast on Depletion

BBC Radio 4 ran a programme called Crude Facts on 19th June at 9:00 pm in which they interviewed Randy Udall, Steve Andrews, Lovins, Nehring and Colin Campbell as well as someone from Greenpeace who claimed that oil was destroying the climate. Udall, Andrews and Campbell drew attention to peak and decline from supply constraints; Nehring, whose earlier work was very good, seemed to extol the exaggerated claims of the USGS, with a great faith in non-conventional oils; and Lovins saw a technological solution in terms of using hydrogen and bio-mass for petrochemicals, and in more efficient vehicles. Overall, the programme was sound and the journalistic "balance", that all programmes demand, did not seriously detract from the stress given to the very serious implications of oil depletion.

131. Middle East Reserves Revised Down

An article in the *Oil & Gas Journal* of June 17th indicates that Middle East reserves have to be revised down, as we have long suspected. The quoted revisions are given in the following table. It mentions that the Saudi government has "mandated" that reserves shall not decline, which evidently may apply more to the reporting than to what is in the reservoir. The ASPO estimate refers to what will be produced from known fields to 2075, a subtle distinction, but one that helps overcome reporting and definitional confusions.

end 2001(Gb)		O&GJ Reported Proved Reserves		ASPO
Country	Past Prod	31.12.01	Revised	
Abu Dhabi	17	92.2	61	59
Iran	53	89.7	69	70
Iraq	27	112.5	91	95
Kuwait	30	94	86	55
Saudi Arabia	91	259.3	160	194

132. The Twin Humps of Eurasia

Russia seems to be acquiring a key role in world depletion, especially if the Middle East itself is constrained by low investment or invasion, recognising also that it has to run ever faster to stand still in its desperate efforts to offset the natural decline of its ageing fields. It has a concentrated geological habitat with most of its oil in a few very large fields, found long ago, leaving much less potential for future discovery.

The Soviet regime was probably quite efficient in exploration. It was able to drill boreholes for scientific information, whereas the Western explorers had to pretend that every wildcat had a good chance of being successful in order to attract the funds to drill it.

Russian geochemists were indeed at the forefront of the research that led to the critical understanding of source-rocks. Accordingly, we may assume

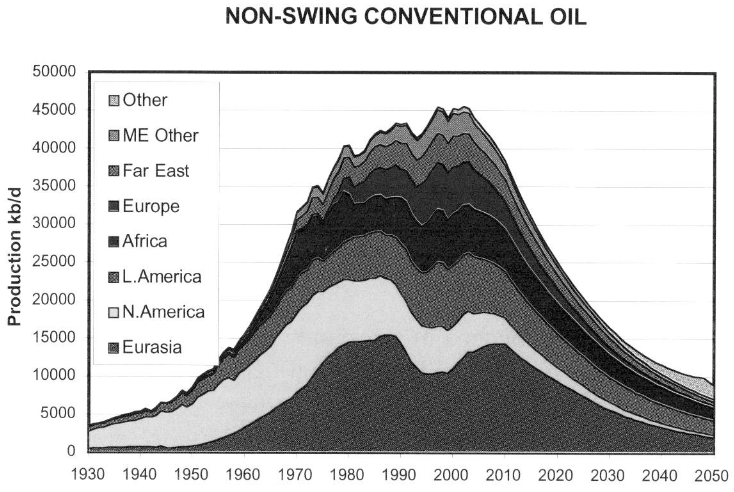

NON-SWING CONVENTIONAL OIL

Figure 36

that they found the main producing basins and many of the giant fields within them, apart perhaps from those in remote Arctic areas, which are here treated as *Non-Conventional*. Production was probably a very different matter being governed by central planning, which at times both over-produced the reservoirs, and at others failed to realise their full potential. In any event, the fall of the Communists led to an abrupt decline in production. It is now being reversed as the situation returns to normal, leading to a second peak around 2010, partly producing what would have already been produced but for the interruption.

Eurasia, as here defined, comprises the former Communist bloc of the Soviet Union, China and eastern Europe. Its twin-humped depletion profile has a large impact on the Non-Swing production as a whole, namely that outside the five main Middle East producers. As modelled, the region as a whole reaches a flat-topped peak from 1997 to 2002 at about 45 Mb/d, and is now set to decline at between 3% and 4% a year. The overall depletion profile is a fair approximation of a classic Hubbert curve. The relatively short span of flush production from Europe, mainly the North Sea, is noteworthy.

133. BP improves its Statistical Review of World Energy
The ASPO Statistical Review of World Oil & Gas pointed out two very misleading items in the BP Review.

- A plot of historical reported reserves, which, by failing to backdate revisions to their discovery, gave the impression that more was being found than was the case.
- A plot showing Reserve to Production Ratio quoted in years, as if it were remotely plausible for production to remain constant for a given number of years and then stop dead overnight.

It is gratifying to find, therefore, that BP has now decided to change, without comment, the first of the offending items in its latest edition, published in June 2002, to show only reported reserves over three years. Let us hope that the second will go in the next edition, so that this widely used publication will begin to provide a valid picture of the actual situation.

134. "Oil Change" – a New Book on Oil Depletion and its Consequences
A new book is to be published in German by DeutscheTaschenbuch Verlag under the auspices of ASPO and the Global Challenges Network, and will by launched in Munich in September. It covers oil and gas depletion, addressing also the consequences and possible solutions.

Ölwechsel! by Colin Campbell, Frauke Liesenborghs, Jörg Schindler and Werner Zittel (assisted by Helga Roth, translator).

Newsletter No. 20, August 2002

135. Country Assessment Series – United Kingdom

The United Kingdom had a strong Neolithic culture, highlighted by the famous astronomical observatory of Stonehenge, long before falling to the Romans in 55 BC. The Roman occupation lasted only a few centuries, but left an indelible mark. It was followed by the dark ages of Viking and Saxon invasions, culminating in the arrival of recycled Danish Vikings from Normandy in 1066, the last military invasion.

General stability brought political and economic progress, including the creation of Parliament, as one of the earlier democratic institutions. The Kingdoms of Wales and Scotland were absorbed into what, in 1801, became the United Kingdom, and Ireland was generally subjugated. Seafarers stimulated trade and exploration throughout the world, paving the way for the British Empire. At its peak in the reign of Queen Victoria, Britain had become the premier world power. Great achievements were recorded in the fields of science, literature and culture.

Britain also led the Industrial Revolution during the 18th Century with mills powered by water to make cloth for export to its colonial markets. The wealth, so created, led to the rapid growth of capitalism, banking, usury, investment and a financial economy. Self-sufficient peasants became wage-earners, consumers and tax payers, many working in gruesome industrial slums. Mechanisation based on iron and steel took many directions. Iron smelting made new demands for energy, at first from firewood, but later from sea-coal, namely detrital coal that had been washed from cliff outcrops and could be collected on beaches. Coal mining followed, at first from shallow pits. The development of steam driven pumps made it possible to deepen the mines below the water table. The pumps were later adapted to provide a steam locomotive for transport, opening the age of the railway that further stimulated trade. The internal combustion engine, using gasoline refined from crude oil, was an evolution of the steam engine with its pistons turning wheels.

Britain successfully resisted and eventually defeated an epoch of French expansion under Napoleon, but during the 19th Century found itself increasingly threatened by a newly united Germany that was overtaking it in industrial prowess and was seeking its own colonial markets. These pressures eventually led to two world wars during the 20th Century, in which the United States intervened to its own advantage, especially in regard to accessing Middle East oil that had previously been a British sphere of influence. Although victorious, Britain was mortally weakened by the wars and voluntarily gave up its Empire, retreating to its island to live in the shade of its former glory. It half-heartedly joined a newly united European community, preferring to retain its particular links with the United States, which eventually replaced the old empires of France and Britain with a new global economic, and now military, hegemony driven by, and for, the dollar.

Massive immigration from the former Empire followed the last War, being permitted at first in a sense of colonial responsibility, but later exploited as a source of cheap labour. The indigenous population aged from falling fertility due to affluence, but the overall population expanded to almost 60 million, with the immigrants and their descendants making up more than ten percent.

Most of Ireland had seceded in 1922, with 26 counties becoming a republic in 1947, leaving a form of civil war to continue in the remaining six counties of northern Ireland. Scotland and Wales are now recovering earlier autonomies with independent legislatures; and various regions within England are heading in the same positive direction. Various immigrant cities have developed, some becoming almost small replicas of Karachi or Kingston, Jamaica. Britain has moved far from its grand imperial past, although echoes remain.

Britain has had a long oil history, both within its own territory, and through the early prominence of its oil companies in the Middle East, Mexico and Venezuela. BP was the flagship with major holdings in Iran, Iraq and Kuwait, while Shell, an Anglo-Dutch enterprise, had a strong position in the Western Hemisphere. BP was once almost a national oil company with a 51% government shareholding and corresponding responsibilities, but Mrs Thatcher disposed of that. Formerly the world's largest vendor of crude oil, it now relies more on merger and acquisition, exemplified by the recent successful take-overs of Arco and Amoco. Its Chairman and Chief Executive now sit on the board of Goldman Sachs, underlining its transition to a financial institution. It claims that BP stands for *Beyond Petroleum*, but cynics are wont to say it stands rather for *Blair Petroleum*, alluding to its impressive political influence (see Evening Standard 19.07.02).

Non-conventional oil shale had been mined in Scotland in the 19th Century, leading to pioneering refinery processes, and minor oilfields had been found onshore in and before the Second World War. But the great thrust came during the 1960s, with the development of the offshore extensions of a prolific belt of gas fields, derived from deeply buried Carboniferous coal measures, that had been discovered in Holland in 1957. Exploration moved northwards in the North Sea to be rewarded by the discovery of Jurassic rifts, containing prolific source rocks, deposited 150 million years ago, which yielded one giant field after another.

Britain entered a phase of socialist government after the War, such that the early stages of its oil boom were dominated by state entities: the British Gas Council and the British National Oil Company. That ended with a reaction to excessive Trade Union demands, especially from the coal workers, leading to an eruption of new capitalism under Mrs Thatcher, who was able to undermine the miners' control of energy by the new indigenous oil supplies that were coming ashore. The state entities, which could have

United Kingdom		
Rates Mb/d		
Consumption	2001	1.65
Production	2001	2.26
	Forecast 2010	1.31
	Forecast 2020	0.72
Discovery 5-year average (Gb)		0.27
Amounts Gb		
Past Production		18.78
Reported *Proved Reserves*		4.93
Estimated Future Production to 2075		
	From Known Fields	11.1
	From New Fields	2.1
	Future Total	13.2
Total Production to 2075		32
Current Depletion Rate		6.1%
Depletion Midpoint Date		1998
Peak Discovery Date		1974
Peak Production Date		1999

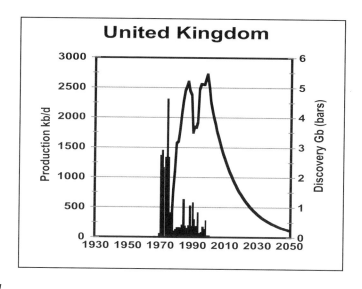

Figure 37

managed long-term depletion to the national interest, were disbanded, and the major international oil companies, along with many small independents, were given every encouragement to deplete the resources as fast as possible.

Production soared as the giant fields were brought on stream with the help of impressive advances in offshore engineering. An early peak was reached in 1987 at 2.6 Mb/d, before production fell as a consequence of a major accident at Occidental Petroleum's substandard Piper Field that called for widespread revisions to operating practices and installations. Production growth later resumed, reflecting a second cycle of smaller discoveries, to reach its overall peak in 1999 at 2.7 Mb/d.

Although the rich deposits of the North Sea dominated production, some other lesser finds were made elsewhere. Lower Jurassic source-rocks gave a solitary large field in Dorset in the otherwise barren English Channel and Western Approaches, and a Carboniferous gas field was found in the Irish Sea. Efforts to find another oil play on the Atlantic margin continue but are likely to be doomed, because the essential prolific Jurassic source-rocks, if present at all, are now too deeply buried to generate oil. The isolated large deposits West of the Shetlands are effectively freak occurrences depending on unique re-migration from earlier accumulations. The scope for gas in this province is more promising, but it will not be cheap.

Britain's brief oil age is in decline. The major companies are withdrawing to be replaced by smaller companies, mopping up small satellites and step-outs, as well as scavenging tail end production from ageing platforms.

Oil production is set to decline at about 6% a year, falling to about half its present level by 2010 and ten percent by 2037. Britain currently consumes 1.65 Mb/d, making it a net exporter. With flat demand, consumption will exceed production around 2007, with the percentage of imports set to rise to 20% in 2010, 50% in 2020, and 95% in 2050. Falling demand from recession would however reduce these import needs.

Gas production is more difficult to forecast due its very different depletion profile. About 3360 G m^3 have been discovered, of which about half have been consumed. Production stands at 106 G m^3; with consumption at 95 G m^3. Making some allowance for new finds, it looks as if production can be maintained at around this level until about 2015, when it is likely to decline at about 20% a year to exhaustion in 2035. Imports will have to rise rapidly from 2015 onwards unless demand can be cut.

It is difficult to imagine the condition of the country fifty years hence. Its own oil and gas will have been substantially exhausted. Failure by the government to recognise natural depletion until too late will have left the country unprepared. Concerns about safety and the environment will have likely delayed the development of nuclear power. Re-commissioning old abandoned coal mines will prove difficult and costly. The growing contribution of solar, wind and tide power will be useful but insufficient.

Soaring energy prices from around 2010 will, it may be supposed, finally lead to falling demand as recession bites in earnest. Even if financial overseas investments, effectively using slave labour in distant lands, ameliorate the internal stresses, it is difficult to avoid the conclusion that the country will find itself unable to support anywhere near its present population, a population, which by then will be, in large measure, of immigrant and mixed extraction. The iron-grip of oil depletion will have left its claw marks.

136. Soaring cost over-runs on Non-conventional oil production

Several reports stress that the costs of expanding *Non-conventional* oil production in Canada are running far over budget. The Edmonton Journal of July 26th for example reports that Shell's Athabaska oilsand project is now expected to cost $5.7 billion compared with a budget of $3.8. Shell Canada's second quarter's earnings fell in parallel to $73 million, down from $314 last year. Suncor's production costs have risen to $16.25 a barrel, far above a targeted $9.

Extraction also makes heavy demands on water, with as much as 26% of Alberta's underground freshwater being used in oil recovery, with even more needed to support expansion (see Platts Oilgram). The government is considering making a charge for this water, which will lift oil extraction costs further. As much as 4 barrels of water are used for every barrel of oil extracted, with conventional waterflood and enhanced recovery making even more demands. Apparently it is not only the cost of the water that is of concern but its availability.

The resources of *Non-conventional* oil may be huge, but they are no substitute for *Conventional* oil in terms of cost and extraction rate. Their entry will therefore have a negligible impact on overall peak.

137. Corporate Confession

Robert Anderson, the former head of Arco, put it bluntly when he said, a few years ago, that the oil industry was "a sunset industry – and the sun is low in the sky". Oil executives today try to put a brave face on the situation choosing their words with care. One reason is to try to encourage new recruits, as their staff is ageing about as fast as the reserves.

Harry Longwell, the Executive Vice-President of Exxon-Mobil, did however come close to confessing to depletion at the Offshore Technology Conference. He said that oil demand would have to rise at 2%, to sustain economic growth before admitting that about half the volume needed to meet such demand in 2010 is not on production today. He estimated the

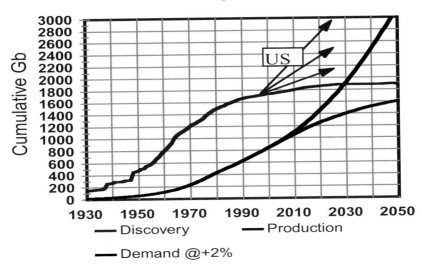

World Discovery & Production

— Discovery   — Production

— Demand @+2%

Figure 38

cost of providing it a trillion dollars or 100 billion a year, much more than is being currently invested. This is an oblique way of saying that supply will not in fact rise at 2% a year. This is made rather plain by the attached graph, although it is not the full picture. The black line plots the discovery of oil and gas liquids, based on Proved & Probable Reserves, with revisions backdated to their discovery. The inflection to decline came in 1964. The trend is extrapolated forward in purple, giving a total of 1900 Gb. The red line depicts past production of conventional crude oil and condensate, projected forward on the assumption of the same 1900 Gb total with flat demand to 2010 followed by decline at the then depletion rate. On this basis, there will still be about 300 Gb left to produce after 2050, so we are evidently not about to run out. The blue line shows a future growth of demand of 2% a year. It is clearly not attainable. Production in oilfields declines slowly due to the physics of the reservoir, so it would be physically impossible to deplete the reserves by 2030.

The black arrows show the implausible High, Low and Mean estimates of discovery between 1995 and 2025 claimed by the USGS. It, even more implausibly, claims that past discovery is under-estimated or under-reported by as much as 10% (Low), 27% (Mean) or 37% (High) even though, by definition, revisions to *Proved & Probable Reserves* are statistically neutral.

Even if this were true, it would not change the shape of the trend with its peak and decline.

It is very evident that a 2% growth in supply is not attainable, which, if Mr Longwell is correct, implies that economic growth is not either.

138. Forecasting Global Oil Supply 2000–2050

The M.King Hubbert Center for Petroleum Supply Studies dedicates its June issue to the above subject, including tables showing realistic reserves and future production forecasts by country.

139. Venezuela

In an editorial in the May issue, the Petroleum Review assesses the predicament faced by Venezuela in the face of depletion. It needs oil revenue to fund a range of much needed government programmes, yet it needs to invest in ever more infill drilling to offset the natural decline of its ageing oilfields. Should it

i)   encourage foreign investment?
ii) allow capacity to shrink by using oil revenue for other purposes?
iii) become an ever more nationalistic member of OPEC, encouraging rigid quota compliance to drive up prices?
iv) have another coup?

Since infill drilling is a losing battle, no matter who pays for it, Option iii) seems the only sensible policy. The US, which evidently supported the failed coup against President Chavez, would no doubt encourage Option iv). A passage from the manuscript of an as yet unpublished penetrating book explains why.

> Soon after US petroleum production had peaked, official policy began emphasizing "free trade" as a global panacea for unemployment, under-development, despotism, and virtually every other economic or political ill. Through its manipulation of the rules of global trade, the US sought to maintain and increase its access to natural resources worldwide. Those rules – written primarily by US-based corporations and encoded in policies of the International Monetary Fund (IMF), the World Bank, and the World Trade Organization (WTO), as well as in treaties like the North American Free Trade Agreement (NAFTA) – essentially said that wherever resources lie, they must be available for sale to the highest bidder; whoever has the money to buy those resources has a legally defensible right to them. According to those rules, the oil of Venezuela belongs to the US every bit as much as if it lay under the soil of Texas or Missouri. Meanwhile technology, or "intellectual property," was regarded as proprietary; thus nations with

prior investments in this strategy were at an advantage, while "underdeveloped" nations were systematically discouraged from adopting it.

140. Improving Oil Company Accounting

The structure of a manufacturing business comprises the plant, which is an asset, and the ongoing business of purchasing raw materials, fabricating products and marketing. Although there are various tricks of treating an operating cost as an asset and in the manner of reporting the details of debt, amortisation etc., it should be well within the grasp of any honest accountant to measure the assets, profit and loss.

But the oil industry is different in one critical respect. In addition to extracting, transporting, refining and marketing, it has to find the oil before it can produce it. What it finds lies far underground where no one can exactly measure it. The oil accountant therefore faces two critical questions: how much has the company found? and when did it find it?

To prevent fraud, the Securities & Exchange Commission set rules whereby companies could report only *Proved Reserves* for financial purposes, which in practice meant the best estimate of what current producing wells would produce in the future. (There is also a sub-category of *Proved Undeveloped*, which we need not explain here). The thrust of this arrangement was to prevent the exaggeration of reserves, with under-reporting being tolerated as prudence. The upward revisions were reported on the date they were determined, which sounds reasonable enough, but in fact greatly distorts the picture.

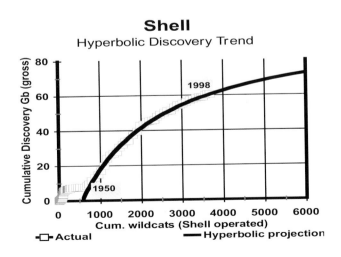

Figure 39

In the aftermath of the scandals of false accounting that rock America, starting appropriately enough with the oil and gas company, Enron, efforts are now being directed to tighten up the rules so as to give the investor a better idea of the value of his stake. One great contribution to this end would be a new requirement for oil companies to report the date of discovery of their oilfields with any revisions to the estimates of their size being attributed to that date. It would soon reveal that companies are replacing reserves less by discovery than by imaginative accounting. Knowledge of the discovery trend would provide the investor with an entirely different image of his assets. If the information showed, as is undoubtedly the case, that the industry has been finding much less than it was producing, the investor might reasonably conclude that pending shortages would drive up the value of the remaining reserves, which might thus become an appreciating asset. In making his dispositions, he might turn to companies that were conserving their reserves rather than producing them at depressed prices, even after taking into account the time-cost of money.

To know the date of discovery is quite as important as to know the amount discovered. The date of discovery of an oilfield is that of the completion of the first successful well drilled upon the prospect responsible for it, even if economic or technological developments over time affect the amount recoverable, and the successive estimates thereof. The attached plot shows the declining discovery rate of Shell, which is probably better than most of its competitors. Reports of percentage reserves replacement as provided in company financial reports are essentially fraudulent.

141. Discovery in 2001 fell

IHS (Petroconsultants) reports that the discovery of oil and gas liquids fell in 2001 to 8.9 Gb. for the world outside North America. It probably means that total world discovery fell from about 24 Gb in the bumper year of 2000 to about 16 Gb, slightly above the average for the last ten years of almost 14 Gb. At a guess, about 6 Gb of it is conventional crude oil (excluding deepwater, polar oil and NGL). It is far below consumption, now running at around 27 Gb.

142. The illusion of Reserve Growth confirmed

A new report by the ever honest and reliable Norwegian Petroleum Directorate (August 2001) includes the attached graph, which confirms the view already emphasised herein about the illusion of reserves growth. Although it is based on what the companies claimed in their applications for licences, which may be very different from what the geologists actually thought, it does give a clear picture.

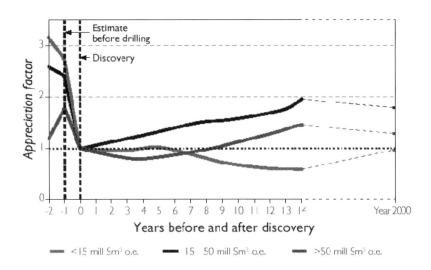

Figure 40

In the case of a large field, explorers are able to report their genuine estimates because the economic threshold is low. The engineers for their part report low initial reserves following discovery, to reflect the initial phases of development only. The reserves are revised upward over time as subsequent phases are added to match the original volumetric estimates of the explorers. In the case of small fields, the explorers are obliged to exaggerate the potential, and the fields do not always live up to the claims made for them.

Both the engineers and the geologists know their job and are perfectly capable of making sound estimates: but what they "sell" to their management or report to government are commonly influenced by non-technical factors. In any event, it serves to dispel the widely held but mistaken view that reserve growth is driven by unforeseen technological developments.

143. New forecast of the imminent peak and decline of European oil

An article by Dr Smith in the August issue of the *Petroleum Review* describes and illustrates the imminent peak and decline of European oil supply, and its vulnerability in a deteriorating world situation. He exposes the misleading nature of "reserve growth" and Reserve to Production Ratio, as well as the failure of the oil industry and the International Energy Agency to properly explain the issue. The arguments, interpretation and data assessment will be very familiar to readers of this newsletter.

144. The Hidden Iraq Agenda

From *The Guardian*, Saturday August 10 2002

Anthony Sampson analyses the roots of America's fear of the Iraqi dictator, and warns that toppling him might cause less stability and more insecurity.

Quote

Is the projected war against Iraq really turning into an oil war, aimed at safeguarding Western energy supplies as much as toppling a dangerous dictator and source of terrorism? Of course no one can doubt the genuine American hatred of Saddam Hussein, but recent developments in Washington suggest oil may loom larger than democracy or human rights in American calculations.

The alarmist briefing to the Pentagon by the Rand Corporation, leaked last week, talked about Saudi Arabia as 'the kernel of evil' and proposed that Washington should have a showdown with its former ally, if necessary seizing its oilfields which have been crucial to America's energy.

And the more anxious oil companies become about the stability of Saudi Arabia, the more they become interested in gaining access to Iraq, site of the world's second biggest oil reserves, which are denied to them. Vice-President Dick Cheney, who has had his own commercial interests in the Middle East, baldly described his objection to Saddam in California last week: 'He sits on top of 10 per cent of the world's oil reserves. He has enormous wealth being generated by that. And left to his own devices, it's the judgment of many of us that in the not too distant future he will acquire nuclear weapons.'

If Saddam were toppled, the Western oil companies led by Exxon expect to have much readier access to those oil reserves, making them less dependent on Saudi oilfields and the future of the Saudi royal family. The US President and Vice-President, both oilmen, cannot be unaware of those interests. Of course Western policies towards Iraq have always been deeply influenced by the need for its oil, though they tried to be discreet about it. The nation of Iraq was invented in 1920, after the First World War. The allies had 'floated to victory on a sea of oil' (as the British Foreign Secretary Lord Curzon put it), but they preferred to conceal their dependence on it: 'When I want oil,' said Clemenceau, the French Prime Minister, 'I go to my grocer.'

But both Clemenceau and Curzon, while they talked about Arab interests and self-determination, knew that what really mattered in Iraq was the oil that was emerging in the North; and the British and

French succeeded in controlling the precious oilfields at Mosul.

Iraqi oil became still more desirable after the oil crisis of 1973 which enabled the Arab producers to hold the world to ransom; and the discovery of huge new oil reserves in the South made Iraq more important as a rival to Saudi Arabia – and Saddam more exasperating as an enemy.

It is true that since the Seventies, as the shortage turned into glut, producing countries have become much more dependent on the global marketplace. Countries which hoped to develop political clout by allocating oil supplies soon found they had to compete to sell their oil wherever they could. And Western companies developed new oilfields nearer home, or in friendlier countries.

But America and continental Europe still depend on uncertain developing countries, mostly Muslim, for much of their energy, and in times of crisis the concern about oil supplies returns. Western oil interests closely influence military and diplomatic policies, and it is no accident that while American companies are competing for access to oil in Central Asia, the US is building up military bases across the region.

In this security context the prospect of a 'terror network' controlling Saudi Arabian oil, which last week's briefing to the Pentagon conjured up, presents the ultimate night mare: a puritanical Islamist regime in Saudi Arabia, and perhaps in other Gulf states, would be prepared to defy the marketplace, with much less need to sell their oil than corrupt monarchies or sheikhdoms. Bin Laden, himself a Saudi, made no secret of his overriding ambition to rid his country of corrupt rulers and return to its austere Islamist roots.

In this scenario Americans would be more determined to get access to oil in Iraq, and the demands to topple Saddam would be reinforced.

There are undoubtedly many different and sometimes conflicting strands behind Washington's attitudes to Iraq. Certainly the public sense of outrage about 11 September, and the fear of terrorism, remains the most potent political force behind the moves against Saddam – reinforced by Israel's dread of Iraq's weaponry.

But there are also the longer-term geopolitical arguments in the Pentagon and the State Department, with commercial pressures behind them, about the need for energy security. And these have become more urgent with the growing worries about the Saudis.

The crucial question remains: would toppling Saddam safeguard Iraq's oil for the West? After all, both previous American Presidents

– Clinton and George Bush Snr – were persuaded not to overthrow Saddam, because the alternative could well be a more dangerous power vacuum. That danger remains. If Iraq were to split into three parts, as many expect, the new oil regions in the South might be become still less reliable, in a region dominated by Shia Muslims who have their own links with the Shia in Iran. And a destabilised Saudi Arabia could make a power vacuum still more dangerous.

The history of oil wars is not encouraging, and oil companies are not necessarily the best judges of national interests. The Anglo-American coup in Iran in 1953, which toppled the radical Mossadeq and brought back the Shah, enabled Western companies to regain control of Iranian oil: but the Iranian people never forgave the intervention, and took their revenge on the Shah in 1979.

The belief that invading Iraq will produce a more stable Middle East, and give the West easy access to its oil wealth, is dangerously simplistic. Westerners live in a world where most of their oil comes from Islam, and their only long-term security in energy depends on accommodating Muslims.

Newsletter No. 21, September 2002

145. County Assessment – Saudi Arabia
This month we take a look, with some trepidation, at Saudi Arabia, potentially the world's largest producer of oil, but one very much subject to artificial politically motivated pressures, which are likely to affect future production in many radical ways. Furthermore, the data on present reserves are particularly unreliable.

Saudi Arabia
It is thought that Cro-Magnon Man, the most advanced early human subspecies, may have evolved in the Middle East only to exterminate poor old Neanderthal in an early example of genocide. In any event, the region has had a very long history. It is curious that some of the world's main religions, Judaism, Christianity and Islam, all had their roots here. They are monotheistic, believing that the Divinity shows his hand on Earth. The Jews, to their misfortune, are still waiting for their prophet; the Christians had theirs but he was nailed to a cross near Jerusalem; and the Moslems had theirs when Muhammed was born in Mecca in Saudi Arabia around 570 AD.

Whereas the Christians give emphasis to the life hereafter, Moslems believe they please their God in their daily lives, seeing the wrath of God if things go badly. They are guided by rather secular priests using the koran, which records the revelations to the Prophet himself. There was a difficulty over the succession, when the Prophet ordained that his mantle should fall to his son-in-law, when the elders preferred the son. This conflict later led to a schism between the Sunni and Shi'ite factions, which still manifests itself throughout the region. Saudi Arabia is Sunni, Iran Shi'ite and Iraq a mixture, which perhaps explains why it needs a strong leader, if it is to remain a single State.

The Moslems were successful, and, from the 7th Century, spread their dominion throughout the Middle East, North Africa and parts of southern Europe, achieving a high level of culture and learning. In the course of this expansion, the power centres shifted from Saudi Arabia to the outposts of empire, especially in Turkey and Egypt, which at different times effectively controlled the homeland, although the desert interior remained under the control of various warlords. One such was Muhammad ibn Saud whose dynasty was founded near what is now Riyadh in the 15th Century

A new religious leader had appeared on the scene in the mid 18th Century in the form of Muhammad ibn Abd al-Wahhab, who founded a fundamentalist sect and supported the Sauds. They were however defeated in warfare with a neighbouring tribe in 1865, being forced to flee temporarily to Kuwait before gradually recovering their lands under a new ibn Saud, a legendary leader of outstanding physical prowess who is said to have sired more than 1000 progeny.

In the epoch leading up to the First World War, most of the Middle East outside Iran belonged to the Ottoman Empire, but enjoyed a degree of autonomy under a rather vague administration. It was a somewhat decadent empire, controlled by its sultan in Istambul, but was propped up to some extent by Britain and France who saw it as a useful buffer to Russian

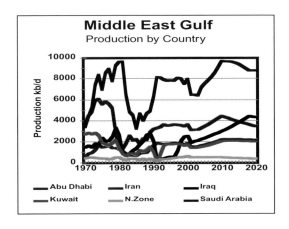

Figure 41

expansion. Its fate was, however, sealed when it decided to side with Germany on the outbreak of the First World War. That led the British to try to foment an Arab rising under the colourful figure of an Oxford academic, Lawrence of Arabia, who donned Arab kit and headed into the interior on a camel. He sponsored the Grand Sherif of Mecca, who held dominion over the western part of the Peninsula including what is now Palestine, reckoning that he was a direct descendent of the Prophet. A British army, assisted by Arab irregulars led by Faisal, the son of the Grand Sherif, marched north to eventually take Palestine and Syria on the promise of the creation of an Arab Kingdom, once the hostilities were over. But the British reneged on their promises at the Treaty of Versailles following the War. This left the door open for the further conquests by ibn Saud, who eventually proclaimed the Kingdom of Saudi Arabia in 1932. It was duly recognised by Britain in return for guarantees for the territorial integrity of the Gulf sheikhdoms. The Grand Sherif left Mecca, when it fell to the Sauds, and retired to Cyprus, thoughtfully taking the Treasury with him. But, the British did try to make amends. They offered one of his sons, the Kingdom of Jordan, and another, the throne of Iraq in the post-war carve up of the Middle East, when it was eventually achieved, despite the conflicting aims of the British Foreign Office, the British India Administration that had a stake in the Middle East, and the French. The boundaries of the Saudi kingdom remain however a little tenuous, especially with the Yemen.

Having few resources, the desert kingdom relied heavily on the income derived from pilgrims visiting the holy shrines of Mecca and Medina, but during The Great Depression of 1930, the flow of pilgrim dried up, leaving the King strapped for cash. Seeking some new enterprise, he turned to a curious disaffected British colonial administrator, by the name of Harry St John Philby, who had established a Ford dealership in the Kingdom and was none other than the father of the well known British double agent, Kim Philby, of Cold War fame or infamy.

In 1932, the Standard Oil Company of California (now Chevron-Texaco) discovered oil in Bahrain, a few miles off the coast of Saudi Arabia, and Philby arranged for an American mining engineer cum archaeologist, by the name of Karl Twitchell, to investigate the possibilities of Saudi Arabia itself. The report was favourable, and after lengthy negotiations, a concession with California Standard was signed in May 1933 in return of a front end payment of £35,000 in gold bars, duly delivered in seven boxes by a P&O liner from London. Neither side realised the immensity of the deal it had struck.

California Standard later brought in Texaco as a partner to help fund Aramco, the operating company, which after some heroic exploration in the best of pioneering traditions made the first discovery in 1940, before hostilities in the Second World War brought operations to a standstill. By 1943, ibn Saud was partly blind and ailing, and Aramco had become concerned about its rights, seeking the support of the US government. As a result, Exxon and Mobil, two other American companies, were brought in to join Aramco, which they eventually did in flagrant disregard to their commitments under the famous red-line agreement that governed the

Saudi Arabia		
Rates Mb/d		
Consumption	2001	1.34
Production	2001	6.47
	Forecast 2010	9.69
	Forecast 2020	8.78
Discovery 5-year average (Gb)		0.66
Amounts Gb		
Past Production		91
Reported *Proved Reserves*		259
Estimated Future Production to 2075		
From Known Fields		194
From New Fields		14
Future Total		209
Total Production to 2075		300
Current Depletion Rate		1.1%
Depletion Midpoint Date		2020
Peak Discovery Date		1948
Peak Production Date		2012

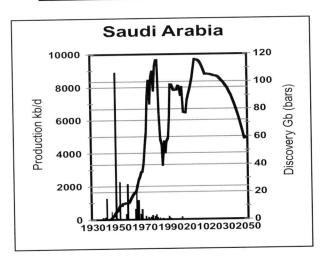

Figure 42

partners in the Iraq Petroleum Company, to which they were both party.

The American influence in Saudi Arabia was cemented when President Roosevelt met the King in February 1945, presumably promising support for his regime in return for access to his oil.

Ghawar, the world's largest oilfield, had already been tentatively identified before the war, but was confirmed in 1948. It should have assured the four companies and their homeland with an ample supply of oil for many years to come, had not it coincided with the creation of the State of Israel after Zionist terrorists had forced the British to surrender their administration of Palestine. A form of civil war has escalated ever since, now arousing new passions, placing world oil supply in jeopardy.

When ibn Saud died in 1953, the kingdom passed to his son, also named Saud who, proving somewhat inadequate, shared power with his brother, Faisal, before being deposed by the family in 1964. Faisal lasted until 1975 when he was assassinated by a disgruntled prince, to be succeeded by his half-brother Khalid, who was in turn followed by yet another half-brother, Fahd, in 1982. It does not sound a very happy royal family, fraught by tensions between the descendants of the many wives of ibn Saud.

The enormous wealth that flowed to the Sauds from their oil revenues made it a curious state. The people were controlled by a Wahabbi religious police under strict laws, whereby adulterers are stoned and thieves subjected to amputation. The population has grown rapidly to over 20 million with an average age of under 20. It includes large numbers of foreign workers and Palestinian refugees, all of whom depend directly and indirectly on oil revenue. Much of the wealth was spent on arms purchases, contrived by enterprising princes for royal fees, such that at times there have been more tanks in the desert than drivers. Deposits were also made in US banks, which unbeknownst to the Sauds, became collateral for soaring US domestic debt.

Relations with the United States have not always been easy, despite the long ties. In 1960, the country joined OPEC to try to regulate the world oil price by proration, following the example of the Texas Railroad Commission, when the US faced a similar challenge. In 1973, it participated with other Arab countries in restricting exports to the United States in response to the latter's support for Israel's military expansion. This triggered the First Oil Shock when oil prices increased five-fold, plunging the world into recession. The country expropriated Aramco in 1979, although still maintaining special relations with its previous owners. But it supported the US and its allies in the Gulf War, permitting US military bases to be established in its territory.

King Fahd is now ailing in Switzerland, and the succession is again in dispute. The crown prince, Abdullah, is evidently now de facto ruler, but Fahd's half brothers, Sultan and Nayef, are contenders. The throne threatens to become a fairly hot seat. It is assailed on one side by an increasingly disaffected population, who resent the Saud's US ties, and, on the other, by the US itself, which is now distancing itself, possibly as a prelude to new military operations in the Middle East, which may lead to control of the Saudi oilfields.

In geological terms, Saudi Arabia covers the western margin of the Middle East basin, and possesses two petroleum systems.

The main system depends of Jurassic source rocks and reservoirs in huge gentle structures, endowed with excellent salt seals that prevented the escape of oil. In fact, there is a single prime structural uplift that straddles the country to which oil from the adjoining basins has migrated over time. The world's largest oilfields Ghawar and its offshore extension Saffaniya, together holding some 130 Gb, lie on this uplift. In detail, these fields are made up of about ten compartments that would have been separate smaller oilfields had the charge of oil not been so great as to cause them to coalesce. Fields outside this prime belt are of about the same size as the individual compartments in the range of 5–10 Gb. As is evident from the discovery plot below, the two super-giant fields of Ghawar and Saffaniya, give a misleading impression of the territory's potential. In geological terms, it is a concentrated habitat with most of its oil in these two super-giant fields, implying in turn that future discovery will deliver much, much less. These two fields are ageing. The southern end of Ghawar has already watered out. A high level of infill drilling is needed to maintain production suggesting that there is in fact very little spare capacity that could be brought on rapidly. Adding significant capacity will take work, much investment and, above all, time.

The second petroleum system depends on deep Silurian source-rocks that have charged patchy and poor reservoirs in the overlying Permo-Carboniferous, yielding mainly gas and condensate, although also some oil on the shallower western extremity of the basin.

Future production is here modelled on the assumption that the world peak of conventional oil production was in 2000 and that recession will hold down demand, such that production is, on average, flat until 2010. By that date, the five Middle East producers may no longer be able to apply their swing role, offsetting decline elsewhere. Each of the swing countries is assumed to increase production until its depletion midpoint, when production declines at the then depletion rate, with the balance being taken up by Saudi Arabia, as shown in the figure. It is obvious that this is a very uncertain forecast, even ignoring the threat of US military action. Saudi reserves are themselves highly suspect, being a state secret. It is almost sure that the reported estimate of 259 Gb is far too high. A recent paper (O&GJ 17.6.02) suggests as little as 160 Gb is "Proved". It is noteworthy that Saudi Arabia reported 167 Gb in 1989, before it had reason to inflate its reserves to protect itself in the "OPEC quota wars". It may be assumed that these were Proved Reserves in a financial sense, being equivalent to about 75% of the actual amount, which would accordingly be about 220 Gb. Production since then has been 33 Gb, reducing the reserves to 184 Gb, to which may be added about 6 Gb for new discovery, giving a total of 194 Gb.

Consumption is reported by the BP Statistical Review as 1.3 Mb/d, which is surprisingly high when compared for example with 1.6 Mb/d in the United Kingdom. If correct, much must be petrochemical feedstock or exported as product.

In conclusion, we may describe Saudi Arabia as an anachronism in the modern world. Its days as a feudal monarchy are almost certainly numbered. Its growing young population, with its large foreign component, who are exposed to the outside world through the antennas of CNN, are increasingly disaffected and disillusioned. While there may be no particular love lost between them and their Palestinian cousins, the universal condemnation of Israel and its US sponsor may galvanise their emotions. Their own understandable despair and resentment may, accordingly, erupt in violent reaction, putting the world's future oil supply in jeopardy, whatever the composition of the future government of the territory. There is a great deal at stake.

Important Postscript

Mo Mowlam, who was until recently a member of Blair's cabinet, has written a compelling article in the Guardian of 5th September 2002, entitled "The Real Goal is the Seizure of the Saudi Oilfields" http://www.guardian.co.uk/Iraq/Story/0,2763,786332,00.html

She argues that the US government must realise that an invasion of Iraq would lead to mayhem throughout the Middle East, leading to the fall of governments, including the Sauds.

She concludes that this is in fact the object of the exercise.

> "Under cover of the war on terrorism, the war to secure oil supplies could be waged. The whole affair has nothing to do with a threat from Iraq – there isn't one. It has nothing to do with the war against terrorism or with morality. Saddam Hussein… is now the distraction for the sleight of hand to protect the West's supply of Oil. And where does this leave the British Government? Are they in on the plan or just part of the smokescreen?"

This comes as no particular surprise, as many commentators have reached the same conclusion, speaking of the new Republic of Aramco that would arise from the ashes of Saudi Arabia, but it is remarkable to find it now coming with the authority of a recent British cabinet minister.

146. USGS Study re-examined

Jean Laherrere has re-examined the USGS study finding inconsistencies and confusion even in what is reported from past production and reserves in the USA itself, quite apart from the flawed forecasts of the future. His graphs tell the story eloquently, showing how the USGS estimates are far out of trend. The Norwegian discovery curve is a firm robust trend pointing to an ultimate discovery of about 34 Gb with a 1000 wildcats, which is very optimistic as it is doubtful if the industry will drill 400 more wildcats to deliver only about 2 Gb. This compares with the NPD estimate of 42 Gb and 48 Gb of the USGS. Norway has had about twenty successive licensing rounds, in each offering the perceived best remaining prospects, which attract drilling commitments. The only chance of these higher estimates being realised is the discovery of an entirely new unforeseen basin in the deepwater or in remote Arctic waters. Pigs might fly.

147. Reserve Reporting

Now that the captains of industry and their accountants are increasingly to be found in handcuffs for exaggerating their assets, we should perhaps ask how the oil industry accounts stack up. After all, Enron, whose management heads the list of scoundrels, was, if not an oil company, at least an energy company.

A major oil company owns refineries, tankers and filling stations but by far the most important of its assets are its reserves of oil and gas, without which the business would soon come to a grinding halt. But these reserves lie far underground where even Arthur Anderson can be excused for uncertain accounting.

But if we were to launch a team of detectives to investigate we might be surprised to find that far from exaggerating their reserves, the oil companies have actually understated them. In fact, when asked about reporting practices, a former managing director of a major British oil company responded:

> "Naturally I did not think it right to claim all of the reserves
> that I inherited. That would not be cricket. I wanted to leave as
> much as possible for my successor"

So, they emerge as white as driven snow, notwithstanding their well known skills in corrupting governments for prime concessions.

To explain this unexpected conclusion, we have to go back to the early days of the United States where individual landowners owned the oil rights. As a result, the ownership of oilfields was highly fragmented not only by area and by depth, as shallow and deep reservoirs in the same field sometimes had different owners. In early years, there was no shortage of Enron-esque tricksters, exaggerating the size of discoveries for promotional purposes, and eventually the Securities & Exchange Commission moved to prevent such fraud by imposing rigorous rules for reserve reporting. In short, the owners of were able to report for financial purposes only the reserves being drained by their current wells, which were termed *Proved*. The reports related only to their particular holding and not to the field as a whole. No one minded if they under-reported the reserves, as the thrust of the rules was to stop fraud by over-reporting.

This long established practice was preserved by the industry as it moved internationally and offshore, most of the companies, being on the US exchanges and subject to SEC rules. They had no reason to complain at the rules, because they found the under-reporting the size of their reserves in this way conferred many benefits. It allowed them to smooth their assets, which would otherwise have fluctuated wildly from occasional discoveries separated by lean years, and it reduced tax in countries operating a

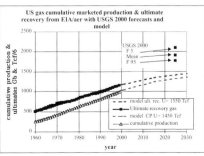

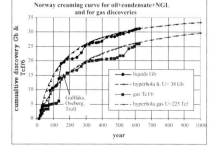

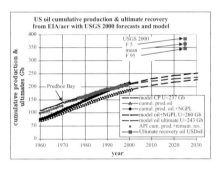

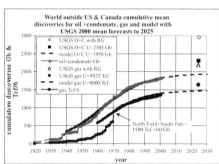

Figure 43

depletion allowance, based on *Proved Reserves*. For most purposes, it was a practical and equitable arrangement. The practice of reporting *Reserve to Production Ratio* in terms of years was a derivative, whereby companies could say that their reserves could sustain current production for a given number of years. What they really meant was that the reserves had been *Proved So Far*, it being implicitly assumed that more could be added as needed by exploration and drilling up the fields. In those days it was not an unreasonable assumption in a world perceived to have near limitless resources which could be tapped at will.

We can, accordingly, forgive Professor Adelman, working with such data, for being misled into making his famous pronouncement:

> *"Minerals are inexhaustible and will never be depleted. A stream of investment creates additions to proved reserves from a very large in-ground inventory. The reserves are constantly being renewed as they are extracted... How much was in the ground at the start and how much will be left at the end are unknown and irrelevant"*

The perception of inexhaustible resources however began to fade when one country after another passed the peaks of discovery and production despite every effort. It is now being replaced by questions of how much is left. Most of the fields are now drilled up to an optimal well-spacing, so little more can be added by new drilling. Advances in technology have also successfully raised the percentage recovered. It follows that *Proved Reserves* have evolved to the point that they cover not just the current wells but the fields as a whole. It means that the companies have less and less left in their under-reported inventories. Some still claim positive reserve replacement, but close inspection shows that it comes more from acquisition than new discovery. It is a perfectly valid financial measurement but does not reflect exploratory success.

This whole business has been greatly confused by the application of probability theory. Naturally all estimates are subject to a degree of uncertainty (or *Probability*) but one range of probability relates to the *Proved Reserves* of current wells, and quite another to what the fields as a whole are expected to deliver. So, it makes little sense to define *Proved Reserves* as having, say, a 95% probability of exceeding the stated value and to say that the field as a whole has a *Mean* probability.

The industry explores the world, drilling many dry holes in the process. The discovery of oil is a transcendental event in terms of adding reserves, and it follows that all the oil ever to be produced from the field in question, under whatever economic and technological conditions as may arise over its life, are logically attributable to the date of the original discovery.

While given a comparatively clean bill of health, oil companies could still improve their public accounts by back-dating their claimed reserves to the discovery date on which they were found. The brokers might recover from the initial shock of discovering that the companies are far from replacing their reserves in any real sense to conclude that what they have left would be an appreciating asset in increasingly short supply.

148. Shell confession

The Sunday Times of August 25th carried a short article entitled *"Oil is running out, says Shell"*, quoting the company as saying "we could be seeing oil shortages by 2025". Whether Shell actually said this in so many words is unsure, but the important point is that the media are beginning to get the message, penetrating the bland words and scenarios.

149. BP shares plunge 6% on falling production

The Times of September 5th carries a revealing article. BP's Chairman, Lord Browne, has evidently enjoyed something like hero status amongst the investment community, which now feels let down when BP announces some technical setbacks, temporarily cutting production.

"He doesn't walk on water" commented one analyst. It demonstrates the short-term nature of the stockmarket, and why the directors of oil companies have to sing to it. Browne's voice has evidently hit a wrong note. Need we wonder that the word *Depletion* is missing from his hymn sheet?

150. Bloomberg accepts the impact of depletion

The penny has dropped for the well-known New York financial institution, Bloomberg, who now alert their clients to the critical impact of oil depletion and the imminent peak in an excellent, well researched, article.
http://www.bloomberg.com/wealth/0902/sep.ft.crude.pdf

151. Biodiversity

Those concerned about the depletion of oil are often described as Pessimists, which is not a compliment, implying some form of moral turpitude. The following article by an eminent biologist however paints an even more depressing picture of a world of solitude after Man has destroyed all other life under flat earth economic principles. But perhaps it carries a message of encouragement. Perhaps those concerned about the depletion of the world's resources of energy will end up as Optimists for without the power of oil Homo Sapiens may after all be less able to secure his own extinction or loneliness.

John Gray is professor of European Thought at the London School of Economics.

Will humanity be left home alone?

John Gray, New Statesman

According to Edward O. Wilson, the greatest living Darwinian thinker, the earth is entering a new evolutionary era. We are on the brink of a great extinction the like of which has not been seen since the end of the Mesozoic Era, 65 million years ago, when the dinosaurs disappeared. Species are vanishing at a rate of a hundred to a thousand times faster than they did before the arrival of humans. On present trends, our children will be practically alone in the world. As Wilson has put it, humanity is leaving the Cenozoic, the age of mammals, and entering the Eremozoic – the era of solitude.

The last mass extinction has not yet been fully explained. Many scientists believe it to have been the result of meteorites whose impact suddenly altered the global climate, but no-one can be sure. In contrast, the cause of the present mass extinction is not in doubt: human expansion. Homo sapiens are gutting the earth of biodiversity.

The lush natural world in which humans evolved is being rapidly transformed into a largely prosthetic environment. Crucially, in any time span that is humanly relevant, this loss of biodiversity is irreversible. True, life on earth recovered its richness after the last great extinction; but only after about 10 million years had passed. Unless something occurs to disrupt the trends under way, all future generations of human beings will live in a world that is more impoverished biologically than it has been for eons.

Given the magnitude of this change, one would expect it to be at the center of public debate. In fact, it is very little discussed. Organizations such as the World Wildlife Fund press on with their invaluable work, and there are occasional reports of the destruction of wilderness; but for the most part, politics and media debates go on as if nothing is happening. There are many reasons for this peculiar state of affairs, including the ingrained human habit of denying danger until its impact is imminent; but the chief reason is that it has become fashionable to deny the reality of overpopulation.

In truth, the root cause of mass extinction is too many people. As Wilson puts it in his book Consilience: "Population growth can justly be called the monster on the land." Yet according to mainstream political parties and most environmental organizations, the despoilation of the environment is mainly the result of flaws in human institutions. If we are entering a desolate world, the reason is not that humans have become too numerous. It is because injustice prevents proper use of the earth's resources. There is no such thing as overpopulation.

Interestingly, this view is not accepted in many of the world's poor countries. China, India, Egypt and Iran all have population programs, as have many other developing nations. Opposition to population control is concentrated in rich parts of the world, notably the US, where the Bush administration pursues a fundamentalist vendetta against international agencies that provide family planning. It is understandable that rich countries should reject the idea of overpopulation. In their use of resources, they are themselves the most overpopulated. Their affluence depends on their appropriating a hugely disproportionate share of the world's non-renewable resources. If they ever face up to that reality, they will have to admit that their affluence is unsustainable.

Another reason for denying the reality of overpopulation is that the growth in human numbers is extremely uneven. In some parts of the world, population is actually declining. This is strikingly true in post-communist Russia. A precipitate fall in public health and living standards has led to a virtually unprecedented population collapse, which is set to accelerate further as an African-style AIDS die-off triggered by the country's enormous numbers of intravenous drug users begins to take hold. In other countries, such as Japan, Italy and Spain, declining fertility is leading to zero or negative population growth. Such examples have given currency to the silly notion that overpopulation is no longer an issue – that, if anything, it is a slowdown in population growth that we should be worrying about.

But while human numbers are falling in some parts of the world, in others they are exploding. The population of the Gulf States will double in around 20 years – against a background of nearly complete dependency on a single, depleting natural resource. Again, despite China's admirable one-child policy, its population will go on growing for much of this century. Globally, the human population will continue to rise for at least a century – even if worldwide fertility falls to replacement level tomorrow. In 1940, there were around two billion humans on the planet. Today, there are about six billion. Even on conservative projections, there will be nearly eight billion by 2050. Eight billion people cannot be maintained without desolating the earth. Today, everyone aspires to live after the fashion of the world's affluent minority. That requires worldwide industrialization – as a result of which the human ecological footprint on the earth will be deeper than it has ever been. If the living standards of rich countries can be replicated worldwide, it is only by making further large inroads into the planetary patrimony of biological wealth.

Rainforests are the last great reservoirs of biodiversity, but they will have to be cleared and turned over to human settlement or food production. What is left of wilderness in the world will be made over to green desert. This is a bleak enough prospect, but what's worse is that it is a path from which there is no turning back. If a human population of this size is to be kept in existence, it must exploit the planet's dwindling resources ever more intensively. In effect, humans will turn the planet into an extension of themselves. When they look about the world, they will find nothing but their own detritus.

There are many who claim to be unfazed by this hideous prospect. Marxists and free-market economists never tire of ridiculing the idea that other living things have intrinsic value. In their view, other species are just means to the satisfaction of human wants, and the earth itself is a site for the realization of human ambitions. These self-professed rationalists are prone to the conceit that theirs is a purely secular view of the world; but in thinking this way about the relationship of humans to the earth, they are in the grip of a religious dogma. The belief that the earth belongs to humans is a residue of theism.

For Christians, humans are unique among animals because they alone are created in the image of God. For the same reason, they are uniquely valuable. It follows that humanity can behave as lord of creation, treating the earth's natural wealth and other animals as tools, mere instruments for the achievement of human purposes.

To my mind, such religious beliefs have caused an immense amount of harm, but at least they are coherent. It is perfectly reasonable to think humans are the only source of value in the scheme of things – so long as you retain the theological framework in which they are held to be categorically different from all other animals. But once you have given up theism, this sort of anthropocentrism makes no sense. Outside the Judaeo-Christian tradition, it is practically unknown. The view of things in which we are separate from the rest of nature and can live with minimal concern for the biosphere is not a conclusion of rational inquiry. It is an inheritance from a single, humanly aberrant religious tradition.

The fashionable belief that there is no such thing as overpopulation is part of an anthropocentric world-view that has nothing to do with science. At the same time, there is more than a hint of anthropocentrism in Wilson's suggestion that we are entering an age of solitude. The idea that, unlike any other animal, humans can take the planet into a new evolutionary era assumes that the earth will patiently submit to their inordinate demands. Yet there is already evidence that human activity is altering the balance of the global climate – and in ways that are unlikely to be comfortable for the human population. The long-term effects of global warming cannot be known with any certainty. But in a worst-case scenario that is being taken increasingly seriously, the greenhouse effect could wipe out densely populated coastal countries such as Bangladesh within the present century, while seriously dislocating food production elsewhere in the world.

The result could be a disaster for billions of people. The idea that we are entering an era of solitude makes sense only if it is assumed that such a world would be stable – and hospitable to humans. Yet we know that the closer an ecosystem comes to being a monoculture, the more fragile it becomes. The world's rainforests are part of the earth's self-regulatory system. As James Lovelock has observed, they sweat to keep us cool. With their disappearance, we will be increasingly at risk. A humanly overcrowded world that has been denuded of much of its biodiversity will be extremely fragile – far more vulnerable to large, destabilizing accidents than the complex biosphere we have inherited. Such a world is too delicate to last for long. There are good reasons for thinking that an era of solitude will not come about at all. Lovelock has written that the human species is now so numerous that it constitutes a serious planetary malady. The earth is suffering from disseminated primatemaia – a plague of people. He sees four possible outcomes of the people plague: "destruction of the invading disease organisms; chronic infection; destruction of the host; or symbiosis, a lasting relationship of mutual benefit to the host and the invader".

The last two can be definitely ruled out. Humankind cannot destroy its planetary host. The earth is much older and stronger than humans will ever be. At the same time, humans will never initiate a relationship of mutually beneficial symbiosis with it. The advance of Homo sapiens has always gone with the destruction of other species and ecological devastation. Of the remaining outcomes, the second – in which over-numerous humans colonize the earth at the cost of weakening the biosphere – corresponds most closely to Wilson's bleak vision. But it is the first that is most likely. The present spike in human numbers will not last.

If it is not forestalled by changes in the planet's climate, we can be pretty sure that Wilson's era of solitude will be derailed by the side effects of human strife. *Resource scarcity is already emerging as a factor aggravating tension in several regions of the world.* In the coming century, it is set to be one of the primary causes of war. A world of eight billion people competing for vital necessities is highly unlikely to be at peace. On the contrary, it is programmed for endemic conflict. New technologies may blunt the edge of scarcity by allowing resources to be extracted and used more efficiently. But their key use will be to secure control over dwindling supplies of oil, natural gas, water and other essential inputs of industrial society.

The internet originated in the military sector. Information technology is at the heart of the revolution in military affairs that is changing the face of war by powering the new generations of computer-guided missiles, unmanned planes and the like. Only a couple of years ago, a host of air-headed publicists was proclaiming the arrival of a weightless world. The reality is just the opposite. The Gulf war was won with computers, and they will be critically important in any attack on Iraq. In that sense, it is true that information technology will be the basis of prosperity in the 21st century. But its main contribution will not be to create a hypermodern, knowledge-driven economy. It will be to enable advanced industrial states to retain control of the most ancient sources of wealth – the world's shrinking supplies of non-renewable resources.

In the past, war has rarely resulted in a long-lasting decline in human numbers. But in a highly globalised world, it could have a new and more devastating impact. With a hugely increased population reliant on far-flung supply networks, large-scale war in the 21st century could do what it has frequently done in the past: trigger food shortages, even famine. Globalization no more engenders world peace than it guarantees an unending boom. It simply magnifies instability.

Summing up his view of the future, Wilson writes: "At best, an environmental bottleneck is coming in the 21st century. It will cause the unfolding of a new kind of driven by environmental change. Or perhaps an unfolding on a global scale of the old kind of history, which saw the collapse of regional civilizations, going back to the earliest in history, in northern Mesopotamia, and subsequently Egypt, then the Mayan and many others scattered across all the inhabited continents except Australia." Wilson's "new kind of history" would involve a worldwide revolution in attitudes and policies. This would include universal access by women to the means of controlling their fertility, abandonment of the belief that there is a natural right to have as many children as you like, and a basic shift in attitudes to the environment in which it is accepted that our fate and that of the rest of life on earth are inseparably linked. These are the minimum conditions for the new kind of history of which Wilson writes. Unfortunately, one has only to list these conditions to see that they are unrealizable. There cannot be a sustainable balance between natural resources and human needs so long as the number of people continues to increase, but a growing population can be seen as a weapon. Many Palestinians and Kurds view having large families as a survival strategy. In a world containing many intractable ethnic conflicts, there is unlikely to be a benign demographic transition to a lower birth rate.

The examples we have of societies in which population has declined in the absence of a big social crisis cannot be replicated worldwide. A policy of zero population growth requires universal availability of contraception and abortion, and limits on the freedom to breed; but the authority that could impose these conditions does not exist. Humans have a long history of mass killing, but have rarely chosen to regulate their numbers intelligently and humanely. If population declines, it will be as a result of war, genocide or the kind of generalized social collapse that has taken place in post-communist Russia. The increase in human population that is under way is unprecedented and unsustainable. It cannot be projected into the future. More than likely, it will be cut short by the classical Malthusian forces of "old history". From a human point of view, this is an extremely discomforting prospect; but at least it dispels the nightmare of an age of solitude. The loss of biodiversity is real, and very often irreversible. But we need not fear a world made desolate by human proliferation. We can rely on Homo sapiens to spare us that fate.

152. Country Assessment Series – Venezuela

Venezuela is a very beautiful country with a diverse terrain. In the South, lie the tropical rain forests of the Orinoco River and the high Roraima hinterland, passing westwards into the grasslands of the Llanos. Several spectacular Andean ranges, capped by Pico Bolivar, at an altitude of 5007m., follow to the north, before giving way to the badlands of Falcon and the deserts of Paraguana, complete with sand dunes and cactus. To the West, lies Maracaibo, an inland shallow sea, while to the East is the Orinoco delta and the Gulf of Paria, which separates Venezuela from Trinidad.

Little is known about its early history, when it was occupied by Amer-indian people. It was sighted by Christopher Columbus in 1498 on his third voyage to the New World. Spanish settlement began in 1520, when Caracas, the capital, was founded in an Andean valley, being administered jointly until 1819 by the Spanish Vice-Royalty of Peru and the Audencia of Santo Domingo. Caracas was the birthplace of Simon Bolivar, known as the Liberator of South America, who, after several years of struggle, achieved independence for Venezuela in 1829, only to die in the following year, a disillusioned man, with his notion of a united Latin America having been destroyed by factional disputes.

The subsequent history has been characterised by revolution, counter-revolution, and dictatorship, interspersed with brief periods of not very successful democratic government. The population amounts to some 25 million, mainly living in the Andean and coastal regions, who are predominantly of mixed European and Negro extraction.

Venezuela has rich natural resources, with substantial iron-ore deposits in the interior, in addition to its ample oil endowment.

The Pitch Lake of Trinidad, which had been known since 1595, attracted early interest to the island's oil potential. The first successful wells were drilled in 1866, only seven years after Colonel Drake's drilled famous discovery in Pennsylvania, which is generally held to be the start of the modern oil industry. The early explorers looked across the limpid waters of the Gulf of Paria to wonder what Venezuela might offer, as it too had a pitch lake. The first well was in fact drilled in 1878 to the south of Lake Maracaibo, but it was not until 1907 that local interest started securing concessions, which eventually passed intro the hands of the foreign companies. They began exploration in earnest in the years preceding the First World War. Shell Oil was one of the pioneers, being introduced to the country by no less than the legendary Armenian oilman, Calouse Gulbenkian, who probably understood how to deal with the Venezuelan dictator, General Gomez. Their pioneering efforts were rewarded when a

Field	Disc.	Gb	Field	Disc.	Gb	Field	Disc.	Gb
Lagunillas*	1926	14	Centro	1957	2	Tejaro	1988	0.75
Bachequero*	1930	8	Lamar	1958	1.75	Ceuta SSE*	1985	0.75
Tia Juana*	1928	5	Ceuta*	1957	1.5	Santa Rosa	1941	0.75
Carito	1988	4.5	Ceuta-Tomo*	1986	1.5	Mene Grande*	1914	0.75
Lama*	1957	4	Lago	1958	1.25	Jobo	1953	0.75
Furrial	1986	3.5	Quiriquire	1928	1	Oficina	1937	0.5
Boscan	1946	2.5	La Paz	1924	1	Mata	1951	0.5
Pueblo Viejo*	1939	2	Cabimas*	1917	0.75	Mara	1945	0.5

* Part of the super-giant Bolivar Coastal Field complex, found in 1917

well on the La Rosa Field in Lake Maracaibo blew out with a flow rate of over 100,000 b/d, having penetrated a highly fractured Cretaceous limestone reservoir. Standard Oil of Indiana (Amoco) also had substantial early rights to Lake Maracaibo, but its Mid-West management in Chicago got cold feet for foreign ventures after expropriations in Mexico in the 1930s, selling out to Exxon in exchange for a block of Exxon stock. Gulf Oil of Pittsburgh was the third principal operator. Venezuela was for many years the jewel in Shell's crown, which by 1932 had made it Britain's largest supplier. The industry went from strength to strength during the inter-war years and into the immediate post war epoch, with production rising from 300,000 barrels a day in 1930 to two million by the mid-1970s.

The expropriation of BP's Iranian interests in 1951 did not pass un-noticed in Caracas, where the government was already in conflict with the companies over the split of oil revenues. It led Perez Alonso, the oil minister, to open discussions with the major Middle East producers, to try to form a world equivalent to the Texas Railroad Commission, which had successfully regulated US over-production to support price. He eventually succeeded with the formation of OPEC in 1960. The government started passing laws qualifying the existing concessions, which paved the way for a full nationalisation in 1976, and the creation of a national company, Petroleos de Venezuela. By now, exploration was at a mature stage, so the main challenge became to develop the extensive heavy oil deposits that had long been known, and to work in the corridors of power at OPEC to obtain the best price.

Venezuela's great oil wealth springs from a happy circumstance, 90 million years ago, when the continents of North and South America began to move apart at a time of intense global warming. Algal growths proliferated, poisoning the seas and giving rise to vast quantities of organic material that sank to the stagnant depth of the opening rifts. With later burial, it formed the *La Luna Formation*, a black shale with large ellipsoidal calcareous, known to early geologists as "wagon wheels". It is one of the world's richest hydrocarbon source rocks, also responsible for oil in Mexico and the Gulf Coast of the USA.

In structural terms, a branch of the Andes divided the country into two provinces: the Maracaibo Basin in the West, and Eastern Venezuela Basin in the East, both of which are prolific oil producers from the same source. The oil has accumulated both in Cretaceous limestones in juxtaposition with the source rocks, and in overlying Tertiary sandstones. The East Venezuela Basin is asymmetrical with a long, gently-dipping, southern flank. Oil has migrated up this flank to shallow depths where it has been weathered and affected by bacterial action, giving rise to the extensive heavy oil and bitumen deposits at depths of 500 to 1500 m along the Orinoco River.

Lake Maracaibo has been subsiding with the extraction of oil, which led the oil companies to build an earth dyke to protect the growing population of the area around Lagunillas, which is sinking below lake-level. Consulting engineers reported that it was safe enough unless there was a major earthquake, when the pebbles in the dyke would flow like marbles. When asked about that risk, they reported that the greatest danger came from

Venezuela – Conventional		
Rates Mb/d		
Consumption	2001	0.49
Production	2001	2.4
	Forecast 2010	2.1
	Forecast 2020	1.8
Discovery 5-year average (Gb)		0.1
Amounts Gb		
Past Production		45
Reported *Proved Reserves*		78
Estimated Future Production to 2075		
From Known Fields		43
From New Fields		6.7
Future Total		50
Total Production to 2075		95
Current Depletion Rate		1.7%
Depletion Midpoint Date		2003
Peak Discovery Date		1941
Peak Production Date		1970

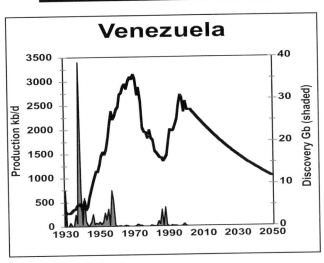

Figure 44

transcurrent faults. The oil companies whereupon sponsored all kinds to research to show that the major faults crossing Maracaibo, which had the hallmarks of lateral movement, had been long dormant, with only minor vertical displacements. They were dismayed when a young geologist published an article on the Santa Marta fault in neighbouring Colombia with a recent lateral offset of 116 km. (Campbell C. J. 1965, *The Santa Marta wrench fault of Colombia and its regional setting*; 4th Carib. Geol. Conf), going to considerable lengths to try to discredit it. Transcurrent faulting in Venezuela has since been established beyond doubt, making this is a catastrophe waiting to happen, but following nationalisation, the foreign oil companies no longer have responsibility for the fate of the many people living below sea-level in Lagunillas.

So far as new discovery is concerned, Venezuela has to rely primarily on tail-end work in the existing producing basins, finding small traps and secondary reservoirs. There is some chance of deep gas-condensate to the south of Lake Maracaibo; in the foothills of the Merida Andes; and in the depths of the East Venezuelan Basin. There is also some possibility of new discovery of mainly gas on the Venezuelan side of the Gulf of Paria and on the shelf to the south, depending on whether the Cretaceous source rocks extend south of the latitude of the Orinoco, which is rather unlikely. Another new opportunity is for gas in rifts off the northern coast, one of which has been found to be productive.

The main potential is for the development of *Non-Conventional* heavy oil along the southern flank of the East Venezuela basin. The country defines heavy oil at a cutoff of 22° API, far above the 17.5° API used in these studies, which distorts both past production and reported reserves. It is here assumed that about 15% of past production was heavy oil, and that the statistics have to be reduced accordingly. It is more difficult to know how to treat reserves. Reported reserves grew from 18 Gb in 1980 to 25 Gb in 1985, when they more than doubled overnight to 56 Gb. It seems that the Eastern Division of the new State oil company decided to include large amounts of long known heavy oil in the reserve base, which had the, not necessarily intended, effect of increasing Venezuela's OPEC quota, causing massive retaliation by other OPEC countries to protect their quota. Venezuela currently reports remaining reserves of 78 Gb, which seems excessive, given that the ultimate recovery of the giants fields, listed below, totals some 60 Gb, and past production amounts to about 45 Gb. Accordingly, the country's remaining reserves of oil lighter than 17.5° API are here generously estimated at 43 Gb.

As already mentioned, the East Venezuela Basin has substantial reserves of Non-Conventional heavy oil, lying at depths of between 500 and 1500m. The traditional method of extraction involved drilling patterns of five closely spaced wells. Steam was then pumped into the peripheral wells to mobilize the heavy oil and drive it to a central producing wells. Long reach horizontal wells are now being used to good effect, managing to extract the oil without steam stimulation. Six projects, involving an investment of some $13 billion, currently produce about 500,000 kb/d. Further efforts may lift production to a plateau of about double that amount by 2005, which could possibly last for 10–20 years, but considerable uncertainties remain as to whether the projects will attract the necessary investment in the face of world recession and political tension.

Venezuela can evidently remain an important exporter, especially to the United States for many years to come, which exposes it to a certain political pressure, as has already been manifested.

The present President, following a long tradition of dictatorial rulers, is Hugo Chavez, an ex-paratrooper who came to power in a landslide popular election in 1998. Venezuela, like many countries, is ruled by a wealthy elite of so-called oligarchs, many of whom, no doubt, shift their wealth overseas, leaving the poor with a minor share of the country's great oil wealth. President Chavez has tried to change this relationship with a decidedly anti-globalist policy, having made well-publicised visits to Cuba and Iraq. He was almost ousted from power earlier this year is a coup, welcomed, if not orchestrated, by the United States, but outwitted the conspirators. No doubt, further attempts to remove him will be made, despite his popular mandate, especially as his influence extends beyond his country. He is a strong voice with the councils of OPEC, whose Secretary-General was until recently also a Venezuelan, Ali Rodriguez-Araque. It is doubtful if Venezuela has any spare capacity at the present time, and that its production is set to decline at about the current depletion rate of 1.75%, a relatively low one. It has every good reason to see OPEC hold or reduce production, making the country's future production ever more valuable. The President has no good reason to meet US demands for cheap oil, and as an ex-paratrooper may not be easily cowed into submission.

Newsletter No. 23, November 2002

153. Recession

In forecasting oil production, it is important to take into account demand as well as supply. Demand was reined in 1979 following the Second Oil Shock, which plunged the world into recession, such that oil production did not rise above its previous level for ten years. The following article by an economist with a major bank suggests that the world is again in deep recession, in part triggered by the high oil prices at the end of the 1990s.

The present Base Case Scenario, as embodied in the above forecast, is that reduced demand will hold the production of *Conventional* oil production about flat until 2010, when supply constraints re-appear. It assumes that production from *Non-Conventional* sources (Deepwater, Polar and Heavy) and NGL will not be constrained. Perhaps it is time to reconsider this assumption.

If all production is constrained by falling demand, then clearly the over-all peak is lower and later, giving a lower rate of subsequent decline. It may help ease the tensions of the transition to the second half of the Age of Oil.

The Costs of Bursting Bubbles
September 22, 2002
By Stephen S. Roach

LONDON – A year after terrorism dealt a seemingly lethal blow to America, talk of resilience and economic recovery is in the air. The nation's inflation-adjusted gross domestic product has risen for four consecutive quarters following a mild downturn in the first nine months of 2001. While the estimated 3.2 percent growth rate over the past year is subdued when compared with the more vigorous rebounds of the past, the hope is that it's a down payment on bigger and better things to come.

But while Sept. 11 was a defining event for America, it was not a defining event for the economy or the financial markets. That role belongs to the stock market bubble of the late 1990s that finally popped in March 2000. There was far more to the excesses of the 1990s, however, than an asset bubble. The bubble expanded high enough, and for long enough, to have infected the behaviour of consumers and businesses alike.

The equity bubble helped to create other bubbles – most notably in the housing market and in consumer spending. Their continued existence poses a serious threat to lasting expansion – and yet, puncturing them raises the grave risk of deflation. This suggests the economy will prove as challenging to America's political leadership as any other issue in the year ahead.

There is good reason to believe that both the property and consumer bubbles will burst in the not-so-distant future. If they do, there is a realistic possibility that the United States, like Japan in the 1990s, will suffer a series of recessionary relapses over the next several years. Yet denial remains deep, just as it was when the Nasdaq composite index was lurching toward 5,000. Few want to believe that this economic expansion may be built on such a shaky foundation.

The evidence in support of a housing bubble is compelling. The 27 percent increase in inflation-adjusted American house prices since 1997 represents the sharpest five-year increase since 1945. This surge is about three times the increase in real housing rents over this period. (The divergence of home prices and rents, which usually move in tandem, is one measure of the speculative element of the housing market.) As their property values rise, hard-pressed consumers have been quick to extract purchasing power from their homes, taking advantage of low interest rates to refinance their property and use the savings to buy cars, furniture, appliances and other luxury goods. Thus the ever-expanding property bubble has become central to the culture of excess that is now driving the United States economy.

The consumer-spending bubble will undoubtedly be the last to pop. Short of savings and long on debt, an aging American population must begin to come to grips with the looming realities of retirement. Yet it must now do so in an era of defined contribution pension plans whose performance has been battered by a devastating bear market in equities. We all know that Americans are addicted to shopping. Yet we also know that, if they want to retire with any kind of financial security, they must increase their savings and rein in their spending. What might cause the consumer-spending bubble to burst? It's hard to say, although several realistic possibilities come to mind – a spike in oil prices, a surge of white-collar layoffs or a collapse of the property bubble. Any one of those developments could send a wake-up call to the American consumer, thereby denying the United States and the broader global economy its main source of support.

But it gets worse. The saga of the post-bubble United States economy doesn't stop with the bursting of the housing and consumer bubbles. Since these events are likely to occur when inflation is already running at a very low rate, they could push the United States into a period of outright deflation – a decline in the nation's overall level of prices for goods and services.

This is a rare and worrisome condition for most economies. The impact of deflation would be most acute for wage earners and debtors. To stay profitable, companies would have to cut jobs or wages, eventually inhibiting consumer purchasing power. And the fixed obligations of indebtedness would have to be paid back in deflated dollars, squeezing overextended borrowers all the more.

America is already on the brink of deflation. Our broadest price gauge, the G.D.P. price index, recorded just a 1 percent annualized increase in the second quarter of 2002. That's the lowest inflation rate in 48 years. Prices of goods and structures – covering nearly half the economy –are already contracting at an annual rate of 0.6 percent. Only in services, where price statistics are notoriously unreliable, are prices still rising.

The hows and whys of America's deflationary perils will long be debated. Two sources seem most likely. First, the bubble-induced boom of business capital spending led to an overhang of new information technologies and other forms of capital equipment in the late 1990s. The result was excess supply, a textbook recipe for lower prices. Also at work are the unmistakable effects of globalization. The modern-day American economy now has a record exposure to global competition. In the second quarter of 2002, America imported a third as many goods as it produced, well in excess of the 20 percent ratio prevailing at the onset of the last recovery in the early 1990s. Significantly, more and more of these goods are coming from highly competitive Asian producers who have much lower cost structures than their American counterparts.

Moreover, with the exception of Korea, every major Asian economy is now in the throes of its own deflation. Consequently, courtesy of ever-expanding trade relations with Asia, America is now buying more and more from countries like China and Japan that are already in deflation. The growing share of these increasingly cheap foreign goods helps drive down prices of products made at home, thereby deepening deflationary pressures.

History tells us that when major asset bubbles burst, deflation is often the result. That was true of the United States in the 1930s and Japan in the 1990s. Most are quick to claim that America is not Japan – that its more flexible, dynamic economy stands in sharp contrast to Japan's economic inertia. But the United States is already a lot closer to the deflationary edge than most concede –and it could go further.

Deflationary risks can never be taken lightly in a post-bubble economy. Yet that's precisely what American investors and policy makers now seem to be doing. If the housing and consumer bubbles pop, then the risk of outright deflation will only increase. It's time to stop pretending this can't happen in the United States.

Stephen S. Roach is chief economist and director of global economics at Morgan Stanley.
http://www.nytimes.com/2002/09/22/opinion/22ROAC.html?ex=1033759619&ei=1&en=e9b4ac3c0318ce31

154. Country Assessment – The United States of America

The New World started drifting away from the Old some 200 million years ago. It was already distant by the time of the arrival of Modern Man, who was able to enter it crossing what is now the Bering Strait some 20,000 years ago, when the sea level was lower in the Ice Age. He found a new continent with a very different animal fauna that had evolved in isolation. Little is known about the early inhabitants who are thought to have numbered some 10 million when new European settlement began in the 15th Century. The Spaniards established a settlement in Florida in 1565, to be followed by various British settlements along the eastern seaboard. France too took a serious interest, founding Quebec in 1608, and controlling much of the Mississippi valley.

Many of the colonists went to the New World to escape from religious persecution at home. European wars in the 18th Century had their effects in the New World, with Britain emerging as the dominant power in 1763, when France surrendered its North American territories. The settlers however soon moved towards independence, not being enthusiastic for various forms of British tax, and after a series of conflicts declared full independence in 1776. A centralised system of government did not come easily, however, as the various settlements, which had evolved into independent states, were reluctant to surrender their autonomy. Constraints on the power of the federal government were established under the Bill of Rights, but have been progressively eroded. The conflicts culminated in a Civil War in 1861–65 between the agrarian South and the industrial North, with slavery being one of the issues. Like most civil wars, it was a vicious affair, costing over 600,000 lives.

There was a great westward migration during the 19th Century, leading to the virtual extermination of the indigenous people; some, it is suspected, being deliberately infected with smallpox as an early example of biological warfare. New waves of immigrants flooded in from over-populated Europe, including particularly Scandinavians, Italians, Jews escaping anti-Semitism, and Irish following a devastating famine in 1845–50.

Texas had been a lightly populated province of Mexico until 1836, when new settlers from the north revolted, declaring it a republic. This prompted a successful war with Mexico, by which the United States acquired Texas, New Mexico and California. Another successful war with Spain followed in 1898, when the United States supported Cuban independence, partly for commercial motives. As a result, it acquired the Philippines, Guam and Puerto Rico, becoming a world power with the imperial aspirations of the day. The sinking of a US warship *Maine*, in Cuba, which was the pretext for the war, was later found to have been due to an explosion on the ship itself, which some think was orchestrated. The US later engineered the secession of Panama from Colombia in order to build the Panama Canal to facilitate trade between the east and west coasts.

The territorial limits of the country eventually stabilised into 48 contiguous states, bordered by Mexico to the south and Canada to the north. Two additional territories were added in 1959: Alaska, which had been purchased from the Russians in 1867, becoming the 49th State; and Hawaii, which had been seized in 1893 over a sugar dispute, becoming the 50th.

The Industrial Revolution of Europe spread to the United States during the 19th Century, as its huge natural resources of iron, coal and, later, oil gave it the essential energy for manufacturing. Floods of new immigrants provided cheap labour. Capitalism took off with a vengeance, throwing up dynasties with extreme wealth, including the houses of Astor, Carnegie, Rockefeller, Morgan, Dupont, to name a few. Wall Street emerged as a premier world financial centre, and the dollar began on its path to world financial domination. These excesses were to some extent countered during the early years of the 20th Century when Theodore Roosevelt brought in the so-called Square Deal with various conservation and regulatory measures, breaking up some of the industrial and financial empires.

The 20th Century history can be summarised into a few key topics. The United States entered the two world wars on the side of Britain and France, its previous colonial masters, and, being spared the ravages of war, replaced them as the dominant economic power in the world. The early inter-war years were marked by an industrial boom, stimulating a speculative bubble on Wall Street, which burst in 1929 bringing on the Great Depression that in large measure lasted until the Second World War gave rise to another boom. It caused great suffering that left a searing memory, deep in the national psyche, leading to greater government involvement in the economy with certain almost socialist attributes. The British and French empires were extinguished by the Second World War, leaving the United States to be countered only by the Soviet Union. These two super-powers glowered at each other for the ensuing 45 years in the Cold War, with several peripheral and rather inconclusive skirmishes, including wars in Korea and Vietnam. The collapse of the Soviets in 1990 left the United States as a solitary super-power, primed for world economic and financial hegemony. Its vast industrial-military complex was in risk of being under-utilised unless new wars or the threat of them should arise. The country's financial dominance was however a mixed blessing, attracting flows of foreign capital that gave rise to possibly unsustainable levels of foreign debt. A critical event was the abandonment of the Gold Standard in 1975, which removed solid foundations for the currency. A post-Cold War speculative bubble burst in 2000 in a situation reminiscent of the crash of 1929.

On September 11th 2001, the World Trade Center in New York, a symbol of globalism, was destroyed by fire, having been hit by two hijacked airliners. The authors and motives of this attack remain somewhat uncertain, but the incident was attributed to Muslim activists. The government declared a general worldwide war of "terror", toppling the government of Afghanistan after a short but brutal bombing campaign. It then turned on Iraq as a

USA – Conventional		
Rates Mb/d		
Consumption	2001	19.63
Production	2001	4.4
	Forecast 2010	2.6
	Forecast 2020	1.4
Discovery 5-year average (Gb)		0.1
Amounts Gb		
Past Production		169
Reported *Proved Reserves*		22
Estimated Future Production to 2075		
From Known Fields		20
From New Fields		6.4
Future Total		26
Past and Future Production		195
Current Depletion Rate		6%
Depletion Midpoint Date		2003
Peak Discovery Date		1930
Peak Production Date		1971

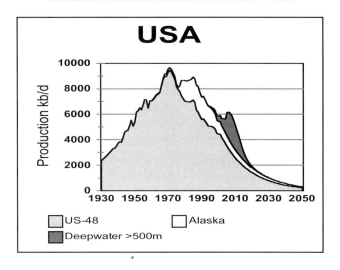

Figure 45

precaution lest it should pose an unspecified threat in the future. It is feared in many quarters that these moves may set in motion a train of events leading to a third world war with devastating consequences. Some have little difficulty in seeing a not very well hidden oil agenda.

The population of the United States amounts to about 290 million. About 75% of the people live in the main urban agglomerations. The black population, being the descendants of former slaves from Africa, number about 30 million, growing at about 15% a year, while those of Hispanic origin amount to about 25 million, and are also growing rapidly. There is, in addition, a large and dynamic Asian community. As in other developed countries, the fertility rate of the long-established segments of the society is below replacement level, but that is more than offset by both massive immigration, much illegal, and the children born to immigrants after arrival. About a third of all births are to unmarried mothers. The country was built of immigrants, who in earlier years soon lost contact with their roots by physical separation, but now easier communications mean that they can retain their close ties. Remittances from Mexican immigrants to their homeland amount to $10 billion annually, being second only to oil as a source of Mexico's foreign exchange. This gives rise to a degree of split loyalty, particularly in the southern states, which is a cause of possible future instability.

The United States, like Britain, operates what can be called a pseudo-democratic system, dominated by two political parties, which select the candidates for election, and heavily control them once elected by denial of the secret ballot. There is widespread political patronage by vested interests. The Presidency has relatively excessive powers and security of tenure over his four-year term of office. The election itself is governed by both direct vote and an Electoral College, furnished by the individual States, under complex rules that are no doubt open to abuse. The Legislature, termed the Congress, consists of a Senate, which represents the individual states, and a House of Representatives, directly elected. A bill passed by Congress is sent to the President who may approve or veto it at his discretion, although the veto can be over-ridden by a two-thirds majority. The country likes to depict itself as "The Land of the Free", forgetting that many other countries enjoy equal or greater freedoms, and may well have a more representative electoral system.

Turning to the country's petroleum endowment, the first point to note is that it is a huge territory of some 9 million km^2, having numerous oil and gas basins. The oil industry had its roots in Pennsylvania, where the self-styled Col. Drake made the first discovery in 1859 in a Devonian sandstone at a depth of 67 feet. It was not in fact a particularly prolific basin, as production had already peaked and begun to decline by 1900. It was however the birthplace of Standard Oil, the mother of Exxon and Chevron, the two surviving US oil giants. Other early oil plays developed in Illinois, Oklahoma and California, and Texas came booming in when a well at Spindletop, near Beaumont, drilled by a one-armed mechanic, blew out on New Year's Day, 1901. Mineral rights in the United States mainly belong to the landowner, which in the early days prompted feverish speculative activity. Discovery peaked in 1930 when the giant East Texas Field was brought in. That itself triggered a slump in oil price from over-production, leading the government to intervene by imposing mandatory cuts in production, managed by the Texas Railroad Commission. It helped delay the corresponding peak of production until 1971. It is an extremely mature area, with future hopes of discovery being confined to small extensions and subsidiary reservoirs in existing producing areas.

The United States dominated world production in earlier years. In 1930, it supplied about 65% of the World's production, but its share has slipped since then to 21% in 1970 and about 7% today. With its burgeoning domestic demand for oil, the country had become a net importer by 1950. Imports began to rise rapidly after peak production, such that they have now passed 60%. The irreversible decline of its production means that even if demand were to be held static, the country would be importing 90% of its needs by 2020. It explains why access to foreign oil has long been officially deemed a vital national interest, justifying military intervention. It might be a tolerable dependency in the case of another commodity such as coffee, because people could learn to drink beer or apple juice, but in the case of oil it effects the very lifeblood of the economy.

The United States also has *Non-Conventional* oil and large amounts of gas, deserving mention. There are a number of heavy oil deposits, such as the Midway Sunset field in California, but they are not readily identifiable in the database. They can continue to support low levels of production for a long time to come. The country also has substantial oil shale deposits, none of which has yet proved to be commercial, but may become so in the future despite giving a negative net energy yield. Alaska offers polar production, being dominated by the giant Prudhoe Bay Field found in 1969, which added approximately 13 Gb of reserves, but is now at a late stage of depletion. Alaska appears to be a concentrated geological habitat, such that new discoveries, including those in areas currently closed for environmental reasons, are likely to be smaller by orders of magnitude. In recent years, exploration has been extended into deepwater areas in the Gulf of Mexico, where some 8 Gb have been found. Production is constrained in this extreme environment but may reach a peak of around 800,000 b/d around 2007 before declining rapidly. It is unlikely that the other deepwater areas have the necessary geology to yield oil, although more exploration is needed to confirm the negative assessment.

The United States has also had a substantial endowment of Natural Gas. It was widely flared in earlier years before a market developed, but is now treated as a prime fuel, especially for electricity generation. Discovery peaked around 1950, giving a corresponding peak in production twenty years later. Gas depletes differently from oil, with production being generally capped below capacity. The resulting plateau of production is now coming to an end, giving rise to higher prices, which prompted a new drilling boom with as many as 16000 wells being completed in 2000, 60% up on the previous average. But the new wells have had to be produced at maximum rate being depleted in a matter of months. Some extra late-stage gas was obtained by tapping the gas caps of oilfields during their dying days. Gas production has increased from 510 G m^3 in 1991 to 555 G m^3 in 2001, a modest increase. Some 25,000 G m^3 have been produced to-date, which is probably about 70% of the total endowment, suggesting that a sharp decline is imminent. The production of natural gas liquids, now running at about 1.9 Mb/d, will fall in parallel with the gas. There are, in addition, large amounts of non-conventional gas in the form of coal-bed methane, and in so-called tight reservoirs, contributing about 10 percent of total supply. Electricity demand is growing, with many gas-fired generators under construction. As a result the United States will have an increasingly desperate need to tap Arctic gas, possibly draining Canada in the process.

It is difficult to avoid the conclusion that the United States faces a dire energy crisis that will radically affect its entire way of life, as indeed the Energy Secretary confirmed before attention was diverted by the events of September 11th. This realisation would at least offer a logical explanation for the massive build-up of military forces in the Middle East, whatever other factors may also be at work. It furthermore adds weight to the expectation of deepening economic recession, expressed by the banker in the preceding article (#186). There is a certain logic in expecting the United States, which led the world into the oil age, to also be the first to experience its decline.

155. A counter view from America

By chance, the following article by Richard Heinberg of New College, Santa Rosa, California was received suggesting that there is a body of opinion in America that fears the direction taken by its government.

MuseLetter #128 / October 2002

Behold Caesar

These days, Julius Caesar and ancient Rome seem to be on the minds of political commentators around the globe. A London

Guardian opinion piece from September 20 was titled "Hail Bush: A New Roman Empire," while Jay Bookman (www.bushwatch.com) explains "The Bush Plan for Empire," and Michael Lind (www.theglobalist.com) asks rhetorically, "Is America the New Roman Empire?"

It was Caesar who transformed the Roman Republic into the Roman Empire. A brilliant general, he waged campaigns throughout modern-day France, Germany, Britain, and Turkey. In 46 BCE, he had himself appointed Imperator for life. Two years later, he was assassinated by a group of conspirators who believed they were striking a blow for the return of the Republic. Thirteen years of civil strife followed. The Republic was finished, but the Roman Empire persisted for another four centuries. Caesar had transformed his world; he was, for a brief time, the most powerful human being in the Western world.

Today the American Republic appears to many pundits to be at a juncture somewhat comparable to the one that Rome confronted in 50 BCE. The analogy is exceedingly imprecise, however: the US is vastly more fearsome than Rome in every respect, possessing weapons no ancient emperor could have dreamed of. Moreover, the American leader, George W. Bush, is far from being a brave and tactically brilliant general, as Caesar was: Bush spent the Vietnam War drinking, snorting coke, and going AWOL from the Texas National Guard. Caesar was also an eloquent orator; the current American leader's abilities in this regard hardly require description.

Nevertheless, Bush has seized leadership of his nation and seems determined both to extend its global influence militarily, and to undermine its democratic institutions, just as surely as his ancient counterpart did. Today, the American administration is preparing to launch a war in the Middle East to advance its imperial ambitions, and is suppressing dissent at home in every way possible. But while Caesar was frank in his war aims – he promised the citizenry colonies, tribute, and slaves – the Bush crowd cloaks its goals in a fog of shifting pretexts.

We are perhaps witnessing a new phase of Pax Americana. But this new order of the world is – for reasons discussed below – destined to persist for far less than four hundred years. And, as was the case with Caesar, victory may come at a high price; though in this instance, it is a price we all will pay.

Rationales for War

War is no small matter for an nation; in the present instance, it is estimated that a new Iraq might cost the US $200 billion or more.

Leaders must have good reasons for such an investment. So far, the US Administration has offered five reasons why Iraq must be attacked. They are as follows:

1. *Iraq is in violation of UN Security Council resolutions.* This is true; Iraq is currently, for example, violating Resolution 687 (01/06/91), establishing UNSCOM; and Resolution 1060 (12/06/96), which was a condemnation of Iraqi refusal to grant inspection access. But these facts do not constitute a believable pretext for war, because Iraq is far from being unique in its violation of UN resolutions. Turkey and Morocco are currently in violation as well. And still another nation in the region, Israel, has refused to comply with literally dozens of UN resolutions, some dating back nearly 50 years. Why single out Iraq?

2. *Iraq has refused UN-mandated arms inspections.* This, of course, is the essence of the particular UN resolutions that Iraq has violated. Arms inspections were mandated by the terms ending the Gulf War of 1991, and inspectors have been absent from Iraq for the past four years. But again, this makes no sense as a pretext for a renewed war. Iraq did comply with inspections up to a point, and evidence suggests that those inspections were working: according to some estimates, 90% to 95% of Iraq's chemical and biological weapons were eliminated, and its nuclear program was almost completely dismantled. When the UN withdrew inspectors in 1998, independent investigations confirmed Iraqi claims that members of the inspection team were "spies" reporting directly to the CIA and to Israeli Mossad. One inspector even left behind a homing device to provide guidance for US bombers, which attacked Iraq in December 1998 during Operation Desert Fox (which, because it played out during the scandal surrounding President Clinton's affair with Monica Lewinsky, was often described as a "wag-the-dog" ruse).

In mid-September, 2002, Iraq agreed unconditionally to the return of weapons inspectors; however, the US responded discouragingly. American secretary of state Powell said that, if UN inspectors attempt to return to Iraq, the US would "move into thwart mode." Before inspectors are allowed back in, the Bush administration is demanding the passage of a new UN resolution that is virtually guaranteed to be unacceptable to Iraq (for example, it calls for the US to have representatives on any inspection team, for the inspection teams to set up *militarily protected* bases and travel corridors in any part of the country they choose, for Iraq to permit unrestricted landing of all aircraft, including unmanned spy planes, and for the US to be able to remove any Iraqi citizen from the country for questioning – all of this effectively dissolving Iraqi sovereignty and amounting to a *de facto* military occupation; if Iraq were to balk at implementing even the smallest detail of the resolution, member states would automatically be entitled to use "all necessary means" to enforce it). The resolution is designed not to make inspections more effective, but to eliminate them and ensure that war ensues.

3. *Saddam Hussein is a brutal dictator who killed his own people.* True enough. But again, as a pretext for war this doesn't make sense. Saddam was just as evil in the 1980s, when he was using poison gas on the Kurds in his northern territories. But then the US approved of him, offering logistical support as well as aid in establishing chemical and biological weapons programs. The US has supported many evil dictators over the years; why attack this particular one now? Is there a sudden crisis of evilness that must be addressed militarily and immediately, even to the point of killing perhaps thousands or tens of thousands of innocent civilians in the process?

4. *Saddam Hussein possesses weapons of mass destruction (WMDs) that pose a threat to his neighbours and to the American people.* But, as documented by the UN and the CIA, Iraq has far less capability in that regard now than in 1990. As noted above, many of Iraq's WMDs were covertly supplied *by the US*. The US itself has vast stores of nuclear weapons, and is the only nation to have used such weapons against a civilian population. Of the countries in the Middle East, Israel has by far the largest inventory of WMDs; yet the US has not proposed that Israel be attacked for that reason. Oddly enough, Iraq's neighbours do not appear concerned about the threat posed to them; indeed, most of them are pleading with the US *not* to attack. And no credible analyst has suggested that, even if Iraq does possess remnant WMDs, its leaders have either the ability or the intent to use them against US citizens, absent a large-scale attack.

5. *Saddam Hussein provides aid to the terrorists who perpetrated the 9/11 attacks on the US.* According to polls, nearly 70% of the American people believe that this is the case, and administration officials have made claims to this effect on several occasions. However, no one has supplied credible evidence for the assertion. Moreover, any such link would be counterintuitive. Osama bin Laden and other radical Islamists detest secular Arab states, of which Iraq is one of the foremost. And secular Arab leaders, in turn, fear and despise the radical Islamists. It was Libya's Muammar Gaddafi – not George Bush or Bill Clinton – who was the first world leader to call for the arrest of bin Laden, in 1994, following terrorist attacks

on his nation. Why would Saddam aid his own sworn enemies? Two other nations in the region have been shown to have much more credible links with al Qaida – Pakistan and Saudi Arabia. Why is Bush not demanding attacks on these countries?

If none of these stated rationales is the true reason for Bush's insistence on war, then the identification of his true motives requires some speculation.

Quest for Empire

Several recent articles, noting the flimsiness of the official war rationale, have discussed possible underlying psychological drives. One writer (Mike Hersh, of Online Journal) tells us that White House insiders privately assert that Bush is "out of control." In prepared speeches, Bush dutifully reads the litany of Saddam's violations and crimes. But in a recent off-the-cuff comment (9/26/02), Bush is reported to have said simply, "This is a guy that tried to kill my dad," referring to a purported failed 1993 assassination plot against ex-president Bush. (The only pieces of evidence ever brought forward for the existence of such a plot were confessions extracted by Kuwaiti torturers; nevertheless, Clinton retaliated with missiles, which hit a residential area and killed eight Iraqi civilians.) Is mere personal revenge the underlying motive for Bush's war?

Revenge may indeed be a contributory factor-at least in the tiny mind of George W. Bush himself. But it is important to remember that many government officials who do not share a personal grudge against Saddam are promoting this war. This is a project that has emerged from a consensus of strategists whose purposes are undoubtedly more sophisticated than the pursuit of a family feud. Since official statements give us almost no insight into the real reasons why the American leadership is determined to pursue an expensive and risky war halfway around the world, one must indulge in a little informed speculation. In what ways might Bush or the people close to him have something to gain from such a war?

When we pursue this line of thought, three clear possible motives quickly come to mind:

1. *Party politics and power*. The American economy is in terrible shape now, with the stock market at levels not seen since 1997, corporate bankruptcies accumulating weekly, and revelations ongoing about corporate accounting fraud at the highest levels. A projected trillion-dollar government budget surplus has become a trillion-dollar deficit in a mere eighteen months. As the bubbles of the exuberant 1990s burst one by one, many economic analysts

believe that the entire world may be teetering on the brink of a depression at least as serious as that of the 1930s. This should be horrific political news for the party in power. However, with Americans' attention riveted by the terrorist attacks of 9/11, Bush and the Republicans have had to endure scant scrutiny. The White House occupant's handlers cannot help but have noticed that terrorism and war do wonders for the leader's poll numbers, while economic headlines do the opposite. An obvious strategy: find ways to dominate the news with fear-inducing, patriotic war talk. David Morris, writing on Alternet, opines that Bush's saber rattling is all about politics, and suggests that, after the November elections, weapons inspectors will return to Iraq and threats of attack will subside.

There's no question that war is good politics, but are there other motives at work that might result in Bush's threats actually being carried out?

2. *Global dominance.* The foreign-policy advisors surrounding Bush all share views typified in a report, "Rebuilding America's Defenses," issued in 2000 by the Project for the New American Century. The report calls for American military dominance of Earth and space, pre-emptive strikes on any potential rival, unquestioning support for Israel, and the ignoring of international opinion in the pursuit of US strategic objectives. Most of the report's authors (including Paul Wolfowitz, deputy defense secretary) are now highly placed administration officials, and the document itself is closely echoed by the official National Security Strategy, released by the administration on September 20. Bush, Cheney, Rumsfeld, Wolfowitz, and the rest appear to view Iraq as a symbolic challenge to US hegemony, Saddam Hussein having survived one US-led attack and over ten years of punishing economic sanctions. The toppling of his regime thus represents a test of the aggressive new American strategic doctrine.

In this view, an attack on Iraq serves an emblematic purpose, sending a message to the rest of the world saying, Defy us at your peril. Yet still something is missing. Why imperil the US economy to project US military might if there is nothing concrete to be gained thereby?

3. *Oil.* Here, perhaps, we get to the real nub of the issue. The US needs oil; its wealth was built on energy resources and on its ability to deploy technologies to use those resources (cars, planes, and industrial machinery). American oil production peaked in 1970 and now the nation imports well over half of what it uses. In order to

maintain its global dominance, the US needs to be able to control global oil prices. However, since the 1970s, the OPEC countries of the Middle East, by virtue of their immense petroleum reserves, have had that power. It is Saudi Arabia, as swing producer, that has opened or closed the spigot to enable economic booms (the mid 1980s and the mid- and late 1990s) or provoke recessions (1973, 2000). Now Saudi Arabia teeters, beset by a growing and youthful population, dwindling per-capita incomes, and simmering Islamist radicalism.

Iraq has reserves second only to those of Saudi Arabia. Because of the war with Iran in the 1980s and sanctions in the 1990s, those reserves are not as fully exploited as those of other nations in the region. This makes Iraq a prize for the taking – a fact not overlooked by Russia and France, which also covet its future oil production. If the US could install a compliant puppet regime in Bagdhad, it could break the back of OPEC, establish its position first in line ahead of Russia and France, and weather any potential upset in Saudi Arabia.

Upon entering office, Dick Cheney, chair of the White House Energy Policy Development Group, commissioned a report on "energy security" from the Baker Institute for Public Policy, a think-tank set up by former US secretary of state James Baker. The report, "Strategic Energy Policy Challenges For The 21st Century," issued in April 2001, concludes: "The United States remains a prisoner of its energy dilemma. Iraq remains a de-stabilizing influence to… the flow of oil to international markets from the Middle East. Saddam Hussein has also demonstrated a willingness to threaten to use the oil weapon and to use his own export program to manipulate oil markets. Therefore the US should conduct an immediate policy review toward Iraq including military, energy, economic and political/ diplomatic assessments."

Cheney, the former CEO of the Texas oil firm Halliburton, was advised principally by Kenneth Lay, the disgraced former chief executive of Enron – the US energy-trading giant that went bankrupt following the revelation of massive accounting fraud. Other advisers included Luis Giusti, a Shell non-executive director; John Manzoni, regional president of BP; and David O'Reilly, chief executive of ChevronTexaco. The Baker report refers to the impact of fuel shortages on voters and recommends a "new and viable US energy policy central to America's domestic economy and to [the] nation's security and foreign policy." It also says that Iraq "turns its taps on and off when it has felt such action was in its strategic interest to do so," adding that there is a "possibility that Saddam

Hussein may remove Iraqi oil from the market for an extended period of time" in order to raise prices. "Unless the United States assumes a leadership role in the formation of new rules of the game," the report warns, "US firms, US consumers and the US government [will be left] in a weaker position."

No doubt all three of these latter factors have converged to galvanize the current Bush policy toward Iraq. In light of these powerful motives, publicly stated concerns about Iraq's violation of UN resolutions and its possession of WMDs pale in significance. The administration has compelling reasons for its attack on Iraq; otherwise it would not invest so much financial and political capital in the effort. It is a shame, however, that those reasons cannot be shared publicly; if they were, an interesting debate might ensue. As it is, politicians and press commentators alike are in the awkward position of having to state plausible-sounding opinions about inherently implausible statements and rationales issuing from the administration. The ensuing charade is painful to witness.

The War's Likely Progress and Consequences

Absurd as its rationales may be, the war itself is a deadly serious prospect. What might happen if efforts to dissuade the Bush administration fail? If the war goes according to plan, it will be over in just a few weeks. An overwhelming air attack will be followed by an invasion of ground troops mopping up Republican Guard resistance in the cities. The Iraqi people themselves will welcome American troops with open arms, delighted to be rid of their tyrant.

Other nations in the region will be cowed into obedience by this show of strength; or, if their regimes display weakness or intransigence, they can be overthrown as needed.

Early in the hostilities, and perhaps prior to their commencement, president Hugo Chavez of Venezuela must be ousted (and killed) so as to terminate his nationalist and leftist influence on OPEC policies and ensure the free flow of oil from his country to the US during the course of the conflict in the Middle East.

Also early in the hostilities, Israel must be expected to take advantage of the exclusive focus of world attention on Iraq by militarily pushing virtually the entire Palestinian population out of the West Bank and Gaza, perhaps into Jordan, thus solving the "Palestinian problem" once and for all.

According to analysts at STRATFOR (the online strategic forecasting service), Dick Cheney and his advisors are working on a long-term plan for post-war Iraq. The currently favoured approach is to unite Iraq and Jordan in a pro-US Hashemite kingdom. The southern Shiite and northern Kurdish areas, where most of Iraq's oil is located, present a dilemma: the former must be prevented from uniting with Iran, the latter from uniting with Kurdish areas in Turkey and agitating for a Kurdish state. Both must be granted some sort of limited autonomy but kept under close US control.

With Iraq's oil resources now accessible to American oil companies, and with Chavez gone from Venezuela, the power of OPEC will have been crushed. Oil prices will fall and the American economy will be saved from ruin (for the time being). American oil companies will grow rich. With large numbers of troops now permanently stationed in the Middle East, the US will have become an overt military empire. That is the outcome if everything goes as expected. Unfortunately, however, a new Iraq war would hardly be the first unprovoked US military adventure, and experience has shown that such adventures *often* don't go according to plan (does the word *Vietnam* ring any bells?). What could go wrong in this instance? One hardly knows where to start.

What if the Iraqi people decide to resist invasion rather than welcoming their American liberators? The campaign could become a house-to-house urban war of attrition with mounting casualties on both sides. At the same time, Saddam Hussein, realizing that he is done for, might well decide to unleash every weapon in his arsenal, with the hope of provoking the widest possible conflagration in the region. The US would then need more than the minimal 200,000 ground troops it is now planning to deploy, and the draft might have to be reinstated. That would in turn provoke more anti-war protests at home, and thus necessitate more government repression. If other states in the region are overthrown by Islamist opposition movements as a result of popular uprisings triggered by the war, efforts by the US to occupy those nations might seriously overextend American forces; then, rather than face defeat on any front, commanders might resort to the use of tactical nuclear weapons. Israel, perhaps finding itself under attack from Arab neighbour states, might itself decide to unleash some of its 200 or so nukes. At the same time, popular outrage throughout the Arab and Muslim world at US actions might result in a dramatic increase in anti-American "terrorism" worldwide. Pakistan, which (unlike Iraq) *does* have functional nuclear weapons, could easily fall to the Islamists; if that were to happen, a nuclear device would probably come to the hands of al Qaida in short order. Not only would the US economy be shattered by high oil prices and the costs of war, but

American cities, and citizens abroad, would be imperiled.

In sum, an outcome in which a years-long World War is triggered, with multiple nuclear weapons being detonated and hundreds of thousands or millions being killed, may be about as likely as that in which everything goes as the war planners hope. All of this to maintain and extend the power of small group of criminal ideologues in Washington, and to keep American motorists fuelled up and mobile for another decade or so.

Who Wants This War?

The potential consequences of the imminent American attack on Iraq are fairly evident to people in most nations around the world – except the people of the US. Here, politicians and pundits alike drone on about the menace of Saddam, while virtually no one dares mention the far greater menace to global peace posed by the geopolitical strategists in the White House. The American people are deeply unaware of their predicament; with the encouragement of television they are – as more than one commentator has put it, and on more than one occasion – "sleepwalking through history." One might get the impression that this is a nation of imbeciles (and this does seem to be the view from the rest of the world); but Americans aren't inherently any more stupid than anyone else. They are being deliberately and systematically dumbed down. Their attention is distracted and manipulated from morning till night by slick PR professionals in both corporate and government offices.

One tool in the arsenal of these professional opinion shapers is the poll. These days we are told that most Americans favour an attack on Iraq, and most think that Mr. Bush is doing a splendid job in leading this brave nation. The polls tend to be deeply disheartening to those who make any attempt whatever to see current events in historical and international context. But one has to view the polls in perspective. What are people actually being asked? Perhaps if questions were rephrased, answers would be more meaningful. What if a random sample were asked, "Do you get your news from alternative sources and think critically about world issues?" The portion of the sample that replied affirmatively might almost exactly correspond with the 40% of the population that is reputed to disapprove of the "president's" job performance. Other possible questions: "Do you watch lots of television and pay minimal attention to civic and world affairs? Are you so absorbed with work and family that you just don't have time to think about much else?" Those who gave an affirmative reply to those questions would, one might well guess, correspond almost identically with the

60% who are said to approve of Bush and his war plans. The latter group is, in effect, saying to pollsters, "Yeah, sure, whatever." ("Do you approve of the way the 'president' is doing his job?" "Yeah, sure, whatever." "Do you want a World War to erupt in the Middle East?" "Whatever.")

Meanwhile the overwhelming majority of letters, phone calls, faxes, and e-mails that have recently poured into the offices of the "president" and members of Congress, as a congressional bill authorizing war was being debated, expressed opposition to an attack. Even senior CIA and Pentagon officials expressed skepticism. Global opinion remains almost unanimously anti-war. It appears that *nobody wants this war except the tiny circle of far-right strategists surrounding Bush.* Yet no one appears able to stand up to these people forcibly enough to stop them. Most of the Democrats in Congress, like Bush, are simply watching the polls and looking toward the November elections; there's no political capital to be made by taking a strong anti-war stand. So the Bushies will probably have their war. And heaven help us all.

Sic Transit Imperium Americanum

George W. Bush aspires to be a Caesar, make no mistake about it. But despite his bellicosity and imperial pretensions, the comparison with Julius utterly fails. Bush jr. perhaps bears more resemblance to some of the feeble and dissolute hereditary emperors of the third century, men whose names are familiar now only to specialist historians.

In reality, the American empire passed its zenith in the late 1960s and early 1970s, as US oil production peaked and the nation squandered its financial wealth on a pointless war in Southeast Asia. Since then, as its petroleum resources and gold reserves have dwindled, the US has been steadily losing ground both politically and economically. Post-peak America is awash in debt, dependent on imports, and mired in corruption. Nations around the world fear its military and watch its television shows, but ridicule its leaders and policies. The far-right ideologues who have hijacked the political and strategic leadership of the country fancy themselves as *establishing* an American empire, whereas they must know in their heart-of-hearts that they are merely presiding over that empire's inevitable twilight. Their chest-thumping patriotic triumphalism would be pathetic if it were not so profoundly perilous. The gambit of an Iraq war is a desperate measure, a floundering attempt to maintain power and authority that are fast slipping away. But, like the flailings of a person caught in quicksand, these efforts can only hasten the undertow. The

US can still destroy, but cannot control the rest of the world. Bush, after all, is just a Caesar wannabe with nukes.

The fall of Rome occurred over several centuries. The fall of imperial America will be much more dramatic and impactful, and much quicker, lasting only decades at the most. What a shame that such a momentous time in the history of the world should be presided over by people who are not only greedy and ruthless (one can almost take that for granted), but talentless and unimaginative as well.

156. New flawed study from the IEA

The IEA maintains its tradition of publishing flawed information on oil supply in its World Energy Outlook of 2002. This is hardly surprising as it allows itself to be advised by the US Geological Survey and Michael Lynch, the well known flat earth economist, and does not, so far as is known, work with the industry database. We may comment on a few highlights.

The IEA defines Conventional Oil to be crude oil, Natural Gas Liquids and natural bitumen. It has supply matching demand, which it estimates will increase at 1.6% a year to 2030.

A comparison with ASPO estimates is given in the following table, after adjusting so far as possible for differing definitions and including processing gains. There is no particular difference to 2010, but thereafter the IEA departs radically to double the ASPO estimate by 2030.

The IEA gives a table of Past Production, Reserves, and Yet-to-Find of respectively 718, 959 and 739 Gb, giving an Ultimate of 2616 Gb. A

	2000	2010	2020	2030
Conventional				
ASPO Conventional	63.8	60.2	46.3	35.7
Deepwater	2.3	7.2	5.0	1
Polar	1.1	2.0	5.9	1.6
NGL	6.0	9.3	10.7	10.7
Total	73	79	68	49
IEA	72.2	83.6	95.8	107
Non Conventional				
IEA	1.1	3.0	5.6	
9.9				
ASPO Heavy	1.4	3.3	4.3	5.0
Processing Gain	1.7	2.2	2.6	3.1
TOTAL – IEA	75.0	88.8	104.0	120.0.
TOTAL – ASPO	76.1	84.5	74.9	57.1

footnote explains that the Reserves apply to 1996 whereas the Yet-to-Find applies to 2000. Although Past Production is described as "to-date", it evidently also relates to 1996. We can see here the footprint of the flawed USGS study, which related to the period 1995–2025. This Ultimate estimate of 2600 Gb, which is not unreasonable for all liquids, gives a simple midpoint of depletion of 1300 Gb. Given that production through 1995 was 718 Gb, the midpoint would be reached when 582 Gb more had been produced, which at 27 Gb a year would be around 2016. Peak is likely to come before midpoint, since the heavy oils will not have much impact. This would not be a wildly inaccurate estimate, only six years from the ASPO estimate, but is inconsistent with the IEA's claim that production would continue to increase to 2030.

The IEA expects that this is to be achieved by obtaining higher recovery rate, thanks to economic and technological factors. It is noteworthy that it refers to rate not percentage recovered. It implies that production would have to fall like a stone after 2030 to respect the ultimate, but the IEA doesn't address that issue as it lies beyond its time-frame.

At the same time, it does confess that "faster depletion will bring forward the time when production peaks", mentioning that the North Sea has done so and is in terminal decline. It also points out that the average age of the fourteen largest fields responsible for one-fifth of world output is more than 43 years, and that the discovery of giant fields has been falling both by size and number.

The IEA gives estimates of the above parameters by country. It is noteworthy that Saudi Arabia is given reserves of 221 Gb as of 1996, which is close enough to the current ASPO estimate of 193 Gb, taking into account production of 16 Gb since 1996. It is far less than the 259 Gb claimed by the country. But then the IEA gives the country a staggering 136 Gb yet to find. Evidently it does not realise that approximately 130 Gb of Saudi Arabia's oil lies in just two fields, Ghawar and Safaniya, found long ago, and that discoveries outside this prime belt have been quite modest, greatly reducing the potential for new discovery.

One can imagine that the economists employed to make these forecasts are well-qualified and intelligent, but lack actual oil company experience. They probably rely on published reserve data, as contained in the BP Statistical Review, and take the Yet-to-Find from the USGS. They cannot be expected to understand the practices of oil company reserve reporting, nor grasp the true impact of technological progress. Without access to properly backdated reserve data from the industry database, they can be forgiven for failing to determine the real discovery trend and its devastating consequences.

Although the report does demonstrate a certain dim awakening to reality, even mentioning the forbidden words "depletion" and "peak", it continues the long tradition of the organisation in providing grossly misleading and erroneous data and forecasts. It is difficult to exaggerate the damage done to its member countries, which are failing to implement sound energy policies as a consequence. They comprise: Austria, Belgium, Canada, Denmark, France, Germany, Greece, Iceland, Ireland, Italy, Luxembourg, Netherlands, Norway, Portugal, Spain, Sweden, Switzerland, Turkey, USA, UK, Japan, Finland, Australia, New Zealand, Mexico, Czech Republic, Hungary, Poland, Korea and Slovakia. Imagine what could be achieved by these countries, if they were to dedicate even a very small fraction of the IEA budget to making a proper study of this critical issue.

157. BP Confesses to Depletion

The official Norwegian Petroleum Diary (No 3, 2002) carries an article by BP prominently depicting in colour the Depletion Plot as given in the frontispiece of the ASPO newsletters. It states that the plot represents the challenge for the company to develop solar energy. This is a marked departure from its normal corporate imagery, which goes out of its way to avoid any mention of the imminent and inevitable peak and decline of oil production.

158. Anniversary and Future

This is the 24th issue of the Newsletter, which means that ASPO has been going for two years. Much has happened during this time to impress us of the transcendental importance of our subject as nations begin to vie for with each other for control of critical oil supplies. Although the Newsletter has erred at times in matters of fact, interpretation, judgement and emphasis, it seems to have been broadly successful in raising awareness. The number of ruffled feathers has been quite small, considering that the ice under the commentary is often thin. New applications to join the distribution list, which now numbers about 150, continue to come in. It is understood that overall readership may exceed about a thousand because of wide onward copying. Even the benighted official institutions begin to shift their ground in our direction. It is noteworthy, for example, that BP has adopted our depletion model, as reproduced at the front of the Newsletter, when contributing to an official Norwegian publication, to demonstrate its serious commitment to solar energy. This indeed is progress.

The membership of ASPO itself has expanded over the two years to become a really representative serious European network. The Uppsala workshop of May 2002 was a watershed that put the organisation on the map. Generous sponsorship gives us hope for a more substantive physical presence. But as we grow, we change. Probably it would be an idea to try to establish not so much a network of key scientists as a confederation of small national committees, each having its own identity within its own environment and resources, but working together in a co-operative manner to further the common objectives.

The "National Committee of France" moves forward with the organisation of the next workshop in Paris for May 2003, which will give an opportunity to plan our future better.

159. Re-visiting a proposed Depletion Protocol

At a Conference in Copenhagen on October 30th, Hermann Scheer, a member of the German Parliament, made the good point that a move towards renewable energies was not so much a desirable option as a necessity, given that the essential feature of *fossil* fuels is their depletion. The pace of transition is really the only point for debate. (See also his excellent book: *The Solar Economy – Renewable Energy for a Sustainable Global Future* [ISBN 1 85383 835 7], which discusses in depth the deeper political, philosophical and economic factors).

So far at ASPO we have concentrated our efforts to determine what the endowment of oil and gas in Nature is in order to model appropriate depletion profiles. We worry if we have got our numbers right, but does it really matter that much? It is surely *"better to be vaguely right, than precisely wrong"*. The impact of any errors in the model is probably small in relation to its overall long-term validity.

There is logic to an earlier suggestion for an international Depletion Protocol whereby countries would agree neither to produce in excess of their current Depletion Rate nor import infringements, but to hope for positive international agreement of any sort is probably unrealistic, given the low and often selfish calibre of government. Even so, great advantages could still accrue to any enlightened country that did unilaterally adopt the principles of such a Protocol. The natural decline of ageing giant fields means that few producers can exceed their Depletion Rate in any event so to adopt the Protocol would simply recognise reality.

For the sake of argument, let us consider the impact on, say, Ireland and Britain, in respectively adopting or rejecting the Protocol. Ireland, having no oil of its own and less military power to secure it elsewhere, could unilaterally accept the conditions of the Protocol, given that it could overcome any impeding EU Directives. In doing so, it would set itself the task of progressively replacing imports by indigenous renewable energies and curbing demand, adapting its economy and life style accordingly. Initially, it might suffer as a consequence, but as tidal rotors around its coasts began to deliver sustainable energy, it would find itself with a progressive competitive advantage in the evolving world situation, as is imposed by Nature. By the end of the Century, Irishmen would be, so to speak, bathing in the sun as they look to the future with confidence, having assured themselves of their own critical supply.

Britain, by contrast, might take an entirely different approach, preferring to burn up the resources as fast as possible under normal economic principles, securing supplies from elsewhere either on the open market or at the gun point of its own or allied forces. Initially, it might be a more successful policy, allowing more popular vehicles to barrel down the highways and more goods to flow from the consumer outlets, to the delight of the voters. But before long, when the country finds that neither the market nor the gun can deliver due to the constraints of Nature, it would discover itself at a disadvantage, being less prepared than the Irishman. Indeed, if advances in technology should now accelerate the rate of oil and gas extraction, as suggested by the IEA, Britain would find itself with very little time to adjust when the need arose, especially as its own production, which has given it a certain false sense of security, is already in steep irrevocable decline.

The argument suggests that it might indeed be an admirable idea for the EU to promote, or at least debate, the proposed Protocol. It would face cries of outrage from the flat earth community and the kleptocrats, but by merely floating the idea, and explaining the rationale, it might encourage some of its more enlightened member-countries to adopt policies that would give them a brighter future.

160. Bush strengthens his grip

Bush and his backers strengthened their grip on the long-suffering American nation in the critical mid-term electoral process, despite the bumper stickers proclaiming "Regime Change Starts at Home"; massive anti-war demonstrations, and an outpouring of Internet and other articles, revealing the backgrounds and interests of the administration (collectively having personal investments of as much as $150 M in oil companies). References include: www.rense.com/general31/thr.htm; *Stupid White Men* by M. Moore ISBN 0-06039245-2, and the works of Thom Hartmann. A leading article in the London Times of 22 Nov. reviewed *Bush at War* by W. Woodward portraying a puzzled, all-powerful President trying to steer between vicious warring cabinet factions, representing lobbies, including the pro-war Israeli lobby.

Blair continues to receive a poor reception across the British press, being compared in the Guardian on November 6th with the symbiotic egret that pecks food from the teeth of a crocodile's jaw. His unquestioning support for Bush is described as "slow political suicide". (see www.monbiot.com). Even the staid Daily Telegraph, with its somewhat imperial tradition, is less than enthusiastic in an article, coinciding with the traditional Remembrance Day ceremonies in London to pay respect to the fallen in two world wars and other conflicts. They took place against a leaden sky and the grim face of Her Majesty the Queen.

161. Country Assessment – Iraq
Last month, we took a look at the United States, so it is perhaps appropriate now to turn to its new enemy.

Iraq
The country covers an area of some 435 000 km2, supporting a population of 20 million. Mountains in the east and north rise to over 3000 m, flanking the fertile valley of the Euphrates and Tigris Rivers, which flow into the Persian Gulf near Kuwait. To the west, lie extensive plains bordering Syria.

Some of the world's greatest ancient civilisations developed in this area. Indeed, the Garden of Eden, where Adam and Eve disported themselves, is supposed to have been located here. Cyrus the Great of Persia conquered the place in 539 BC, before it fell to Alexander the Great in 331 BC. Greek and later renewed Persian dominion followed until it was overrun by Muslim Arabs in the 7th Century. It was later subject to Mongol invasions, and the attentions of Persian and Turkish rulers, before the Ottomans established firm dominion in the 17th Century, operating eventually through three local administrations (vilayets) having a fair degree of delegated authority. Various nomadic Arab tribes were never fully integrated, and the Kurds in the north, being descendants of the ancient Medes, have long sought their independence.

The area began to attract the conflicting commercial and political attentions of Britain and Germany during the latter part of the 19th Century. Britain, as a trading sea-power, was interested in the coastal areas, including what is now Kuwait; also establishing a shipping company on the Tigris to serve the interior. Germany, being a land-power, proposed building a railway from Berlin to Baghdad, recognising its importance in a military context. The Middle East itself seems to have been of limited interest to Britain, but had strategic importance as a bastion against Russian expansion threatening communications with India, the jewel in its imperial crown.

Oil had been known in the area since antiquity, being used as a form of mortar in the construction of Babylon. New interest developed in the early years of the 20th Century, when engineers came across oil seepages in the course of surveying the concession granted by the Ottoman Sultan for the proposed German railway. The Sultan called in a young Armenian oilman, by the name of Calouste Gulbenkian, to investigate, launching him on what became his life's work to develop Iraq's oil. To this end, he established the

Turkish Petroleum Company in 1912. It was owned by the Deutsche Bank (25%), which controlled the previous railway concession that conveyed the mineral rights, Shell (25%) and the Turkish National Bank (50%). The latter had been set up by British financial interests, with Gulbenkian holding 30%. The British government then intervened to secure a holding for what is now BP, reducing Gulbenkian's share to 5%.

The rights to the concession were confirmed on June 28th 1914, a few days before the outbreak of the First World War, in which Turkey sided with Germany, with whom it already had close links. The importance of oil became evident during the war, and France and Britain, followed by the United States, began to discuss the eventual carve up of the Middle East while hostilities were still in progress. It was already perceived to hold much of the world's endowment.

Negotiations began in earnest in the peace treaties that followed the war, eventually giving Britain mandated administrative control of the territory. It was declared a Kingdom, with the crown being placed on the head of Prince Feisal, the son of the Grand Sharif of Mecca, Britain's premier ally in the war, who had been promised an Arab Kingdom in return for his contribution. In fact, Feisal had first been put on the throne of Syria, but was recycled when that country came into the more republican French sphere of interest. It was agreed that Iraq's oil, which had become a central issue, would be produced by what became the Iraq Petroleum Company (IPC) with the following shareholding:

Shell (Anglo-Dutch)	23.75%
BP (British)	23.75% (previously Anglo-Persian & Anglo-Iranian)
CFP (French)	23.75% (now TotalFinaElf)
Exxon (US)	11.875% (now Exxon-Mobil)
Mobil (US)	11.875% (now Exxon-Mobil)
Gulbenkian (Independent)	5%

The companies also agreed not to compete with each other throughout most of the previous Turkish Empire, including Saudi Arabia: Exxon and Mobil reneging on the agreement when they joined Aramco in Saudi Arabia in the 1930s.

Exploration soon commenced to be richly rewarded with the discovery of the Kirkuk Field in 1927, holding about 16 Gb of oil in a large surface structure, obvious for miles around in the northern, Kurdish, part of the country. Production rose gradually to the Second World War, reaching 100 kb/d by 1947. It was not, accordingly, a particularly important exporter to that point.

The post-war epoch was characterised by growing nationalism throughout the region, being encouraged when the United States opposed an Anglo-French military strike to prevent Egypt sequestering the Suez

Canal in 1956. Most of the producing countries nationalised the holdings of the foreign oil companies over the ensuing years: Iraq doing so in 1972. Exploration continued successfully, testing the prospects already identified by the IPC. As many as fifty oilfields have been found, of which about half are giant fields, together holding some 90 Gb. Of that, about 50 Gb lie in just three fields: Rumaila (1953); Kirkuk (1927); and East Baghdad (1976). Production to-date from all fields amounts to almost 30 Gb, leaving about 60 Gb for the future plus whatever new exploration might turn up. It seems very clear from the size distribution of the fields that the bulk of Iraq's oil has already been found, with many of the smaller discoveries still awaiting development.

Saddam Hussein was born in 1937, making him 65 years of age. He joined the Ba'athist Party in 1957, which was an Arab version of Communist style dictatorship. In the following year, the then King, Feisal II, was beheaded in a coup led by a Colonel Kassim, who was backed by Egypt. He in turn fell in another coup that brought the Ba'athists to power in 1968, appointing Saddam Hussein President in 1979. As described above, the country was a somewhat artificial construction, comprising Kurds, who have long sought independence, in the north; Shi'ites with links to Iran in the south; and Sunnis around Baghdad, the capital. Evidently, it takes a strong leader to hold these disparate groups together as a nation. It previously had a substantial, well-integrated Jewish community, Baghdad having been one of the great centres of Judaic culture in the 5th Century, but it was driven out by popular outrage on the creation of the State of Israel.

In 1974, heavy fighting broke out between government forces and Kurdish separatists, who were being backed by Iran, but the dispute was settled when Iran withdrew its support in return for resolution of a long-standing boundary dispute, related to the key Shatt al-Arab estuary of the Tigris-Euphrates river system, Iraq's main trade route. But tensions with Iran erupted again on the fall of the Shah in 1979 when unrest among the Iranian Kurds spilled over into Iraq. It soon developed into a full-scale war, which dragged on for almost eight long years with colossal loss of life to both sides. Although nominally neutral, the United States backed Iraq during this conflict, still smarting from an incident in which American citizens were taken hostage in Tehran, following the fall of the Shah. During the late 1980s, the United States supplied Iraq with substantial bank credits and technology to rebuild its military strength. The Soviets too developed close ties, furnishing credit and weapons.

Meanwhile, President Reagan and Mrs Thatcher resolved to try to bring down the Soviet regime, ending the policy of co-existence. According to the book, *Victory*, by Peter Schweizer, the first step was to rearm the Afghans to

Iraq – Conventional		
Rates Mb/d		
Consumption	2001	0.4
Production	2001	2.0
	Forecast 2010	3.0
	Forecast 2020	4.5
Discovery 5-year average (Gb)		0.06
Amounts Gb		
Past Production		27
Reported *Proved Reserves*		113
Estimated Future Production to 2075		
	From Known Fields	95
	From New Fields	13
	Future Total	108
	Past and Future Production	135
Current Depletion Rate		0.7%
Depletion Midpoint Date		2021
Peak Discovery Date		1948
Peak Production Date		2013

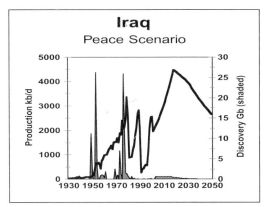

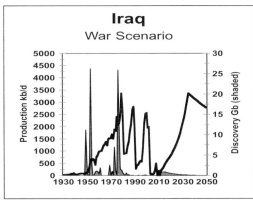

Figure 46

end the Soviet occupation, and undermine its military credibility. This was achieved with the help of King Fahd of Saudi Arabia, who funded the covert purchase of arms in Egypt for shipment to none other than Osama bin-Laden, who was backing the Taliban with CIA support.

The next step was to persuade King Fahd to step up production to undermine the price of oil. The Soviets relied on oil exports for foreign exchange, which they now needed in greater amounts to buy equipment with which counter the new US "star wars" initiative. It was a successful strategy, which contributed to the fall of the Soviets, but was achieved at a cost, as the low price of oil was bankrupting not only King Fahd but the Texan oil constituents of Mr Bush Sr.

While all this was going on, Kuwait arbitrarily increased its reported reserves by 50% in 1985 although nothing particular had changed in its reservoirs. It did so in order to raise its OPEC production quota, which was based on reserves. It also began pumping from the southern end of the Rumaila field that straddles the ill-defined border with Iraq. The latter complained bitterly both about what amounted to the theft of its oil across the border, and the loss of oil revenue, as prices fell consequent upon Kuwait's failure to observe its contractual OPEC agreement.

Now, US strategy moved to strengthen the price of oil, dispatching an emissary, Henry Shuyler, to encourage its ally, Saddam Hussein, to intervene in the councils of OPEC to enforce quota agreements sufficiently to achieve that end. It was recognised that words might not be enough to concentrate the minds of the OPEC ministers. Exactly what was proposed is not known, but it seems clear that a border incident to stop Kuwait producing from the southern end of the shared oilfield was contemplated. This interpretation is confirmed by the words of April Glaspie, the US ambassador to Baghdad, who, on the eve of the invasion of Kuwait, made a statement to the effect that boundary disputes between Arab countries were of no concern to the United States. It was clearly an authorised statement, being released simultaneously in Washington under the signature of James Baker, the Secretary of State.

However, Saddam Hussein, possibly misunderstanding a wink and nod, did not stop with a border incident, mounting a successful full-scale invasion of Kuwait on August 2nd 1990. April Glaspie, on being woken by journalists with the news, reportedly responded *"Oh My God, they haven't taken the whole place, have they?"*, which hints of collusion or at least fore-knowledge.

US policy now changed to condemn its former ally. A series of UN resolutions called for Iraq to withdraw from Kuwait by January 15th 1991, leading to a US aerial bombardment, when it failed to comply. Ground forces, led by General Schwarzkopf, crossed the frontier, killing tens of thousands of Iraqis and destroying most of its military capability, before being ordered to halt at the gates of Baghdad when a cease-fire was agreed. The dissident Shi'ites in the south and the Kurds in the north saw this as their moment to rise, but were successfully suppressed by remnant government forces. Hundreds of thousands of Kurdish refugees fled into neighbouring Turkey and Iran, where they were not exactly welcome.

The United Nations then imposed a trade embargo on Iraq, making it effectively swing oil producer of last recourse, which provided a useful mechanism for stabilising the world price of oil at no cost to anyone but Iraq. In common with many countries, Iraq had made certain progress in developing modern nuclear, chemical and biological weapons, but by 1998, UN inspectors had reported that virtually all such facilities had been destroyed. The embargo remained, however, although it was partly relaxed for "humanitarian" reasons when the price of oil rose to uncomfortable levels. Several European, Russian, Chinese and other companies have signed agreements to develop oilfields as soon as the embargo is lifted, committing over $1.7 trillion to do so. US officials have stated that such agreements would be nullified by the overthrow of the present government, paving the way for the entry of the major US companies. Whether Blair's support for Bush would be rewarded with a stake for BP remains to be seen.

The United States is now moving against Iraq on the grounds that it might be re-developing its weapons such that they could pose a threat to unspecified targets in the future. Little concrete evidence has yet been furnished, but the United States claims to be entitled to move on the so-called precautionary principle in its new generalised "war on terror". There has been no love lost between Saddam Hussein and Osama bin-Laden, who is said to have been behind acts of terror, so evidently a different strain of terror is attributed to Iraq.

Future Scenarios

The United Nations has now passed a Resolution calling for the re-entry of the famous inspectors, to which Iraq has agreed. At the time of writing (Nov 20, 2002), it appears that some of the tension has gone out of the situation following the US elections when President Bush and his backers succeeded in strengthening their grip on the country, partly on the strength of a certain war fever.

Scenario of Peace

As it is impossible to begin to forecast the actual turn of events, we will have to be content with a fanciful fairy tale scenario. Let's imagine that a certain Baron Werner von Wittel, a historian from Munich University, was working on archives in the Istanbul Museum, and when clearing away the ancient parchments at the end of the day, his eye was caught by a document that had slipped from the pile. Closer inspection showed it to be a hitherto unknown contract between Abdul Hamid, the last of the Ottoman Sultans, and Calouste Gulbenkian, his heirs and assigns, dated 1897. It conveyed irrevocable rights in perpetuity to the oil lands of the Vilayet of Mosul and surrounding territories, being countersigned by three descendants of the Prophet; the Ambassadors of France, Great Britain and the United States; and the Chief Rabbi of Jerusalem. It carried a special proviso establishing legal protection against regime change, such being guaranteed by the co-signatory governments.

The Baron, realising its possible current significance, passed a surreptitious copy to his Ambassador. A search of the international phone directory located a Mr Gulbenkian in Lisbon who was invited to present himself. Some frantic diplomatic activity ensued.

Next day, the sunburnt shrewd face of the Armenian appeared at the Embassy gates. After some pleasantries confirming that he was indeed the great grandson of the signatory to the contract, he asked to make some international phone calls. The first was placed to the Oval Office where he quickly struck a deal for Texan oil interests to have exclusive rights to Iraqi oil, with an overriding royalty to the President's favourite charity. The next was to Saddam Hussein in Baghdad, who readily accepted a proposed 20% royalty to the Iraqi State and an override to his family charity, not wishing to be up-staged by the US leader.

The UN inspectors rushed in to find that the suspect biological laboratory had been developing a flu vaccine, and that the suspect radio-active materials had been purchased for an ageing Russian nuclear electricity generator in Basra. Within a few weeks, Kofi Annan was able to sign a clean bill of health.

The Gulbenkian Oil Company was incorporated in Bermuda with obscure nominee shareholders; and Western contractors were brought in to undertake the rehabilitation of the industry. The new management decided to establish three divisions: the Foothills; Central; and Western Divisions. Bankers arrived, offering unsecured bonds to finance the operation. The financial press was enthusiastic for what was identified as one of the most straightforward and fair oil deals struck in recent years. It was free of the hidden subsidies taken as operating cost against taxable income; the politicians had been paid off reasonably; and the national interests of Iraq and the United States protected. The latter was particularly pleased to be able to draw on Iraqi supplies at cost plus 25%. The slim figure of the Armenian was seen being carried shoulder high through Wall Street from one champagne lunch to another.

Now began the serious work of developing Iraq's remaining oil in a highly technical manner, free of political or speculative pressures. The following ten-year work programme was drawn up for the monumental task in hand, recognising that it was costly and slow onshore work in tough terrain.

1. Exploration.

Establishment of a staff of 30 geologists and support

- Regional compilation of existing reports, data, samples
- Photogeological evaluation
- Geochemical laboratory
- Field work to identify and map source rocks using modern methods
- Seismic surveys: five crews to provide modern coverage
- 30 Geophysicists with computing support to interpret the data

The objective was to develop a steady stream of prospects to be tested by five exploration rigs, which were to be kept in continuous operation.

2. Reservoir Engineering

Establishment of a staff of 30 reservoir engineers to evaluate past well data on producing fields, optimise reservoir performance, introduce pressure maintenance and waterflood programmes as appropriate. It was found that several reservoirs had suffered serious damage from over-production in recent years.

3. Operations

- Operate ten workover rigs, repairing and reconfiguring existing wells;
- Operate ten rigs dedicated to infill drilling and new field development

4. Support Services and Construction

- Warehousing, purchasing, housing, personnel
- Road construction to remote locations
- Helicopter and air services
- Transport and maintenance
- Pipeline construction: maintenance and repair
- Loading terminals: repair, expansion and operation

The scale of the work soon became evident. A remarkable feature was the good co-operation from the Iraqi workforce, including professionals with invaluable local knowledge. Although the true production potential was not determinable for several years, a valid preliminary assessment gradually did emerge as follows:

1. Foothills Division

It transpired that the geology of the Zagros Foothills had already been well determined by the old Iraq Petroleum Company. All the major structures of interest had been identified; and most had been tested. What remained to do was to evaluate deeper plays in the frontal thrust-belt, in part relying on secondary source and reservoir objectives in complex structural conditions. It called for intensive geophysical work.

2. Central Division

This was the prime area of interest, but its geology was complicated by the superimposition of two structural frameworks: the original sedimentary basin, running northward from the Gulf, having been cut but late-stage northeasterly transverse movements that broke it up into alternating structural highs and lows. The lows were found to be generally gas-prone or non-prospective, although having some scope for secondary source and reservoir potential. All the larger prospects on the highs had been identified long ago; and most had been tested.

3. *Western Division*

This area is characterised by very low structural relief offering subtle, but possibly large, stratigraphic traps. A deeper gas-prone play, relying on Silurian sources charging patchy, poor-quality Permian reservoirs, is also in range in this area. While generally poorly prospective, it might nevertheless turn up a few positive surprises.

In Summary

There can be little doubt that Iraq has the potential to produce much more oil. Exactly how much will be known only after a comprehensive evaluation and new information. However, it would be reasonable to assume on today's evidence that about a total of 125 Gb will have been discovered by 2010, with about another 10 Gb to come in after that. Some 30 Gb have been produced to-date. Production stands at about 2 Mb/d, the amount being uncertain because of smuggled exports through Turkey, Syria, Jordan and Iran, which appear unseen in the statistics of those countries. It turns out that there is very little instantly available spare capacity. Under optimal, unconstrained operating conditions, it would be reasonable to expect production to rise to about 3 Mb/d by 2010, reaching a peak of 4.5 Mb/d around 2020. By then, it might be able to supply about one-quarter of US needs, assuming its consumption did not rise greatly in the future, under very advantageous terms, far below the then current world prices.

The scenario assumes general political stability, with any moves to Kurdish separatism being brutally suppressed by Turkey, Iraq and Iran, now with a UN mandate and US help. Such repression would be nothing new having been practised by both the current government and Britain during the 1920s, when it had found it necessary to call in the Royal Air Force strafe Kurdish positions in its effort to maintain order and underpin the IPC concession.

A Scenario of War

If, on the other hand, the country is subject to military attack, it would be hard pressed to maintain even present production in the face of death, devastation and prolonged conflict, with the oil installations being subject to continuing relatively easy acts of sabotage. The silver lining would be that there would be more left for the survivors of the apocalypse. It is too awful a scenario to contemplate in detail.

Summary

It is difficult to present the standard data table that normally accompanies these evaluations because the available information is so unreliable, but the following is based on the current model, for what it is worth, approximating with the Peace Scenario. Two plots give the production profiles under the two alternative scenarios, but are no more than illustrative. The country has no known Non-Conventional potential but probably plenty of gas for the future.

(It is worth mentioning in passing that the USGS, which may be misleading the US government, proposes much higher Mean estimates, having come upon an outdated consultant's map of Iraq that has been floating around the industry for many years. It shows notional prospect leads, several of which were subsequently tested, and deserves no particular credence.)

162. BBC Open University Video

The BBC has produced an admirable video for the Open University on energy, and fossil fuel depletion, featuring C. J. Campbell and the former Chairman of Shell, who surprisingly did not seriously disagree with the premise of oil decline in the not too distant future. It was produced by Anne-Marie Gallon and carries a reference FOUT657R T206/VC1B1-01. (BBC, Perry Building, Walton Hall, Milton Keynes,

Bucks, MK7 6BH, England). ASPO members could well find it useful for teaching or illustrative purposes.

163. New book on depletion and sustainable energy

A new book by members of ASPO has been published in Germany as follows:
Title: Ölwechsel!
Subtitle: Das Ende des Erdölzeitalters und die Weichenstellung für die Zukunft
Authors: Colin J. Campbell, Frauke Liesenborghs, Jörg Schindler, Werner Zittel
Editor: Global Challenges Network (www.gcn.org)
Publisher: Deutscher Taschenbuch Verlag (dtv) Series: dtv-premium, No. 24321 ISBN 3-423-24321-4

164. Words of Wisdom

Attention is drawn to the works of John Attarian, an economist from Michigan, who explains with great lucidity the devastating consequences of outdated flat-earth economic principles, seeing the depletion of oil as one of the mechanisms that will impose change to a more sustainable life-style. Some of his work can be found on www.thesocialcontract.com

165. The USGS seeks to perpetuate its mistake

We can readily forgive the USGS for its flawed new study of the potential for oil discovery, which counters its own excellent work over the past thirty years. It is indeed a difficult subject calling for judgement and experience. It is less easy to forgive them for trying to perpetuate their mistake, which is doing great damage by misleading foreign governments and international agencies.

A Reply by C. J. Campbell

to

"Global Petroleum Reserves – A View to the Future"
by Thomas S. Ahlbrandt and J.McCabe,,United States Geological Survey
Published in Geotimes, November 2002

Ahlbrandt and McCabe have written an elegant article choosing their words with extreme care to present what seems to be an authoritative account of the world's oil and gas situation, based on a study made by the United States Geological Survey in 2000. But a closer look shows it to be a thoroughly flawed study that has done incalculable damage by misleading international agencies and governments.

The study was in fact a marked departure from earlier sound evaluations made by the USGS over a thirty years period under its previous project director, the late C.H. Masters. He showed that he understood the situation well, using great skill to deliver the message, albeit at times between the lines, as he recognised its sensitive nature.

Neither of the authors claim practical oil experience. That is betrayed by their mindset, which is more appropriate to the mining geologist for whom resource concentration is as important as occurrence. They say they speak to a Mr Green of Exxon, but we do not know what he tells them or with what motive: another spokesman in the same company reportedly made the succinct comment: *"you get what you pay for, and that came free"*.

It is an old trick for the politician to answer a question that is not asked. No one need be seriously concerned about when the last drop of oil will be produced. What matters – and matters greatly – is the date when the growth of past production gives way to decline from resource constraints. This is the transcendental issue, given the world's dependence on abundant oil-based energy, furnishing 40% of all traded energy and 90% of transport fuel, essential to trade. The USA itself experienced the discontinuity in 1971, and the same pattern of growth to decline has been repeated from one country to another around the world. A recent example is the United Kingdom, with production peaking in 1999, twenty-seven years after peak discovery. Production inevitably has to mirror earlier discovery after a time lag. The world peak of discovery was in 1964, so it should surprise no one that a corresponding peak in production is now imminent. It is so self-evident, even if our eyes are too blinkered to see it.

The authors present the comforting notion of a resource pyramid, implying that the World can seamlessly move to more difficult and expensive sources of oil and gas when the need arises. This is the case in eastern Venezuela and western Canada where huge deposits of degraded oil flank the basins, but they are exceptional. There is a polarity about oil that the authors fail to grasp: it is either present in profitable abundance or not there at all, due ultimately to the fact that it is a liquid concentrated by Nature in a few places having the right geology. They speak of "crustal abundance" when a glance at the oil map shows clusters of oilfields separated wide barren tracts.

They give emphasis to "reserve growth" as a new element, although recognised and dismissed by their predecessor, yet fail to point out that the text of the study itself expresses grave reservations. "Growth" is in fact more an artefact of reporting practices than a technological or economic dynamic. In short, reserves described as *Proved* for financial purposes refer to what has been confirmed so far by drilling, saying little about the full size of the field concerned. Clearly, it was absurd to apply, as the study did, the experience of the old onshore fields of the USA, with their special commercial, legal and reporting environment, to the offshore or international spheres, where very different conditions obtain.

The authors speak of their impressive probabilistic methods, which in the study allowed them to quote estimates to three decimal places. In, for example, the famous case of little known NE Greenland, the study states with a straight face that there is a 95% subjective probability of more than zero, namely at least one barrel, and a 5% probability of more than 111.815 Gb (billion barrels). A *Mean* value of 47.148 Gb is then computed from this range, being incorporated in the global assessment. Can we really give much credence to the suggestion that this remote place, which has so far failed to attract the interest of the industry, holds almost as much, or indeed more, than the North Sea, the largest new province to be found since the Second World War? Could this be pseudo-science at its best?

Turning to the actual estimates, the authors state that the sum of past production, reserves, reserve growth and undiscovered comes to about three trillion barrels, but then claim that the peak of production will not arise before the mid-Century. Experience shows that the onset of decline comes at, or before, the midpoint of depletion, due largely to the immutable physics of the reservoir that impose a gradual decline on production towards exhaustion. Depletion Midpoint on their estimates will come when 1500 Gb have been produced, which will be around 2020 at present production rates, or sooner if demand should rise. A mid-Century peak implies a precipitate fall thereafter, which is implausible. But this line of reasoning does not paint the full picture, because it fails to distinguish the different categories of oil. There is, clearly, a huge difference between a Middle East free flowing well and digging up a tar-sand in Canada with a shovel. As the authors themselves state, there will be an increasing reliance on heavy oils, low on their resource pyramid, which are slow to produce and will not contribute significantly until after global peak for obvious reasons. The USGS study did not forecast production itself, but simply indicated the amounts to be found over the 30-year study period ending in 2025. But the internal evidence, flawed as it is, indicates a peak long before the mid-Century. If that were not enough, we can now compare the actual results with forecast over the first six yours of the study period. The indicated average annual discovery is 24 Gb, whereas the actual has been less than half that amount. This is doubly damning because it would be normal to expect the larger fields to be found first as the past record amply confirms.

What the USGS failed to do was to extrapolate past discovery trends in the world's mature basins, containing most of its oil and gas, having properly backdated reserve revisions to the discovery of the respective fields. It is axiomatic that a field is found by the first successful borehole drilled into it, even if its size is not exactly known at the outset. Had the USGS done that, it would have had the benefit of the considerable experience of the oil industry working in the real world, which is likely to give a better view of the future than abstract geological assessment couched in subjective probability ranking.

The authors accuse those, who draw attention to the manifest failure of the study, of having hidden agendas, introducing the colourful but unhelpful designations of Cornucopian and Malthusian, when all we seek is a realistic assessment of this critical issue.

The article reviews two specific areas: the Caspian and Iraq. Is it a coincidence that the United States earlier this year attacked Afghanistan, which borders the Caspian, and now turns its guns on Iraq?

We can forgive its authors for having got it wrong as it is a difficult subject, calling for long years of experience, as marshalled by their predecessor, but to perpetuate the error with persuasive language and specious argument verges on the culpable.

166. Petroleum and People

A paper by C. J. Campbell under the above title appears in *Population and Environment* v.24/2 (Nov 2002), reviewing the impact of oil depletion on society in general.

167. Dire UK Gas situation

Papers from the UK gas industry reveal that gas production peaked in 2000, declining at 2% while demand increases in like amount, such that the UK will be importing 50% of its needs in ten years. The position for poor Ireland, which relies on UK gas, will be even worse. Little thought seems to have been given to exactly where the imports will come from, with misplaced reliance on the famous flat earth "Open Market" to deliver.

Part 3

The Graphs Of Jean Laherrère

Jean Laherrère's first career was with the Total Oil Company, based in France. His early years were spent working as a geophysicist in Algeria, where he was involved in the discovery of the giant Hassi Massaoud field. Later, he undertook exploration assignments in Australia and Canada before returning to Paris as Deputy Exploration Manager.

A new career unfolded in "retirement" when he concentrated on an evaluation of the world's reserves of oil and gas and their depletion. One of his particular achievements was the discovery of the parabolic fractal, which showed that objects in a natural domain plot as a parabola when size is set against rank on a log-log scale. It is particularly relevant to oilfield distributions because once the larger fields have been found, which usually happens early, they set the parameters of the parabola which may be extrapolated to the smallest field. The difference between what has been found to-date and the theoretical curve is what is yet-to-find.

Another considerable contribution was the recognition of discovery cycles, based on properly backdated reserve revisions, which could be convincingly correlated with corresponding production cycles after a time-lag. It provides a strong tool for forecasting future production from past discovery.

He has a particular interest in producing graphic representations of the many difficult relationships. A selection from a large collection that have been published or used in conferences is reproduced below. They are often complex as they evaluate diverse factors and relationships, but the captions explain their essential meaning.

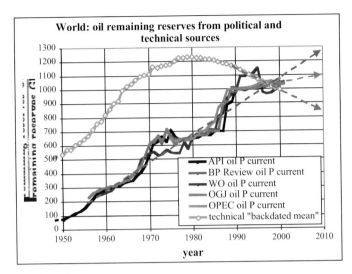

Graph 1. Reserve Reporting

This graph compares published data on so-called *Proved Reserves*, with revisions taken when announced, and *Proved & Probable Reserves*, with properly back-dated revisions, as reported in the industry database. Stated simply, Proved Reserves refer to estimated future production from an oilfield's current stage of development as reported for financial purposes, with a tendency to under-report, whereas *Proved & Probable Reserves* refer to estimates of all future production. Many analysts have been misled into concluding that more is being found than is the case.

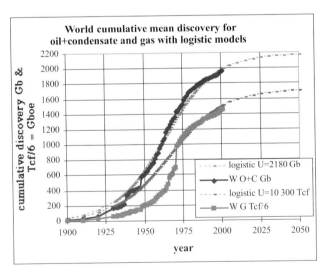

Graph 2. Extrapolation of discovery

This graph extrapolates the discovery trends of the World's conventional oil and gas. It is evident that discovery is flattening. The extrapolation shows that exploration will have virtually ended before 2050, by which time a total about 2200 Gb (billion barrels) of oil and condensate (liquids that condense naturally from gas) and 10 300 Tcf (trillion cubic feet) of gas will have been found. (Note tarsands and related heavy oils are not included).

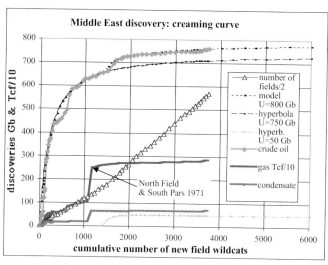

Graph 3. Discovery in the Middle East

This graph plots the discovery of oil and gas in the Middle East against the number of exploration boreholes ("wildcats"), showing also the number of fields. It is noteworthy that most of its oil was found with the first 1000 wildcats, suggesting that the future potential is very reduced. An ultimate recovery of 700–800 Gb is indicated. Much of the gas lies in a single field, found in 1971. It straddles the Qatar — Iran boundary, being known in the respective countries as North Field and South Pars. The Iranian extension is commonly attributed to a later date, which is a mistake because the field as a whole was found in 1971.

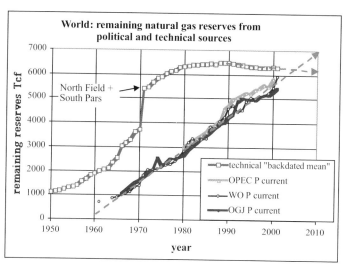

Graph 4. Gas Reporting

This graph compares published *Proved Reserves* with *Proved & Probable Reserves* from the industry database. The quoted published sources are OPEC, World Oil and the Oil & Gas Journal. See also the comments made for Graph 1.

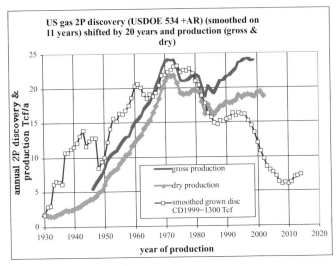

Graph 5. US Gas discovery and production

This relates the discovery of gas in the United States with production after a time shift of twenty years. Discovery peaked in the 1950s and has since fallen to about the same level as it was in 1925. It follows that production is set to decline steeply. The plot also compares gross production with so called dry production after the removal of dissolved liquids.

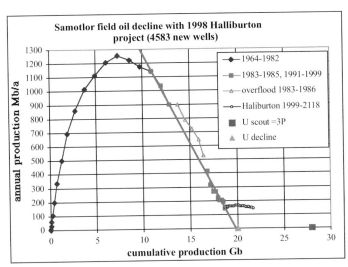

Graph 6. The Decline of Production in the Samotlor Field of Russia

This plots annual against cumulative production whereby the decline rate plots as a straight line. Extrapolating the decline shows that the field is capable of producing not more than about 20 Gb (billion barrels). It shows that the official estimate of 28 Gb cannot in practice be reached. It demonstrates that reserves were generally exaggerated under the Soviet classification, which did not take into account technical and economic constraints.

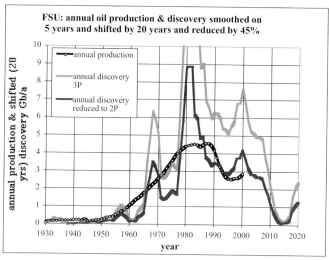

Graph 7. Oil Discovery and Production in the Former Soviet Union

This shows smoothed annual discovery (2P), having been corrected from the exaggerated official estimates (3P) with the corresponding production after a time-shift of twenty years. Discovery peaked in the early 1960s, with the corresponding peak of production following in the early 1980s. Both exploration and production collapsed on the fall of the Soviets but both are now recovering, which will give a second peak at some time in the future.

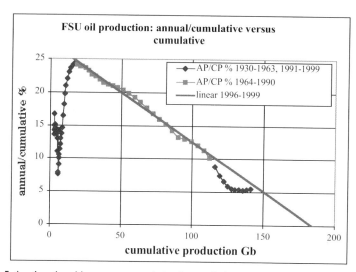

Graph 8. Estimating the ultimate recovery of the Former Soviet Union

This illustrates a mathematical technique whereby the ultimate recovery may be calculated from the ratio of annual production to the total of all past production against cumulative production. Extrapolating the trend of points to an ultimate recovery of about 180 Gb (billion barrels).

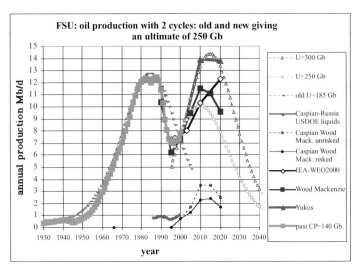

Graph 9. Estimates of future production from the Former Soviet Union

This graph evaluates the production profile of the Former Soviet Union under two cycles, assuming an ultimate recovery of 250 Gb (billion barrels), somewhat higher than that indicated by the plot in Graph 8. This model would give a second peak at 10 Mb/d in 2010. Also shown are the higher estimates from other authorities, which are based mainly on announced plans and aspirations, without paying due attention to the implications in terms of ultimate recovery.

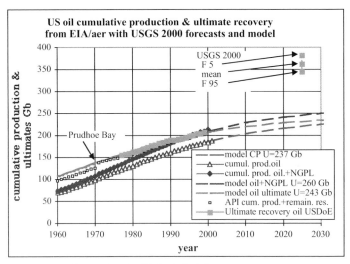

Graph 10. Implausible oil forecast by the US Geological Survey

This plot shows the record of past discovery in the United States for crude oil and crude oil combined with natural gas liquids from gas plants (NGPL), duly extrapolated to show what can be expected from new discovery to 2025. Superimposed on the graph is the range of estimates of the total to be found by 2025 made by the US Geological Survey in 2000. It is evident that they bear no relationship to the past trend and can be confidently dismissed as utterly implausible.

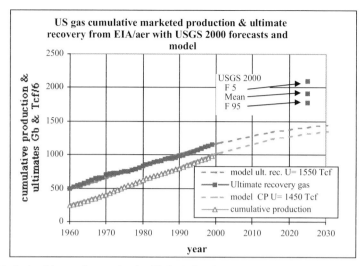

Graph 11. Implausible gas forecast by the US Geological Survey
This shows a corresponding plot for gas, indicating that the USGS estimates are implausible in relation to the trend of past discovery.

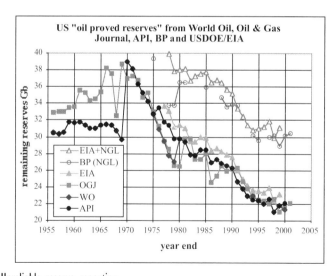

Graph 12. Unreliable reserve reporting
This plot shows the wide range in the reported Proved Reserves for the USA from different publications.

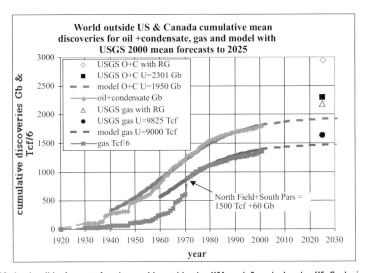

Graph 13. Implausible forecast for the world outside the USA and Canada by the US Geological Survey
The graph is a comparable plot showing the extrapolation of the past discovery of oil & condensate and gas, showing that the USGS forecast is far out of trend.

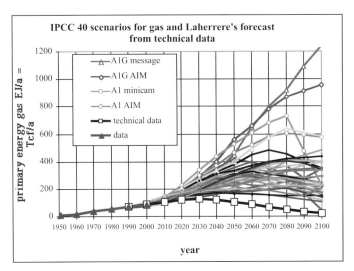

Graph 14. Implausible models used in the climate change debate
The International Panel of Climate Change has issued a large number of scenarios of hydrocarbon emissions, all greatly exaggerating the amounts available in the face of depletion, indicated in black.

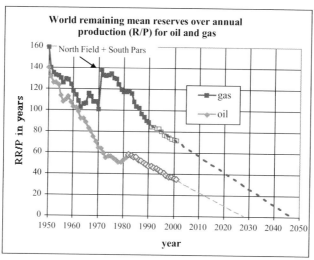

Graph 15. Extrapolation of Reserve to Production Ratio (R/P)

Security of supply is commonly quoted in terms of R/P ratio quoted in years, as if it were remotely plausible that current production could be maintained for a given number of years and then stop over night, when all oil fields are observed to decline gradually towards the end of their lives. This plot shows that R/P ratio for oil has declined from 140 years in 1950 to 35 years in 2000, which by extrapolation implies exhaustion by 2030, a ridiculous proposition.

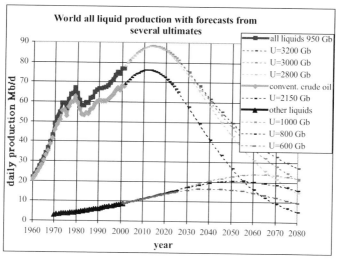

Graph 16. Oil production modelled on alternative resource assumptions

This graph shows the production of all liquid hydrocarbons modelled on alternative assumptions of the total recovery at 2800 Gb, 3000 Gb and 3200 Gb. giving a peak around 2015.

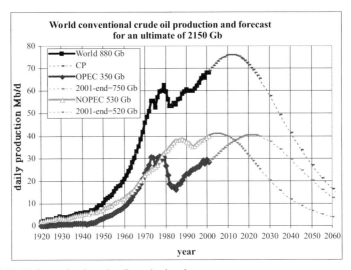

Graph 17. World Conventional crude oil production forecast
This graph shows past production of conventional oil for OPEC, Non-OPEC and the World as a whole, with forecasts of future production based on an assumed ultimate recovery of 2150 Gb.

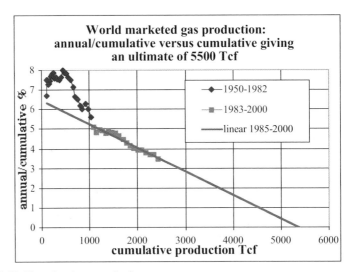

Graph 18. World marketed gas production
This graph shows cumulative production compared with the percentage of annual to cumulative production, which by extrapolation suggests an ultimate recovery of 5500 Tcf.

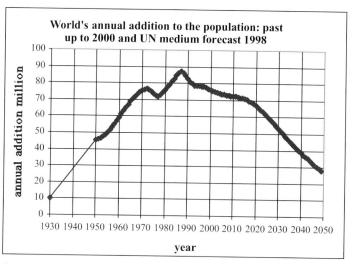

Graph 19. Annual growth in population

The annual increase in population was about 1 million during the 19th Century, rising to 10 million for the first half of the 20th Century. It reached 45 million in 1950 and peaked at around 87 million in 1987, before falling to 77 million in 2000. It appears that a symmetrical curve is building, such that past growth will be matched by future decline. The 1998 UN forecast of 30 million in 2050 may be too high.

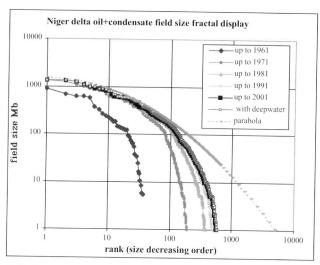

Graph 20. Parabolic Fractal

This graph shows the evolution of the size distribution of fields in the Niger Delta, with rank plotted against size on a log-log scale. Few larger fields were added after 1971, meaning that that segment of the full distribution was complete. A fractal law of self-similarity states that a complete segment defines the total. It in turn provides the parameters for the parabola, by which to model the full distribution and determine what is left to find, mainly in ever-smaller fields.

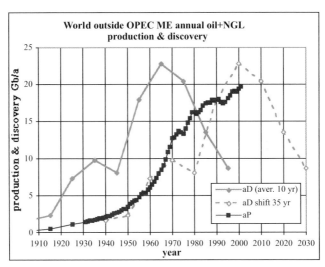

Graph 21. Annual discovery and production for the World outside the Middle East OPEC countries
This shows a moving average of discovery of crude oil and Natural Gas liquids (with a ten-year period) compared with production after a time shift of 35 years. The peak of discovery was in the mid-1960s, suggesting that production is now set to decline steeply.

Part 4

Aspo Statistical Review of World Oil and Gas

Many analysts use the BP Statistical Review of World Energy, wrongly believing that the oil and gas statistics have at least the tacit blessing of a knowledgeable oil company. They do not, being simply reproduced from the Oil & Gas Journal, which in turn compiles information from governments without being in a position to verify its validity.

To remedy this situation, ASPO itself endeavours to put out at least more reliable information. It has chosen a similar title to raise awareness of the issue in the hope that more authoritative data may come into the public domain.

Although there are no particular technical or scientific difficulties in determining the size of an oil field or in extrapolating the discovery trends, we may be sure that the estimates are wrong, given the ambiguous definitions and lax reporting practices. That is not in doubt. The question is by how much? It is an important question deserving urgent attention, given the world's dependence on oil supply.

Aspo Statistical Review of World Oil and Gas

Explanations and Definitions

Conventional Oil Production
Effective Date – end 2002
This table gives the current best estimate of future production of conventional oil by country and region. The following explanations will help decipher its meaning.

Regular Oil (also termed *Conventional*) is here defined to exclude the following categories:

> Oil from coal and "shale" (actually immature source-rock)
> Bitumen (defined by viscosity)
> Extra-Heavy Oil (<10° API)
> Heavy Oil (10–17.5° API)
> Deepwater Oil (>500m)
> Polar Oil
> Natural Gas liquids extracted from plants – NGL(P)

(Condensate which condenses naturally at surface conditions is treated with crude oil, with which it is commonly metered.)

The estimates refer to production through 2075 to avoid having to worry about tail end production and any enhanced recovery practices that may succeed in extracting more oil from fields approaching the end of their lives. The amounts added after 2075 are likely to be too small to have any material impact on peak production, which is the prime subject under investigation.

The source of production is indicated. The amounts coming from Known Fields is distinguished from the amounts expected from yet-to-be-discovered New Fields. Total production to 2075 as well as All Future production is indicated.

The explanation of the columns in Table 1 are as follows:

Column A   List of producing countries with total production in excess of 500 Mb. The regions comprise the following:
W. Europe
ME Gulf   – Abu Dhabi, Iran, Iraq, Kuwait, Neutral Zone, Saudi Arabia
Eurasia   – the former communist bloc of East Europe, FSU and China
N.America – Canada and USA
L. America – South America and Mexico
Africa
Europe   – Norway to Italy
Far East   – Pakistan to Australia and Vietnam (excluding China)
ME Other   – Middle East countries other then ME Gulf, including Turkey

	Other
	Unforeseen – unforeseen future production to yield a rounded total
Column B	Annual production in average thousands of barrels per day
Column C	Annual production in billion barrels
Column D	Total production to end 2002 ("*Cumulative Production*")
Column E	Average percentage increase or decrease over past five years
Column F	"*Proved Reserves*" as reported by the Oil & Gas Journal
Column G	"*Proved Reserves*" as reported by World Oil
Column H	Adjustment to remove any identifiable Non-Conventional oil and NGL(P), as well as the cumulative production of any period of unchanged reports, which normally reflect a failure by the country concerned to update its reserve estimates. It is implausible to imagine that new discovery or reserve revision would exactly match production, so in the absence of other information the reported reserves are reduced by the production of the period in question
Column I	The percentage of *Proved Reserves*, as reported by the Oil & Gas Journal, to the assessed *Proved & Probable* (or *Mean Probability*) *Reserves*, which are deemed to equate to future production from known fields to 2075
Column J	Future production to 2075 from known fields ("*Reserves*")
Column K	Sum of Columns D and J ("*Discovered-to-Date*")
Column L	Estimated production to 2075 from new fields to be found in the future ("*Yet-to-Find*")
Column M	Total future production to 2075, being the sum of Columns J and L ("*Yet-to-Produce*")
Column N	Total production to 2075 ("Ultimate")
Column O	Percentage of Total discovered (Col. K/N)
Column P	Depletion rate, being current production (Col. C) as a percent of future production (Col. M)
Column Q	Date of the midpoint of depletion, when Col D = Col N/2
Column R	Date of peak production, actual if past, forecast if in the future

Further Parameters

The following definitions refer to Table 2

Column S	Discovery in fields holding more than 500 Mb of total production to 2075
Column T	Time lag in years between peak discovery and peak production as assessed
Column U	Period of unchanged reserves as reported by the Oil & Gas Journal in years
Column V	Peak annual production as a percent of total production to 2075
Column W	Number of producing wells as reported by the Oil & Gas Journal
Column X	Production per producing well
Column Y	Reserves per producing well
Column Z	Number of wildcats to-end 2002 (wildcats are boreholes drilled to find new fields)
Column AA	Year when maximum number of wildcats were drilled
Column AB	Estimated number of future wildcats
Column AC	Past average discovery per wildcat
Column AD	Average discovery per wildcat over past five years
Column AE	Implied future discovery per wildcat
Column AF	Assessed gas reserves
Column AG	Yield of condensate Gb/Tcf
Column AH	Condensate reserves

Resource Based Production Forecast

Table 3 gives the forecast of future production by country assuming that production declines as the current or midpoint depletion rate, whichever comes first, with some short term adjustments as appropriate to local circumstances. The model assumes that the production of conventional oil is flat to 2010, due to demand constraints from recession. The ME Gulf region is assumed to act as a swing producer to 2010 offsetting the natural or modelled decline elsewhere. Production in the individual swing countries is modelled such that each produces in proportion to its share of the regional total until it peaks at its depletion midpoint, with the balance being made up by Saudi Arabia. The swing role ends in 2010, when the countries concerned are no longer able in practice to offset decline elsewhere.

This is no more than scenario. There are alternative possibilities around peak, but the range of option diminishes in the future as natural depletion, driven by falling discovery rate and the immutable physics of the reservoir, exerts an increasing grip on supply.

Although demand has been consistently rising at about 2% a year for many years, that need not continue if the economy declines in the face of the excessive financial manipulation of the past or if a new world war, driven in part for access to foreign oil, should break out. The population too may decline from war, disease, loss of essential energy supplies, as well as falling fertility due to changed behavioural patterns arising from past affluence.

These are human reactions that are hard to predict, but the depletion of oil and gas, formed in the geological past, is dictated by Nature.

		World — REGULAR OIL PRODUCTION TO 2075 — 2002 Unit:Gb (billion barrels). Excluding: (1) oil from coal, shale, bitumen, (2) heavy, deepwater & polar oil, (3) plant NGL																
		KNOWN FIELDS								**Future**	**Total**	**NEW FIELDS**	**FUTURE** Known & New Fields	**TOTAL**	Revised 9/Jan/03			
		Present		Past		*Reported Reserves*									% Disc	Dep. Rate	MP Dep	Peak Prod
		kb/d	Gb/a		5yr	World	O&GJ	Adjust	%									
	Country	2002	2002	Total	Trend	Oil			Rept'd									
	A	B	C	D	E	F	G	H	I	J	K	L	M	N	O	P	Q	R
1	Saudi Arabia	7380	2.69	94.4	-2%	259.3	259.3	0.0	135%	192.0	286	13.6	206	300	95%	1.3%	2020	2013
2	Russia	7385	2.70	121	4%	53.9	60.0	0.0	95%	63.2	185	15.4	79	200	92%	3.1%	1992	1987
3	US-48	4239	1.55	171	-2%	21.3	22.4	-9.0	65%	20.7	191	3.7	24	195	98%	6.0%	1971	1971
4	Iraq	2030	0.74	27.4	-1%	115.0	112.5	-4.2	120%	90.2	118	17.4	108	135	87%	0.7%	2032	2019
5	Iran	3450	1.26	54.4	-1%	99.1	89.7	-5.2	125%	67.6	122	8.1	76	130	94%	1.6%	2010	1974
6	Venezuela	2415	0.88	46.4	-2%	50.2	77.8	-30.0	110%	43.5	90	5.1	49	95	95%	1.8%	2003	1970
7	Kuwait	1600	0.58	30.8	-2%	96.5	94.0	-6.7	165%	52.9	84	6.3	59	90	93%	1.0%	2022	1971
8	Abu Dhabi	1690	0.62	18.0	-2%	61.9	92.2	-8.1	150%	56.1	74	3.9	60	78	95%	1.0%	2028	2014
9	China	3400	1.24	27.3	1%	29.5	18.3	0.0	70%	26.1	53	3.6	30	57	94%	3.9%	2002	2002
10	Libya	1300	0.47	22.9	-1%	30.0	29.5	-1.5	97%	28.9	52	3.2	32	55	94%	1.5%	2011	1970
11	Nigeria	1930	0.70	22.4	-2%	30.0	24.0	-7.5	55%	30.0	52	2.6	33	55	95%	2.1%	2009	1979
12	Mexico	3180	1.16	30.0	1%	23.1	12.6	0.0	60%	21.0	51	4.0	25	55	93%	4.4%	2000	2002
13	Kazakhstan	800	0.29	5.7	11%	-	9.0	0.0	30%	30.0	36	4.3	34	40	89%	0.7%	2028	2028
14	Norway	3150	1.15	16.3	1%	10.3	10.3	0.0	70%	14.7	31	2.0	17	33	94%	6.4%	2003	2001
15	UK	2250	0.82	19.6	-2%	4.6	4.7	0.0	45%	10.5	30	1.9	12	32	94%	6.2%	1998	1999
16	Indonesia	1120	0.41	19.8	-3%	9.2	5.0	-0.4	50%	9.2	29	2.0	11	31	94%	3.5%	1993	1977
17	Algeria	850	0.31	12.1	1%	17.0	9.2	-3.2	50%	12.0	24	3.9	16	28	86%	1.9%	2008	1978
18	Canada	1028.4	0.38	18.7	-2%	5.6	180.0	-174.8	100%	5.2	24	1.1	6	25	96%	5.6%	1986	1973
19	Azerbaijan	300	0.11	8.01	6%	-	7.0	0.0	55%	12.7	21	2.3	15	23	90%	0.7%	2014	2014
20	N.Zone	535	0.20	6.63	-1%	4.7	5.0	-2.0	55%	5.51	12	3.4	9	16	78%	2.2%	2008	2004
21	Oman	895	0.33	6.98	-1%	5.9	5.5	-1.0	60%	7.48	14	0.54	8	15	96%	3.9%	2003	2003
22	Qatar	640	0.23	6.76	2%	13.8	15.2	-10.2	90%	5.53	12	0.72	6	13	94%	3.6%	2001	2000
23	Egypt	750	0.27	8.67	-3%	2.4	3.7	0.0	105%	3.52	12	0.80	4	13	94%	6.0%	1995	1995
24	India	663	0.24	5.58	0%	3.8	5.4	0.0	105%	5.11	11	1.31	6	12	89%	3.6%	2004	1997
25	Argentina	750	0.27	8.26	-2%	2.9	2.9	0.0	90%	3.20	11	0.54	4	12	96%	6.8%	1994	1998
26	Australia	633	0.23	5.77	4%	3.8	3.5	-0.2	65%	5.03	11	1.20	6	12	90%	3.6%	2003	2000
27	Malaysia	760	0.28	5.31	1%	4.5	3.0	-0.3	70%	3.89	9.2	0.80	5	10	92%	5.6%	2002	2002
28	Colombia	583	0.21	5.74	-5%	1.9	1.8	0.0	50%	3.68	9.4	0.57	4	10	94%	4.8%	1999	1999
29	Angola	700	0.26	4.55	-3%	6.0	5.4	-10.6	-100%	5.21	9.8	0.24	5	10	98%	4.3%	2004	1998
30	Ecuador	398	0.15	3.29	1%	2.6	4.6	0.0	105%	4.41	7.7	0.80	5	8.50	91%	2.7%	2007	2004
31	Romania	118	0.04	5.71	-2%	1.2	1.0	0.0	75%	1.22	6.9	0.58	2	7.50	92%	2.4%	1973	1976
32	Brasil	300	0.11	4.50	-7%	8.6	8.3	-12.0	-500%	0.74	5.2	0.76	1	6.00	87%	6.8%	1991	1997
33	Syria	490	0.18	3.78	-2%	2.3	2.5	-1.8	40%	1.73	5.5	0.48	2	6.00	92%	7.5%	1998	1995
34	Turkmenistan	180	0.07	2.88	7%	-	0.5	-0.2	25%	1.48	4.4	1.64	3	6.00	73%	1.8%	2003	1973
35	Dubai	200	0.07	3.71	-7%	0.9	4.0	-2.0	250%	0.82	4.5	0.22	1	4.75	95%	6.6%	1990	1991
36	Brunei	185	0.07	2.99	6%	1.2	1.4	-0.7	45%	1.42	4.4	0.09	2	4.50	98%	4.0%	1989	1978
37	Trinidad	127	0.05	3.20	1%	0.7	0.7	0.0	60%	1.12	4.3	0.19	1	4.50	96%	3.4%	1983	1978
38	Gabon	294	0.11	2.86	-3%	2.4	2.5	-0.6	125%	1.53	4.4	0.11	2	4.50	98%	6.1%	1997	1996
39	Ukraine	72.474	0.03	2.63	-4%	-	0.4	0.0	50%	0.74	3.4	0.63	1	4.00	84%	1.9%	1984	1970
40	Peru	90.885	0.03	2.33	-4%	0.9	0.3	-0.6	-30%	1.03	3.4	0.39	1	3.75	90%	2.3%	1991	1983
41	Yemen	350	0.13	1.62	-1%	2.4	4.0	-1.3	200%	1.37	3.0	0.52	2	3.50	85%	6.4%	2003	1999
42	Vietnam	304	0.11	0.88	12%	2.2	0.6	-0.5	5%	1.81	2.7	0.55	2	3.25	83%	4.4%	2009	2005
43	Denmark	365	0.13	1.33	10%	1.1	1.3	0.0	95%	0.96	2.3	0.71	2	3.00	76%	7.4%	2003	2002
44	Uzbekistan	150.15	0.05	1.00	-2%	-	0.6	-0.2	30%	1.43	2.4	0.57	2	3.00	81%	2.5%	2009	2009
45	Congo	250	0.09	1.52	1%	1.6	1.5	-1.2	35%	0.87	2.4	0.36	1.23	2.75	87%	6.9%	2000	2001
46	Germany	71.7	0.03	1.93	5%	0.3	0.3	0.0	95%	0.36	2.3	0.11	0.47	2.40	95%	5.2%	1976	1966
47	Italy	87	0.03	0.88	-4%	0.6	0.6	0.0	55%	1.13	2.0	0.24	1.37	2.25	89%	2.3%	2010	1997
48	Tunisia	71	0.03	1.20	-2%	0.5	0.3	-0.1	30%	0.66	1.9	0.34	1.00	2.20	84%	2.5%	1998	1981
49	Sudan	210	0.08	0.24	145%	0.7	0.6	-0.1	40%	1.22	1.5	0.55	1.76	2.00	73%	4.2%	2009	2005
50	Thailand	130	0.05	0.42	15%	0.6	0.6	0.0	70%	0.83	1.3	0.24	1.08	1.50	84%	4.2%	2008	2005
51	Bahrain	31.51	0.01	0.99	-3%	-	0.1	0.0	90%	0.13	1.1	0.39	0.51	1.50	74%	2.2%	1985	1990
52	Cameroon	69	0.03	1.03	-9%	-	0.4	-0.7	-100%	0.26	1.3	0.05	0.32	1.35	96%	7.4%	1992	1986
53	Bolivia	31.5	0.01	0.42	3%	0.5	0.4	0.0	65%	0.66	1.1	0.27	0.93	1.35	80%	1.2%	2016	2016
54	Netherlands	42	0.02	0.82	-5%	0.1	0.1	0.0	40%	0.27	1.1	0.16	0.43	1.25	87%	3.5%	1991	1989
55	Turkey	47	0.02	0.83	-5%	0.3	0.3	0.0	130%	0.23	1.1	0.14	0.37	1.20	88%	4.4%	1992	1991
56	Croatia	21	0.01	0.48	-6%	0.1	0.1	0.0	20%	0.34	0.8	0.18	0.52	1.00	82%	1.5%	2003	1988
57	Austria	18.4	0.01	0.77	-2%	0.9	0.1	0.0	65%	0.11	0.9	0.07	0.18	0.95	93%	3.7%	1971	1955
58	France	26.2	0.01	0.72	-5%	0.1	0.1	0.0	90%	0.16	0.9	0.06	0.23	0.95	94%	4.1%	1987	1988
59	Hungary	21.7	0.01	0.67	-3%	0.1	0.1	0.0	70%	0.15	0.8	0.13	0.28	0.95	86%	3.0%	1984	1982
60	Papua	46	0.02	0.34	-9%	-	0.2	0.0	65%	0.37	0.7	0.19	0.56	0.90	79%	2.9%	2008	1993
61	Pakistan	60	0.02	0.46	2%	0.3	0.3	0.0	100%	0.31	0.8	0.55	0.44	0.90	86%	4.7%	1998	1992
62	Albania	6.2	0.00	0.53	-2%	0.0	0.2	0.0	65%	0.19	0.7	0.08	0.27	0.80	90%	0.8%	1986	1983
63	Sharjah	44	0.02	0.47	-7%	-	1.5	0.0	900%	0.16	0.6	0.16	0.33	0.80	80%	4.7%	1998	1998
64	Chile	7	0.00	0.42	-4%	0.0	0.2	0.0	400%	0.03	0.5	0.05	0.08	0.50	90%	3.1%	1979	1982
	REGIONS																	
1	ME Gulf	15316	5.59	231	-3%	636.5	652.7	-26.2	135%	464	695	53	517	749	93%	1.1%	2021	2014
2	Eurasia	12455	4.55	176	3%	84.7	97.1	-0.5	70%	138	314	29	167	343	91%	2.5%	2000	1987
3	N.America	5267.4	1.92	189	-2%	27.0	202.5	-183.8	72%	26	215	5	31	220	98%	0.0%	1973	1972
4	L.America	7882.4	2.88	105	-1%	91.3	109.7	-42.7	84%	79	184	13	92	197	94%	3.0%	2000	1998
5	Africa	6424	2.34	77	-1%	91.8	77.1	-25.4	61%	84	162	12	96	174	93%	2.4%	2006	1997
6	Europe	6010.3	2.19	42	0%	18.0	17.6	0.0	63%	28	71	5	33	76	93%	6.2%	2000	2000
7	Far East	3898.4	1.42	42	1%	91.3	20.0	-2.1	64%	28	70	6	34	76	91%	4.0%	1999	2000
8	ME. Other	2697.5	0.98	25	-3%	25.6	33.1	-16.3	96%	17	43	3	21	46	93%	4.6%	2000	1998
9	Other	509	0.19	3	6%	-	3.1	0.0	50%	6	10	0	7	10	95%	2.7%	2011	1978
10	Unforeseen											10	10	10				
	Non-Swing	45143	16.5	665	0%	364	560	-270.9	0%	407	1071	6	487	1152	93%	3.3%	1998	1997
	WORLD	60460	22.1	896	-1%	1000	1212.8	-297.1	105%	871	1767	133	1004	1900	93%	2.2%	2004	2000

Note: ME Gulf = Abu Dhabi, Iran, Iraq, Kuwait, NZ, Saudi Arabia ; Eurasia = China & former communist bloc; N.America = Canada & USA; Other= small & future producers; "Unforeseen" = rounding item. Non-Swing = World less ME Gulf. Heavy = <17.5oAPI. Deepwater = >500m

World							FURTHER PARAMETERS									2002
		Revised		12/Jan/03												
	Disc in Giants	Peak Lag	Static Res. Report yrs	Pk/Ult	Prod. Wells	Prod. /well kb/d	Res. /well Mb	Wildcats To-date	Peak Date	Future	Discovery/Wildcat Past Av. Mb	Last 5yr Av. Mb	Future Av. Mb	Gas Reserves Tcf Assessed	Cond. Yield Gb/Tcf	Cond. Reserves Gb
Country	S	T	U	V	W	X	Y	Z	AA	AB	AC	AD	AE	AF	AG	AH
1 Saudi Arabia	85%	65	0	42%	1560	4.93	123.1	145	1968	44	1975	430.3	308.6	313	0.037	11.6
2 Russia	60%	27	0	41%	41192	0.16	1.5	8238	1988	556	22	11.3	27.7	1600	0.011	18.0
3 US-48	26%	41	10	51%	239754	0.02	0.1	393322	1981	12073	0	0.0	0.3	177	-	-
4 Iraq	68%	91	5	35%	1685	1.40	53.5	145	1978	204	811	24.6	85.3	92	1.975	181.0
5 Iran	74%	13	4	16%	1120	3.30	60.3	352	1975	121	346	395.9	66.7	956	0.026	25.0
6 Venezuela	58%	29	0	23%	15395	0.17	2.8	2210	1985	96	41	36.8	53.7	174	0.003	0.5
7 Kuwait	91%	33	11	15%	790	2.17	67.0	52	1963	12	1610	0.0	521.8	47	0.025	1.2
8 Abu Dhabi	82%	50	12	35%	1200	1.53	46.7	178	1970	26	416	0.0	151.8	123	0.032	3.9
9 China	41%	43	0	50%	72255	0.05	0.4	2279	1998	1134	23	5.1	3.2	89	0.012	1.1
10 Libya	56%	46	5	10%	1470	0.93	19.7	1717	1963	305	30	4.7	10.6	77	0.009	0.7
11 Nigeria	16%	124	4	14%	2586	0.81	11.6	1261	1992	199	42	56.5	13.1	155	0.021	3.3
12 Mexico	55%	25	0	55%	2991	1.05	7.0	867	1990	312	59	0.7	12.7	44	0.085	3.7
13 Kazakhstan	52%	28	0	50%	11676	0.06	2.6	348	1996	532	103	419.6	8.1	100	0.057	5.7
14 Norway	40%	22	2	46%	833	3.90	17.6	670	1991	135	46	13.3	15.0	129	0.006	0.8
15 UK	40%	25	0	53%	1387	1.68	7.6	2740	1990	259	11	5.7	7.3	62	0.026	1.6
16 Indonesia	38%	32	1	24%	6373	0.19	1.4	3355	1974	514	9	2.9	3.9	168	0.015	2.5
17 Algeria	74%	160	11	19%	1312	0.64	9.1	1129	1962	230	21	1.1	16.7	139	0.050	6.9
18 Canada	46%	15	10	27%	54061	0.02	0.1	18760	1978	2606	1	0.0	0.4	63	-	-
19 Azerbaijan	35%	143	0	50%	2102	0.14	6.1	134	1988	100	155	0.0	22.8	40	0.032	1.3
20 N.Zone	84%	53	11	19%	578	0.98	9.5	36	1962	14	337	0.0	240.2	10	0.010	0.1
21 Oman	6%	31	2	49%	2298	0.42	3.3	606	1984	141	24	10.3	3.8	37	0.021	0.8
22 Qatar	52%	60	1	48%	417	1.61	13.3	88	1998	7	140	0.0	102.5	363	0.073	26.5
23 Egypt	35%	35	0	51%	1258	0.60	2.8	1381	1985	78	9	1.2	10.3	58	0.023	1.4
24 India	25%	23	0	37%	3300	0.20	1.5	1063	1991	307	10	4.3	4.3	27	0.016	0.4
25 Argentina	26%	38	0	60%	14328	0.05	0.2	3804	1985	297	3	0.1	1.8	41	0.023	1.0
26 Australia	22%	36	1	44%	1417	0.45	3.5	4115	1985	648	3	4.4	1.8	164	0.025	4.1
27 Malaysia	11%	29	1	53%	788	0.94	4.9	661	1970	198	14	5.6	4.0	83	0.011	0.9
28 Colombia	39%	7	0	51%	7641	0.08	0.5	1462	1970	149	6	1.3	3.9	13	0.070	0.9
29 Angola	13%	2	6	35%	561	1.24	9.3	613	1983	149	16	39.3	1.6	20	0.001	0.0
30 Ecuador	26%	35	0	42%	1044	0.39	4.2	291	1993	64	26	3.9	12.5	2	0.000	0.0
31 Romania	11%	119	1	54%	6000	0.02	0.2	525	1997	71	13	0.0	8.2	14	0.074	1.0
32 Brasil	0%	22	0	54%	11983	0.03	0.1	3096	1991	677	2	1.9	2.8	16	0.006	0.1
33 Syria	33%	29	9	40%	132	3.93	13.1	279	1992	91	20	1.3	5.3	6	0.018	0.1
34 Turkmenistan	34%	9	3	23%	2460	0.07	0.6	406	1986	101	11	1.7	16.3	101	0.005	0.5
35 Dubai	0%	21	16	52%	200	1.35	4.1	47	1984	2	96	0.0	109.3	4	0.020	0.1
36 Brunei	70%	49	12	36%	779	0.23	1.8	151	1975	7	29	0.5	12.6	11	0.036	0.4
37 Trinidad	13%	19	1	41%	3911	0.03	0.3	318	1994	64	14	17.3	2.9	29	0.015	0.4
38 Gabon	40%	1	5	48%	375	0.80	4.1	620	1991	45	7	3.1	2.5	1	0.016	0.0
39 Ukraine	0%	8	3	22%	1353	0.05	0.5	148	1990	54	23	0.6	11.8	26	0.023	0.6
40 Peru	13%	122	1	29%	4915	0.02	0.2	411	1983	71	8	2.4	5.5	17	0.059	1.0
41 Yemen	14%	21	13	35%	302	1.16	4.5	273	1992	99	11	2.8	5.2	18	0.018	0.3
42 Vietnam	15%	30	4	37%	28	10.89	64.8	165	1994	46	20	36.9	12.1	15	0.017	0.3
43 Denmark	0%	31	0	44%	213	1.64	4.5	168	1985	34	14	14.8	21.0	5	0.000	0.0
44 Uzbekistan	0%	17	3	51%	2190	0.07	0.7	266	1991	134	9	0.2	4.2	51	0.027	1.4
45 Congo	0%	3	8	52%	445	0.60	2.0	183	1992	46	13	18.7	7.9	2	0.046	0.1
46 Germany	0%	14	0	29%	1138	0.06	0.3	2110	1958	46	1	0.1	2.5	19	0.007	0.1
47 Italy	0%	16	0	32%	208	0.31	5.4	2539	1962	155	1	0.8	1.6	8	0.342	2.7
48 Tunisia	36%	3	4	21%	211	0.33	3.1	442	1981	96	4	1.5	3.6	1	0.181	0.1
49 Sudan	0%	3	1	27%	9	22.22	135.1	124	2001	99	12	9.5	5.5	1	0.123	0.1
50 Thailand	0%	24	0	38%	749	0.15	1.1	248	1983	85	5	5.1	2.9	21	0.032	0.6
51 Bahrain	79%	58	1	55%	496	0.06	0.3	17	1983	2	66	0.0	193.2	3	0.029	0.1
52 Cameroon	0%	4	15	23%	255	0.31	1.0	171	1977	7	8	3.3	7.7	3	0.004	0.0
53 Bolivia	0%	50	1	49%	328	0.09	2.0	359	1975	70	3	7.2	3.8	30	0.026	0.8
54 Netherlands	0%	8	0	45%	196	0.14	1.4	1092	1985	85	1	0.0	1.9	69	0.018	1.2
55 Turkey	0%	22	0	47%	846	0.06	0.3	1027	1975	155	1	0.1	0.9	1	0.004	0.0
56 Croatia	0%	38	3	32%	723	0.03	0.5	151	1985	96	5	0.0	1.8	3	0.002	0.0
57 Austria	54%	8	2	18%	950	0.02	0.1	629	1977	75	1	0.1	0.9	1	0.003	0.0
58 France	0%	30	0	52%	403	0.07	0.4	1939	1959	21	0	0.0	2.9	1	0.018	0.0
59 Hungary	0%	18	0	45%	944	0.03	0.2	933	1964	167	1	1.0	0.8	3	0.106	0.4
60 Papua	0%	6	0	7%	39	1.46	9.5	165	1990	12	4	0.0	15.8	14	0.013	0.2
61 Pakistan	0%	9	0	29%	250	0.24	1.2	432	1993	99	2	0.9	1.3	28	0.004	0.1
62 Albania	0%	55	12	42%	2275	0.00	0.1	42	1987	72	17	1.1	1.1	1	0.000	0.0
63 Sharjah	0%	18	3	50%	49	0.98	3.4	8	1994	7	80	0.0	23.3	0	0.000	0.0
64 Chile	0%	22	7	57%	315	0.02	0.1	523	1974	14	1	0.0	3.4	6	0.013	0.1
REGIONS																
1 ME Gulf	80%	66		43%	6933	2.58	67.0	908	1976	420	766	245.5	126.3	1540	0.145	222.8
2 Eurasia	51%	23		34%	143170	0.08	1.0	13568	1987	3027	23	22.0	9.7	2029	0.015	29.9
3 N.America	28%	42		59%	293815	0.02	0.1	412082	1981	14970	1	0.0	0.3	241	0.000	0.0
4 L.America	48%	21		47%	62851	0.13	1.3	13341	1982	2066	14	2.9	6.7	373	0.023	8.5
5 Africa	40%	36		15%	8482	0.78	9.9	7610	1963	1254	21	13.1	9.7	456	0.028	12.6
6 Europe	35%	26		50%	5328	1.15	5.3	11887	1986	810	6	5.6	6.5	293	0.022	6.6
7 Far East	29%	33		51%	13723	0.29	2.0	10324	1985	2117	7	4.4	3.1	530	0.018	9.6
8 ME. Other	25%	33		46%	4740	0.61	3.7	2345	1991	505	18	3.4	6.3	431	0.064	27.8
Other	0%	22		9%												
Unforeseen	0%	22		0%												
Non-Swing	40%	41		51%	532109	0.09	0.8	473387	1962	22378	2	2.6	3.6	4353	0.000	1.0
WORLD	56%	41		51%	539042	0.12	1.6	474295	1962	22793	4	3.6	5.8	5894	0.007	38.5

RESOURCE BASED PRODUCTION FORECAST

End 2002 — Revised 18/Jan/03 — Base Case Scenario

Regular Oil by Country (Mb/d)

	2000	2005	2010	2020	2050
Saudi Arabia	8.00	6.61	8.98	8.96	6.57
Russia	6.33	8.55	9.43	4.85	0.66
US-48	4.45	3.53	2.59	1.40	0.22
Iran	3.68	3.99	4.73	3.61	1.61
China	3.24	2.99	2.42	1.58	0.44
Norway	3.21	2.73	1.91	0.94	0.11
Mexico	3.01	2.78	2.21	1.41	0.36
Venezuela	2.57	2.32	2.12	1.77	1.02
Iraq	2.57	2.35	3.00	4.55	2.66
UK	2.51	1.86	1.35	0.71	0.10
Nigeria	2.03	1.92	1.87	1.46	0.69
Abu Dhabi	1.90	1.96	2.50	2.25	1.34
Kuwait	1.77	1.85	2.36	2.10	1.28
Libya	1.41	1.37	1.37	1.17	0.68
Indonesia	1.27	1.01	0.84	0.59	0.20
Canada	1.08	0.62	0.51	0.34	0.10
Oman	0.93	0.83	0.67	0.44	0.13
Egypt	0.81	0.62	0.46	0.25	0.04
Algeria	0.81	0.85	0.81	0.65	0.34
Argentina	0.75	0.61	0.43	0.21	0.03
Angola	0.74	0.68	0.52	0.31	0.07
Australia	0.70	0.58	0.48	0.33	0.11
Malaysia	0.69	0.64	0.48	0.27	0.05
Qatar	0.69	0.57	0.48	0.33	0.11
Colombia	0.69	0.50	0.39	0.24	0.06
Kazakhstan	0.68	0.93	1.18	1.71	0.91
India	0.65	0.63	0.52	0.35	0.11
N.Zone	0.63	0.55	0.52	0.40	0.19
Syria	0.52	0.39	0.26	0.12	0.01
Ecuador	0.40	0.41	0.38	0.27	0.10
Denmark	0.36	0.28	0.20	0.09	0.01
Brasil	0.36	0.24	0.17	0.08	0.01
Yemen	0.35	0.31	0.22	0.11	0.01
Gabon	0.33	0.24	0.18	0.09	0.01
Vietnam	0.30	0.30	0.28	0.14	0.02
Dubai	0.28	0.16	0.12	0.06	0.01
Azerbaijan	0.28	0.68	0.82	0.70	0.33
Congo	0.27	0.20	0.14	0.07	0.01
Sudan	0.19	0.27	0.25	0.10	0.01
Brunei	0.18	0.16	0.13	0.08	0.02
Uzbekistan	0.16	0.17	0.20	0.12	0.03
Turkmenistan	0.14	0.18	0.16	0.13	0.06
Romania	0.12	0.11	0.10	0.08	0.04
Trinidad	0.12	0.11	0.10	0.07	0.02
Thailand	0.11	0.14	0.12	0.06	0.01
Peru	0.10	0.08	0.08	0.06	0.03
Italy	0.09	0.08	0.08	0.07	0.03
Cameroon	0.09	0.05	0.04	0.02	0.00
Tunisia	0.08	0.07	0.06	0.04	0.02
Ukraine	0.07	0.07	0.06	0.05	0.03
Papua	0.07	0.05	0.05	0.03	0.01
Germany	0.06	0.06	0.05	0.03	0.01
Turkey	0.06	0.04	0.03	0.02	0.01
Sharjah	0.05	0.04	0.03	0.02	0.00
Pakistan	0.04	0.05	0.04	0.03	0.01
Bahrain	0.03	0.03	0.03	0.02	0.01
Hungary	0.03	0.02	0.02	0.01	0.01
Netherlands	0.03	0.04	0.03	0.02	0.01
France	0.03	0.02	0.02	0.01	0.00
Bolivia	0.03	0.04	0.05	0.05	0.02
Croatia	0.02	0.03	0.02	0.02	0.01
Austria	0.02	0.02	0.01	0.01	0.00
Chile	0.01	0.01	0.01	0.00	0.00
Albania	0.01	0.01	0.03	0.03	0.00

Regular Oil by Region (Mb/d)

	2000	2005	2010	2020	2050
ME Gulf	18.5	17.4	22.1	21.9	13.3
Eurasia	11.1	13.7	14.4	9.3	2.5
N.America	5.5	4.4	3.2	1.8	0.3
L.America	8.0	7.1	5.9	4.2	1.6
Africa	6.7	6.3	5.7	4.2	1.9
Europe	6.3	5.1	3.6	1.9	0.3
Far East	4.0	3.6	2.9	1.9	0.5
ME. Other	2.9	2.4	1.8	1.1	0.3
Other	0.5	0.5	0.5	0.4	0.1
Unforeseen		0.0	0.1	0.1	0.7
Non-Swing	45	43	38	25	8
WORLD	**64**	**60**	**60**	**47**	**22**

Excluding:tar;heavy;deepwater;polar oil & PNGL

Other Hydrocarbons

Oil (Mb/d)

	2000	2005	2010	2020	2050
Heavy Oils	1.5	2.6	3.3	4.3	5.7
Canada	1.0	1.3	2.0	2.8	3.9
Venezuela I	0.5	0.5	0.5	0.6	1.1
Venezuela II	0.0	0.6	0.7	0.7	0.2
Other	0.1	0.2	0.2	0.3	0.5
Deepwater	1.2	7.1	8.8	4.7	0.1
G. Mexico	0.4	1.9	2.7	0.7	0.0
Brasil	0.8	2.0	2.0	1.5	0.0
Angola	0.0	1.7	1.9	0.7	0.0
Nigeria	0.0	0.8	1.2	0.9	0.0
Other	0.0	0.7	1.0	1.0	0.1
Polar	1.1	1.2	1.8	5.7	0.1
Alaska	1.0	0.8	0.5	0.2	0.1
Other	0.1	0.4	1.3	5.5	0.0
Other	0.0	0.8	1.5	2.0	1.5
Subtotal	3.8	11.6	15.4	16.7	7.4

GAS & GAS LIQUIDS

Gas (at 6Tcf = 1 Gboe)

	2000	2005	2010	2020	2050
Gas	39	45	53	59	36
Non-con gas	2	2	2	4	10
Subtotal	41	47	55	64	46

Gas Liquids

	2000	2005	2010	2020	2050
NGL	6	8	9	11	6

ALL HYDROCARBONS

	2000	2005	2010	2020	2050
Gas	41	47	55	64	46
Liquids	73	80	85	74	35
Processing Gain	2	2	2	1	1
Total	116	129	142	139	82

BALANCE

with a notional 1.5% annual demand growth

Liquids Mb/d

	2000	2005	2010	2020	2050
Supply	75	82	87	76	36
Demand	75	81	87	101	158
Balance	0	1	0	-25	-122

NOTES

Regular Oil excludes
Oil from coal & "shale"; bitumen; Extra-Heavy Oil;
Heavy Oil (<17 API); Deepwater (>500m) & Polar
Oil and NGL from gas plants
Base Case Scenario assumes flat Regular
production to 2010, when ME Gulf can no longer
in practice offset natural decline elsewhere
ME. Gulf is Abu Dhabi, Iran, Iraq, Kuwait, NZ and
Saudi Arabia, with 37% of world supply by 2010
Venezuela I = ordinary heavy
Venezuela II = 4 Extra-Heavy oil projects

Africa
Conventional Oil

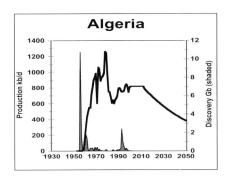

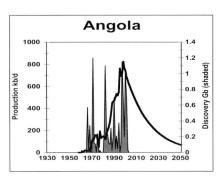

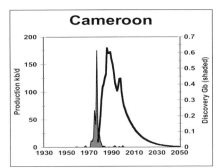

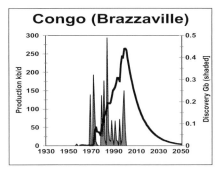

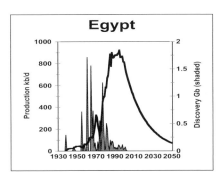

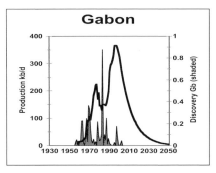

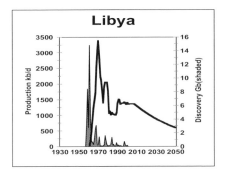

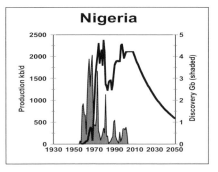

Africa
Conventional Oil

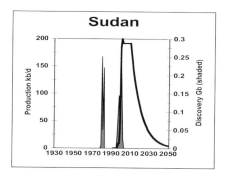

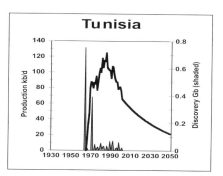

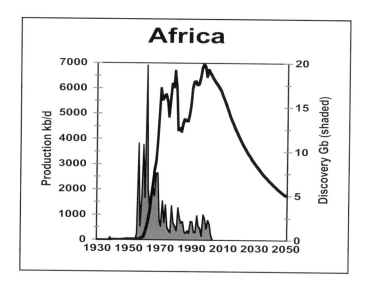

Eurasia
Conventional Oil

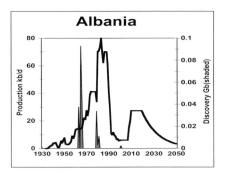

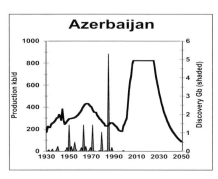

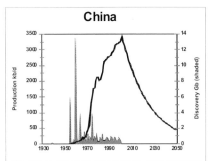

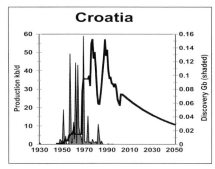

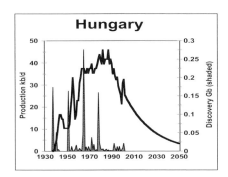

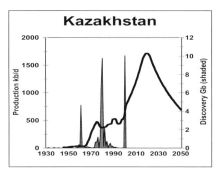

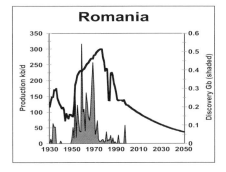

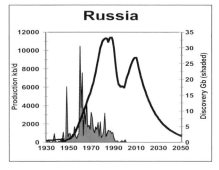

Eurasia
Conventional Oil

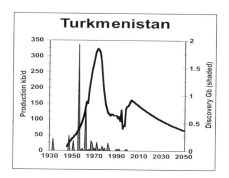

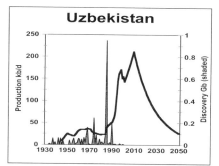

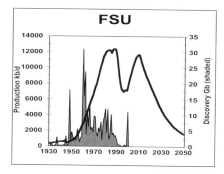

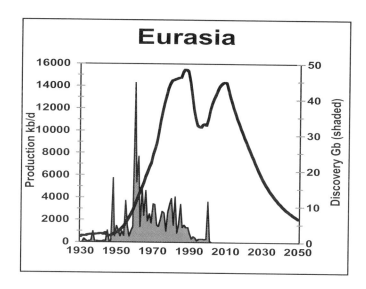

Europe
Conventional Oil

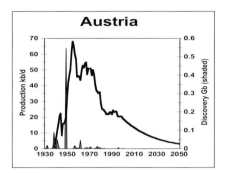

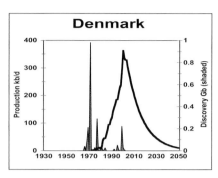

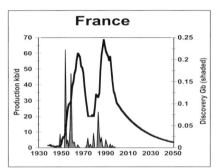

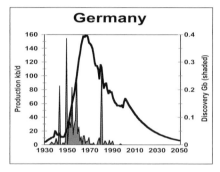

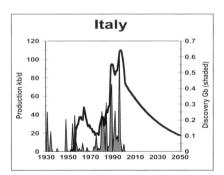

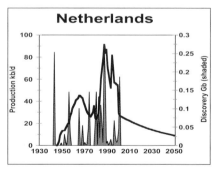

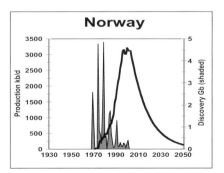

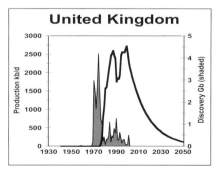

Europe
Conventional Oil

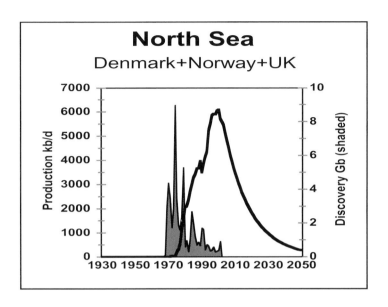

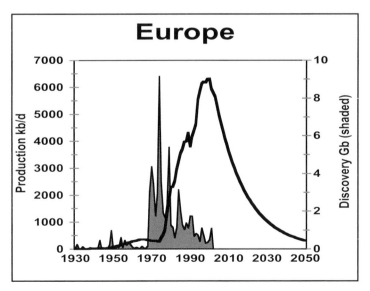

Far East
Conventional Oil

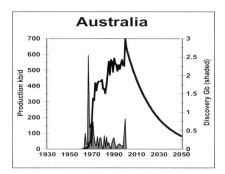

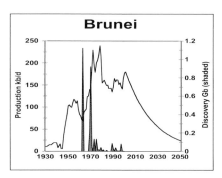

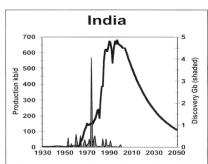

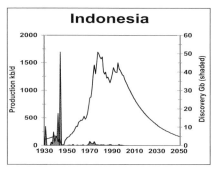

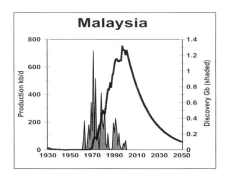

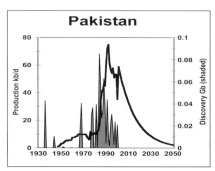

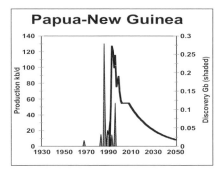

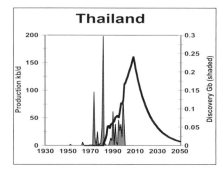

Far East
Conventional Oil

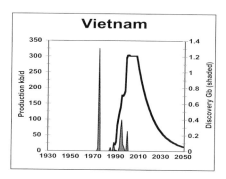

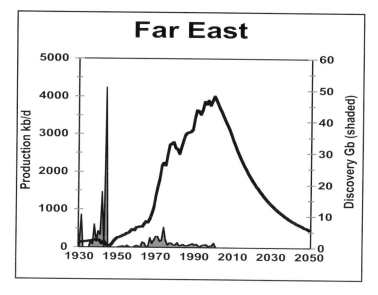

Latin America
Conventional Oil

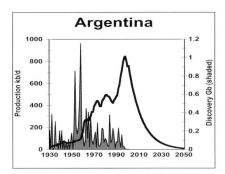

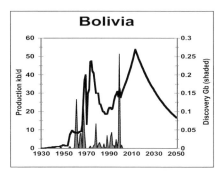

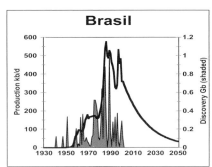

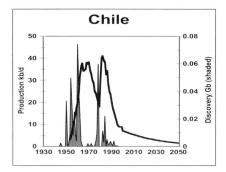

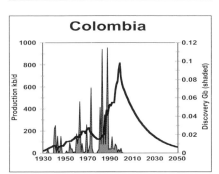

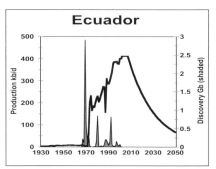

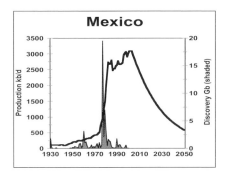

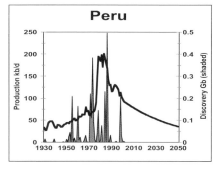

Latin America
Conventional Oil

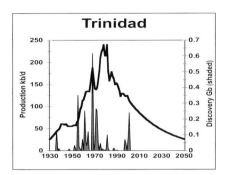

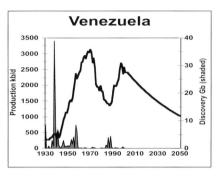

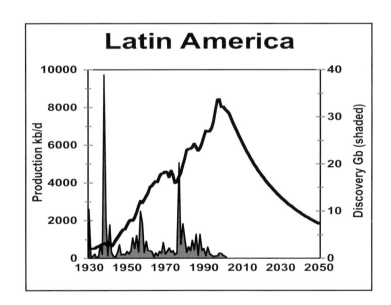

Middle East (Other)
Conventional Oil

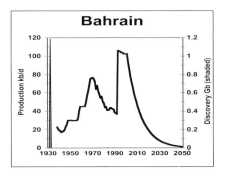

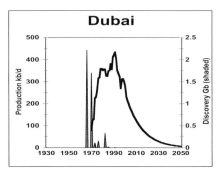

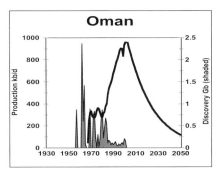

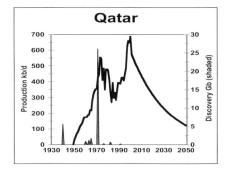

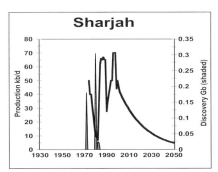

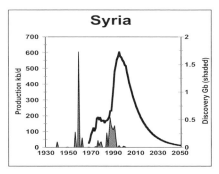

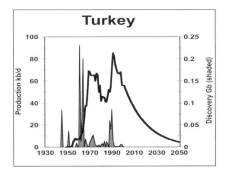

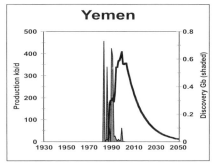

Middle East (Other)
Conventional Oil

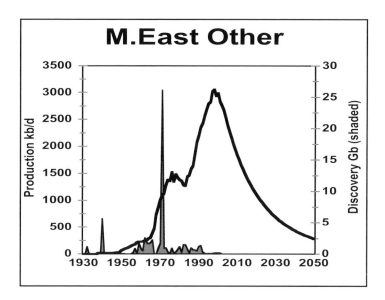

Middle East Gulf
Conventional Oil

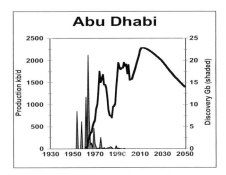

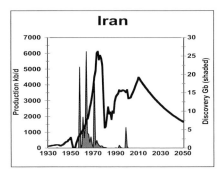

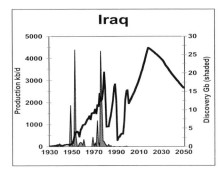

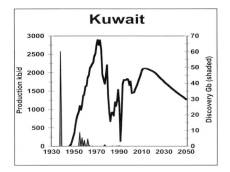

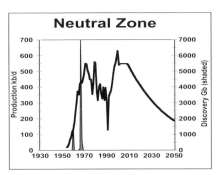

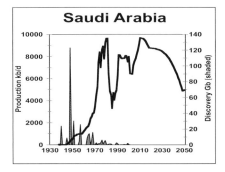

Middle East Gulf
Conventional Oil

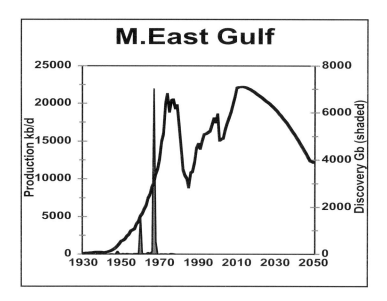

North America
Conventional Oil

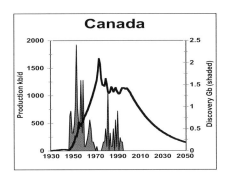

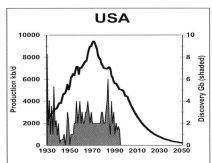

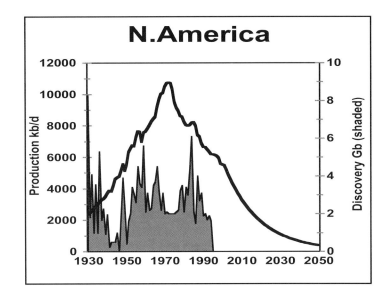

Part 5

Synthesis
by
C. J. Campbell

From the Dawn of Time

The simple limpet, *Patella*, has lived little changed for over 500 million years of recorded history, having found a sustainable place in the environment, clinging to a rock as the waves washed over him. Other species were more adventurous, finding ways to exploit more efficiently the particular niches in which they found themselves. As they did so, their numbers grew, as a genetic momentum perfected their adaptation. They were highly successful so long as their niche lasted, but when the environment changed for geological or climatic reasons, they died out as victims of their very success. They did not manage to evolve backwards to the more sustainable simple stock from which they came. In this respect, Darwin did not get it exactly right when he explained evolution in terms of the survival of the fittest. Evidently, the fittest over the short-term were not the same as the fittest over the long-term.

Man (and in using the term I imply no disrespect for the fair sex) swung down from the trees about four million years ago to evolve into a discrete species that became ever more adapted to its environment. He passed through several subspecies to arrive at a rather primitive version of Modern Man about two million years ago, when he started making use of crude stone implements. Next followed a large brained descendant, known as Cro-Magnon, only about 200,000 years ago, who probably indulged in a degree of genocide, ridding the World of his more primitive cousins. He came into his own about 50,000 years ago in Europe with the retreat of the last Ice Age.

Cro-Magnon started as a hunter, fisherman and cave dweller but gradually learned to herd cattle and grow crops, which in turn led to settled primitive agriculture. It gave rise to a more political social structure to control the future of a static community. About nine thousand years ago in Egypt, his descendants learned how to smelt gold and copper, perhaps largely for ornaments, before a metallurgical breakthrough, only four thousand years ago, gave an alloy of copper and tin for tougher bronze weapons and tools.

This marked the beginning of truly modern man. If we were to compare the fossil record of life on Earth since the Cambrian with a single day, Modern Man appeared only at twenty seconds to midnight, making him by all means a recent arrival.

He proliferated, organising himself into societies, which eventually made great cultural advances, building cities with cathedrals to sophisticated Gods. But he did not lose his jungle instincts, being given to frequent warfare, as a particular society sought to dominate and enslave its neighbours. Empires waxed and waned, but overall, the human population lived a sustainable life within the resources of the planet that were at its disposal.

A second metallurgical breakthrough came about one thousand years ago when our ancestors found out how to use charcoal and bellows to raise the temperature sufficiently to smelt iron, which could be worked into vastly stronger weapons and implements, followed later by a still stronger material, steel.

The last chapter opened only two hundred years ago, or on the stroke of midnight, with the coming of the Industrial Revolution, having its origins in Britain, where energy from millstreams was harnessed by the water wheel to drive looms to weave cloth.

Gold and silver had been used as a medium of exchange for centuries, but the advent of the Industrial Revolution led to a new capitalism, as the mill owners accumulated wealth by the use of machinery, costing less than human energy. The burst of new capital stimulated expansion and the search for new markets, which in turn prompted the growth of empires, notably those of Britain, France and Russia. In parallel with that came a greatly increased use of debt and credit, implying a confidence in the system and its progress, sometimes underpinned by military force. It in turn brought an expansion of usury that fed new money into the system, fuelling further economic growth.

Charcoal was at first used for smelting iron, leading to deforestation in the industrial areas. Sea-coal, being lumps of coal washed from outcrops that could be collected from beaches, had been known for a long time, but the increased demand for fuel led to mining coal in shallow pits. The pressure to deepen the mines below the water-table led to the development of steam-driven pumps. An ingenious inventor then saw how the pump could be adapted by feeding steam into the cylinders, causing the pistons to turn a wheel. The steam engine resulted, which was soon adapted to drive a locomotive ushering in the age of rail, which expanded trade and travel greatly. Sail gave way to steam, opening up world trade. Before long, a further technological development brought the fuel directly into the cylinder in the form of the internal combustion engine. At first, it used benzene distilled from coal, but soon turned to petroleum refined from crude oil.

These developments were already well advanced by the end of the 19th Century. The British Empire ruled a world. The new energy freed people from drudgery, helping to make possible great achievements in science, medicine, literature and general culture. The amazing pace of progress made a deep impression, leading people to believe that Man was the master of his environment in a world of near limitless resources to be bent to his will. These notions were enshrined in the new subject of Economics, which enunciated ineluctable laws of supply and demand. The world was seen as a market place such that if the price of wheat should rise, the farmer would grow more in the next sowing, returning the system to an everlasting equilibrium. Financial management became more sophisticated. Religious teachings further consecrated the special position of Man in the Universe, being perceived to be closer to God than the birds and the beasts of the field. This, together with the new subject of eugenics, led to notions that certain races were better than others, and that wars were an essential part of evolution.

The population began to soar to unprecedented heights.

The First Half of the Age of Oil

Oil has been known from Antiquity, when people collected it from natural seepages or dug shallow wells for it. The burning bush of biblical fame was very probably a gas seepage, while the Zoroastrians worshipped the eternal flames of Baku on the Caspian, where hydrocarbon source-rocks caught fire.

The modern oil industry had its roots in the middle of the 19th Century on the shores of the Caspian and in Pennsylvania. The most famous, but not the first, early well was that drilled in 1859 by the self-styled Colonel Drake at Titusville, Pennsylvania, striking oil at a depth of 67 feet in a Devonian sandstone. The equipment for drilling was already well developed, having been used for drilling wells for salt brine, needed to preserve meat in the days before refrigeration.

Drake's well stimulated a boom, as an army of adventurers and speculators descended on Pennsylvania to develop "rock oil". It was a cheap replacement for whale oil that had been used for lighting but was now becoming scarce and expensive, thanks to the depletion of whale stocks through over-hunting. Primitive stills to refine the product were soon erected, and before long, kerosene was being exported to Europe in sailing ships.

But the boom was soon followed by bust, demonstrating a characteristic feature of the oil industry. Rapid exploitation of this highly profitable new form of energy led to glut, followed by a collapse in price. Many fortunes were lost as well as made. John D. Rockefeller, one of the oil pioneers, determined to bring order to the industry with the creation of Standard Oil. The word *Standard* conveyed the notion of a consistent high-quality product, as poorly refined kerosene had often contained volatile components, causing explosions. He was primarily a refiner and marketer, who managed to control the industry by establishing a stranglehold through preferential rebates from the railroads, which in those days transported most of the oil. He was much reviled by the independent producers, and his near monopoly was eventually broken up by the government in 1911 under anti-trust legislation. Many of the world's larger oil companies, including Exxon, Mobil, Chevron, Amoco and Arco, had their origins in Standard Oil.

Meanwhile, oil was being found and produced in other parts of the world: from Borneo and Sumatra, the birthplace of Shell Oil, to the Middle East where BP brought in the first major discovery in Iran in 1906. Other important finds were made in the Caspian, Latin America, and Texas, the birthplace of Texaco and Gulf Oil, two other major companies.

The first automobile, developed by Carl Benz in Germany, took to the road in 1881, followed in 1904 by the first tractor to plough its furrow. The Wright brothers lifted off in the first aircraft driven by petroleum in 1903. In short, oil was beginning to fuel the world by 1914 on the outbreak of the First World War.

Wars

Wars have been a recurring feature since the birth of Mankind, as one society sought to establish dominion over another. But relatively speaking, they were rather local affairs, incidental to the lives of the great mass of people. One army, led by its king or baron, would meet another on a set field of battle to resolve a conflict.

That changed in 1914 with the advent of total war. It affected entire nations, becoming more prolonged and vicious with the new weaponry provided by steel and high explosive. A critical factor was the new mobility provided by rail and steam, which allowed reinforcements and ammunition to be delivered rapidly. In effect, it was the first industrial war, having its roots in industrial conflict and competition.

Britain's industrial prowess, driven by coal, had been gaining strength over the previous century. Under the forces of capitalism, it needed the expansion and growth of a mercantile empire, backed by political force to ensure the sanctity of contract. The pound sterling was the world currency, with the Bank of England *promising* to back its money with gold. But a recently unified Germany was overtaking Britain in industrial output with its great iron and steel industry. It needed its own empire, as well as a source of raw materials, especially coal and iron. While there were many catalysts that triggered the outbreak of war, the underlying tensions had been building for a long time. They related to industrialisation and the growth of population.

The United States entered the war in 1917, decisively turning the scales in what had become a bloody stalemate, as exhausted armies faced each other under appalling conditions in the fields of Flanders. President Wilson brokered a peace under the slogan "Peace without Victory". The last thing he needed were resurgent British and French empires to compete with his own new industrial dominion. His terms allowed the Germans to delude themselves into believing that they had agreed to end fighting without being defeated. The French, for their part, pointed the finger of guilt for the war squarely at Germany, claiming reparations for the huge losses they had sustained. They were still smarting from the reparations that they had been obliged to pay at the end of the Franco-Prussian War less than fifty years before, which included the loss of Alsace-Lorraine, the largest iron-ore deposit in Europe. It had made German industrialisation possible, explaining why the first act of war on August 10th 1914 was in this critical territory. The French were also resentful that on the last days of the war the retreating Germans had flooded the coal mines of northern France to hinder its recovery. Iron and coal were what it was all about.

The defeat of the Ottoman Empire, which had sided with Germany in the war, put control of the Middle East into the hands of Britain, France and the United States. They established spheres of influence, drawing up new political frontiers, which gave birth to the important oil producing countries of Iraq, Kuwait and Saudi Arabia.

The Second World War was, in many respects, a continuation of the first, as Germany sought to reassert itself, aiming primarily to build an empire in eastern Europe and Russia, then suffering under the Communist yoke. It perceived itself to be in need of *lebensraum*, or room-to-live, having developed the notion that it was a superior race, which would somehow be a civilising influence for the world – a view not widely shared. The war was now fought with even more sophisticated industrial weaponry, fuelled by petroleum.

Japan, an ally of Germany, which had developed as another industrial power in the Far East, took the opportunity of the fall of France to move on French territories in Indochina. The United States retaliated by prohibiting the export to Japan of steel, scrap iron, and oil. The German advances into

Russia in 1941 gave its ally the confidence to try to secure its essential oil supply from the Borneo and Sumatra, which were then colonies of the defeated Dutch. As a pre-emptive strike to that end, it attacked the US naval base at Pearl Harbour in Hawaii in December 1941, bringing the United States into the war. The conflict was in large measure about control of oil, the essential energy supply on which modern countries depended.

Oil was critical in the Second World War, fuelling the tank battles and the massive air raids in which cities were flattened and fire-stormed. Indeed, oil was one of the factors that contributed to the eventual defeat of the Germans. The United States was a major oil producing country, able to supply Britain, and Russia had its own sources, but the Germans for their part had to rely on oil from conquered Romania and the synthetic oil that they managed to make from coal. The battlefields of Normandy were strewn with dead and dying horses, which were still widely used by the German army to save oil.

At the time of writing, the Middle East is under threat of invasion by the United States, as it, in its turn, faces a growing need for access to foreign oil: its indigenous supplies having been in decline for more than thirty years. As in the previous wars, there have been pretexts and catalysts but the underlying cause is the same: the control of natural resources. Oil is certainly one of the most critical, being responsible for 40 percent of all traded energy and 90 percent of transport fuel.

Population

We have mentioned the word *lebensraum*, or room-to-live, upon which Hitler, who suffered the pangs of hunger in his youth, sought to justify his country's expansion. Today, it is called Carrying Capacity, as scientists study the number of people that a planet of finite dimensions and resources, with its thin skin of atmosphere, can support.

As the following plot makes clear, the world's population stood steady at below 500 million for most of the last two millennia, with a slight dip in 14th Century from the Black Death. But then it doubled during the first half of the 19th Century, as industrialisation, based on coal, brought economic expansion. The world was still a large place, although the massive emigration from Europe to the empty spaces of the New World speaks of desperation, as there was not enough land to support the growing population at home under the farming methods of the day. This is underlined by the fact that it was viable for Europe to import soil nutrients, derived from the excrement of seabirds in Chile and Peru, in sailing ships. Synthetic fertilisers came later when Norway harnessed hydro-electric power to extract nitrogen from the air.

The arrival of oil in the middle of the 19th Century made it possible for

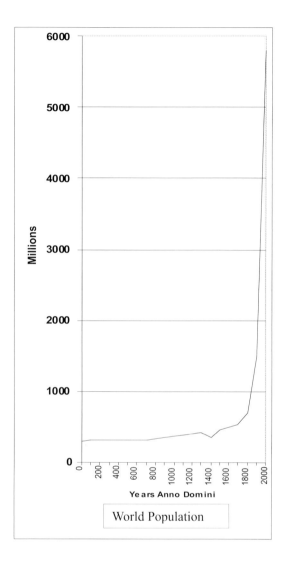

World Population

the population to expand six-fold, exactly in parallel with the growth of oil. It was as if everyone had an un-fed and barely paid team of slaves to do his manual work for him. Mechanised farming, combined with irrigation, itself largely relying on oil-driven pumps, led to a huge increase in food supply. Newly engineered plant types gave increased yields, but had a voracious appetite for synthetic nutrients and water.

Two world wars, together with the massive civilian exterminations of the Soviet Union and Germany, led to the violent death of an estimated 54 million people, but the loss had no perceptible impact on the overall trend. *Homo Sapiens*, having tapped the world's fossil energy supplies of coal, oil and later natural gas, became immensely successful. But it has been a success at the expense of other species. They are being wiped out by human destruction of essential environments at a rate equalling those in the geological record when massive volcanic eruptions blotted out the sun for centuries or asteroid impacts shook the planet.

Economic growth has led to new wealth, which was not evenly distributed. The United States emerged supreme from the Second World War, when a weakened Britain and France voluntarily surrendered their empires. The new economic empire, built with, by and for the dollar, replaced them, spreading its unseen financial tentacles throughout the world. There was a great disparity, not only between the rich and poor nations of the world, but within the rich countries themselves. Manufacturing was progressively transferred to the poorer countries so as to benefit from what almost amounted to slave labour, while the profits, partly in the form of mysterious credit, flowed home to the wealthy nations, which became ultra-consumers. Hairdressers in affluent capitals served their clients, both arriving in large cars. Property values soared as neo-palaces were built for the new executive kleptocracy, and as hitherto humble people developed new aspirations. Conditions in many poor countries deteriorated, especially in the growing urban agglomerations. Political tensions arose in many places. Colombia is not the only country to face civil war as a consequence of globalism, caused in its case by a trade in narcotics made from its indigenous coca leaf.

But the growth in population was achieved more by rising longevity than increased fertility. Average life expectancy in 1950 was just below 40 years, but by 1991 had risen to over 60. It was highest in the wealthy nations, but was countered there by declining birth-rates, consequent upon the emancipation of women, who preferred paid employment to raising families at home. The fertility rate in much of Europe and the developed world is now running at far below the replacement rate of 2.1 children per woman.

As the indigenous populations age and wither, their place is being taken by new immigrants, who are being welcomed as cheap labour, despite the ethnic tensions they bring. Within a generation or so, a typical child in, for example, Spain will grow up lacking brothers, sisters, cousins, aunts and uncles, but enjoying the inherited wealth of his antecedents. He will be served by migrants from North Africa, who have successfully crossed the Straits of Gibraltar on flimsy boats. This curious outcome is seemingly consistent with economic principles. But in due time, the immigrants will tire of serving their master with ever more paella, and demand their fair share of the fruits of their labours, as have many slaves before them.

Falling fertility is not however being experienced in the Middle East, where a combination of unearned oil revenue and traditional family patterns is causing a population explosion. The resulting youthful generation, however, faces a difficult future, depending on the uncertain disbursement of the proceeds of oil revenues by essentially feudal governments, over which it has no control. As the population expands, the share of the patrimony has to decline, as indeed will the patrimony itself as oil depletion grips even these countries in the years ahead. There is not much gainful employment to be had in the barren deserts.

The Age of Information

In 1894, Guglielmo Marconi succeeded in transmitting a message over the distance of a mile by means of electromagnetic radiation. He was later to span the Atlantic from transmitters on the west coast of Ireland. Speech transmission followed in 1906, and the first commercial radio station started broadcasting news and entertainment ten years later. By 1926, five million homes in America were equipped with radio. Television followed, taking off after the Second World War.

Today, with the help of satellite dishes, the entire world is linked to a continual stream of imagery pumped relentlessly from television screen and radio loudspeaker. Many people are subjected to it throughout their waking hours either deliberately or as background in their home or place of work.

In earlier years, certain national broadcasting systems, such for example as the State-controlled BBC, did endeavour to broadcast uplifting material in the fields of music, drama, debate and objective news, delivered by announcers with authoritative accents. Those days are long over, as polls and ratings, being a surrogate for market forces, were applied to provide the diet that the listeners and viewers were perceived to want, irrespective of its intrinsic worth.

There can be little doubt that the attitudes of the modern world are formed by television imagery, which is almost universal, with everyone from the Borneo native in his longhouse to the Wall Street executive being subjected to variants of the same stuff. The imagery itself commonly depicts an unreal life of affluence, sex and violence, not to be found in the real world of most people's experience. World catastrophes occupy a fleeting moment on the screen before the cameras switch to the next item. In addition to the news and feature programmes, the airwaves carry the full blast of advertising, which has drawn well-hewn skills of persuasion to condition us to a world of rampant consumerism. Indeed, at times we are urged, almost as a matter of moral duty, to go out and spend, which is perceived to be a contribution to our economic prosperity, when in reality it simply hastens our demise from resource depletion.

This enormous power to influence has been used in some places for direct political ends, and even where it has not been abused, it has served to standardise viewpoints, conditioning people everywhere to the same general perception of the world in which they live. Everyone has come to accept a form of virtual life, whereby at the click of a switch they can be transported to see a close-up of the eye of a penguin in Antarctica, or the hair style of the latest star as she belts a tennis ball in Flushing Meadow. The image of President Saddam of Iraq at his cabinet table in Baghdad flicks to a virtually identical picture at President Bush at his. From a casual glance it is hard to distinguish one from the other. If we have come to lose track of the real world in which we live, with its natural blessings and constraints, we have only our addiction to television to blame.

The advent of the computer, and especially the personal computer, during the latter years of the 20th Century was another remarkable development. It has made possible electronic communications throughout the world. Whereas in the past, news and commentary were filtered by editors, now it pulsates around the world free of any constraint. Although censorship is avoided, the scope for the promulgation of false information or flawed interpretation is greatly increased. But, on balance, most would see it as a benefit and achievement.

Whatever the strengths and weaknesses of our condition, the fact remains that it was, in large measure, an abundant supply of cheap oil-based energy that put this brief 150-year span of history so far out of context with all that had preceded it.

How much oil was found to make all this possible?

Since we have already touched on the early days, we can take up the story of oil as peace returned at the end of the Second World War. By 1950, the United States was providing 52% of the ten million barrels a day produced worldwide, as it reaped the harvest of its own discovery that had reached a peak twenty years earlier. Only16% of the world's oil came from the five key countries bordering the Persian Gulf, although some mammoth fields had already been found there. In Iran, BP retained its monolithic position under the concessions granted to it in 1903. In Kuwait, the same company shared with Gulf Oil of the United States the country's oil wealth, dominated by the huge Burgan Field, that had been found in 1938. In Saudi Arabia, Aramco, which was owned by a group of American companies led by Chevron, was looking forward to producing the world's largest field at Ghawar that had been confirmed two years before. In Iraq, BP, Shell, Exxon, Mobil, CFP of France and Mr Gulbenkian, with his legendary 5%, were developing major resources that had been found over the preceding twenty-five years. The Soviet Union was likewise systematically exploring its colossal territories, finding the more prolific basins and most of the giant fields within them.

With the return of peace, the world's oil was under the firm control of seven major international companies which brought it from the wellhead to the filling station through a central chain of command, little different in structure from that which had controlled the war effort. The total discovery to the mid-century amounted to about 500 billion barrels, of which only about 65 billion had been consumed. Clearly, there was plenty left for the future, but already, some finite limits were being observed. The science of exploration had advanced to the point of understanding the essential geological requirements. It was recognised that the main prospective areas onshore had already been identified. Estimates, based largely on the yield of rock volumes, suggested that what had been found represented between one-half and one-third of the world's ultimate endowment.

Then, in 1951, the hegemony of the major oil companies was interrupted by political events in Iran leading to the expropriation of BP's local subsidiary as a longstanding conflict over the share of oil revenues rose to a head. The company's directors in London could not believe that the British government, which owned 51% of the company, and had recently emerged victorious from a long war, would stand idly by and watch the loss of the jewel in the crown. But a war-weary Britain under its new socialist government had no stomach for foreign conflict. The message was not lost on other governments, encouraging, first, Colonel Nasser of Egypt to sequester the Suez Canal and, later, the main oil producing countries, which followed suit, expropriating foreign oil interests during the 1970s.

Despite this devastating loss, BP was nothing if not resilient, and soon set about finding new sources of supply in a world that still had them. It went into Libya; it teamed up with Shell in Nigeria; and it brought in a major new discovery in Alaska in the very backyard of its American competitors. It later managed to return to Iran but was forced to share its once exclusive position with several American companies.

Attention turned to the offshore. Already deviated wells had been drilled from the shore to tap extensions of known productive belts, but now wells were drilled from platforms erected on the seabed or from anchored barges in shallow water. Later, came the semi-submersible rig, floating on pontoons beneath the wave-base, which opened up the continental shelves of the world to routine exploration. The effort was rewarded by the discovery of offshore oil in many areas, highlighted by the North Sea, which began to deliver a crop of giant fields in the late 1960s.

By 1970, 1200 billion barrels had been found, over double the amount recorded in 1950, of which almost 250 billion had been consumed. In 1950, only 13% of what had been found had been used, but by 1970 the percentage had risen to 21%. Consumption had been rising rapidly at an average of about 7% a year, leading some responsible oil companies to warn that the rate of increase could not be sustained. Production in the United

States, which had once supplied the world, reached a peak and began its irreversible decline, reflecting the earlier peak of discovery forty years before.

As discussed further below, in 1973, certain Arab producing countries, concerned about Israeli military expansion, briefly suspended exports to the United States, triggering the First Oil Shock, when prices rose five-fold, never to return to previous levels. It was followed by a second shock in 1979 when panic buying, occasioned by the fall of the Shah of Iran, drove prices sky high. The world was plunged into recession, causing the demand for oil to fall, such that production did not return to its previous level for almost ten years.

Meanwhile, exploration and production continued throughout the world, with a great surge of tax-driven drilling in the early 1980s, but the results were modest, with most new discovery being in smaller fields in the established basins. In the 1990s, the industry turned to the last frontier offered by prospects in extremely deep water, bringing in a series of giant fields in the Gulf of Mexico and along the margins of the South Atlantic, where rather uniquely favourable geological conditions obtain. It is axiomatic that no one would look for oil in such an extreme environment if there were anywhere else easier.

By the year 2000, as much as 1750 billion barrels had been found, of which about 850 had been consumed. This represents 48%, vastly up on the corresponding factors of 21% in 1970 and 13% in 1950. Looking back, it has become evident that the peak of world discovery was in 1964. The world is now finding about one barrel for every three it consumes, depending precisely on what categories of oil are being measured. The world has been well and truly explored by an industry deliberately searching out the biggest and best prospects. Exploration was, in practice, barely affected by economic constraints because the companies enjoyed the facility in many countries of being able to offset its cost against taxable income, such that it was commonly spending "10-cent dollars". Great advances in geological understanding had identified the critical source-rocks, and new high-resolution seismic surveys showed every nook and cranny. Modern engineering improved performance all round.

As many as sixty-five estimates of the world's ultimate endowment of conventional oil over as many years have averaged just under two trillion barrels, depending again on exactly which categories are being measured. Most of them refer to *Conventional* oil, which has supplied most oil to-date and will dominate all supply far into the future. Although there are large additional deposits of *Non-Conventional* oil, such as tar-sands and oil shales, their extraction is too slow and costly for them to have much impact on the overall peak of production.

So, taking a broad brush, and not worrying about the finer points of definition which tend to confuse the issue, we can say that by the end of 2002, the world will have consumed almost half the amount of oil it will ever produce. It means that the growth of production that has marked the past 150 years must soon give way to decline.

The Second Half of the Age of Oil

The foregoing discussion has emphasised what an extraordinary chapter in history the last 150 years has been. There have been huge developments in every aspect of life, but the important change was, arguably, the appearance of a new and abundant supply of energy to turn the wheels of industry and trade. Much of it came from oil. Whereas coal had to be excavated with pick and shovel, oil, once found, simply flowed from the wellhead to offer a cheap, efficient and convenient fuel.

The growth of the past must now be mirrored by the decline of the future. The turning point does not necessarily come exactly at the halfway point, but it cannot be long delayed without imposing an implausible and very undesirable precipitate fall.

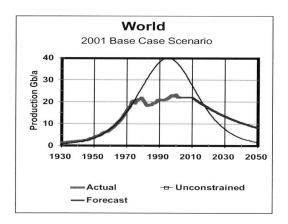

Figure 47

This plot shows the theoretical symmetrical bell curve reflecting an unfettered environment of exploration, production and consumption. There was a close match until the oil shocks of the 1970s, when high prices curbed demand, which had the effect of delaying and lowering the actual peak, also making the subsequent decline less steep than it would otherwise have been. Additional amounts of *non-conventional* oil will further lessen the decline rate. While we do indeed face the reflection of past growth, the mirror is a little warped around peak.

A Polemic Forecast

The foregoing discussion tried to place the peak of oil in a historical context, relating it to the evolving condition of Man and the extraordinary growth of population. But in doing so, it was necessary to depart from the strict analysis of oil depletion on which some scientific expertise could be claimed. The gulf between science and speculation widens as we turn to evaluate the possible future course of events.

The scientist is trained to a rigorous discipline of observation and logical deduction, and is reluctant to indulge in speculation about political developments for fear that to do so would undermine the credibility of his scientific arguments. On the other hand, there is no reason to suppose that a scientist's view of political situations should be any less relevant than those formulated by politicians, political analysts and commentators, who daily mould public opinion. There is a subject called political science, but whether the whims of human behaviour can be subjected to scientific analysis is perhaps unsure. At least, the scientist can claim objectivity in his assessment, albeit as an amateur.

It is for this reason that the latter part of this discussion has been entitled a *Polemic Forecast*. It does not claim any scientific basis, but simply offers a sight of the future that seems to flow logically from past and current events, many of which are intimately related to the world supply and consumption of oil. Indeed, it is precisely this close link that imposes the need to address these subjects: sensitive, difficult and uncertain as they are.

The expressed view of the future is that seen through a pair of eyes that has also looked backwards to identify the role of oil in the evolution of Mankind to this point. We can be sure that this vision will prove wide of the mark. That is not in doubt: the question to ask is "by how much?"

Few would deny that the world faces, at this very moment, a serious threat that a new war is about to break out for reasons not unrelated to oil supply. In the summer of 1914, few people could believe that the assassination of an Austrian archduke by a consumptive Serbian student would plunge Europe into four years of bitter war, from which it never truly recovered. Nor could people in the summer of 1939 believe that another devastating war was about to break, despite all the compelling evidence. People are always reluctant to think the unthinkable, and yet as we gathered the harvest beneath the summer skies of 2002, we faced a new military build-up that may divide the world, releasing pent-up passions with incalculable consequences. The starting point of this discussion has to be the State of Israel, and its conflict with the Arab people in a region, which is endowed with approximately half the World's remaining oil. It is the flash-point.

The State of Israel

The Kingdom of Israel came into existence around 1000 BC, when Saul and his son, David, established a regime along the Jordan valley. It covered a strip of territory about 100 miles wide, extending from the Euphrates River to the head of the Red Sea, over parts of what are now Israel, Palestine, Syria, Jordan, Lebanon, and Egypt. Most of it was inland, being separated from the Mediterranean by a Phoenician coastline and Philestia on the site of present-day Gaza. It was a monotheistic state, in which the daily norms of communal life were sanctified as being divinely inspired. The people came to believe that they were uniquely favoured by God, who had revealed himself to them. This gave them a certain uncompromising confidence in their actions, which remains a national characteristic.

The early history of the state was one of conflict, advance, retreat and internal discord between the several factions, some with religious overtones as each sect competed for a better claim to divine inspiration. Christianity was in fact to emerge in this way from a Judaic tradition.

The Jordan valley appears to have been a place of continual ferment. In 47 BC, it became part of the Roman Empire after one of the factions sought its support to settle a local conflict. But the tensions continued with strife between Jews and Romans, Jews and Christians and between various factions of Jews themselves. Over many years, Jews had left the troubled homeland to settle in other countries, but they retained their religious beliefs and separate identity. By AD 135, the Romans had had enough of the place, and decided to settle the discord once and for all by raising Jerusalem to the ground, and killing or selling its inhabitants into slavery. The loss of the homeland had a deep psychological impact on the many overseas communities, although most had long since lost any close link. The notion of recovering the homeland soon took root, and has festered ever since, being to some extent seen to be God's will.

The immigrant Jewish communities naturally concentrated in towns in their adopted countries where they became traders, business people and above all bankers. The Jewish banker began to accumulate wealth and exercise influence. He was however somewhat of an outsider, and in the smaller communities of the day, people could better perceive usury as a sort of financial cancer, whereby one man's gain had to be another's loss. It was deemed to be a sin by the Catholic Church, which inspired a degree of anti-Semitism.

In the 19th Century, the Jewish community in Europe had begun to integrate, providing many outstanding figures in science, literature, music, and politics, including amongst many others: Albert Einstein; Benjamin Disraeli, the British Prime Minister; and Karl Marx, not to forget the famous Rothschild bankers. Some sought total integration, while others preferred to elevate their very Jewishness to a new-found state of respect. But their success triggered a reaction from those who felt themselves to be in some way threatened, and anti-Semitic movements developed particularly in Russia, France and Germany. Many Jews emigrated to the United States in the hope of a new life away from the ancient prejudices of Europe, soon to secure a strong stake in the financial community, which in turn gave them a corresponding political influence.

These anti-Semitic pressures led to a revival of the notion of a homeland under a new Zionist political movement, founded in 1896 by an Austrian, named Theodor Herzl.

The First World War offered the protagonists a chance to implement their plans. Chaim Weizmann, a Russian-born chemist and Zionist activist from his student days, came to Britain in 1904 to teach at a university, before becoming director of the Admiralty laboratories in 1916, when he succeeded in developing a new method for synthesising acetone, a key ingredient of explosives. In this way, he gained access to Arthur Balfour, the Foreign Minister, whose primary mission at the time was to bring the United States into the war on Britain's side, despite the large numbers of German immigrants in its country. Weizmann, displaying the skills of a trader, succeeded in persuading Balfour that the Jewish Community in the United States had sufficient influence to win support for Britain, which it would exercise in return for the promise of the establishment of a Jewish homeland in Palestine, in the event that Britain should emerge victorious. Lloyd George, the Prime Minister, thought it was worth a try, and agreed the so-called Balfour Declaration of 1916 to this end. Publication was delayed until America did enter the war a year later, suggesting that there was indeed a link.

A British army, supported by Arab irregulars, captured Palestine and neighbouring territories from the Turks during the last two years of the war. British control continued after the war, being confirmed in the Versailles Treaty of 1922 when it was given a League of Nations mandate to administer the territory under conditions incorporating the Balfour Declaration to facilitate a "Jewish national home". The term "home" was understood to mean a good deal less than "State". Britain, also having obligations to its Arab allies, divided the mandate into two areas, designating the lands west of the Jordan River as Palestine, and those to the east as Transjordan. The latter obtained qualified independence in 1928, under King Abdullah ibn Hussein, the son of Britain's main ally during the war, namely the Grand Sherif of Mecca, who claimed direct descent from the Prophet. Britain had promised him an Arab Kingdom in return for his support. So far as Palestine was concerned, Britain tried to restrict Jewish immigration to a level that could be assimilated, but the numbers rose uncomfortably in the ensuing years, especially after the establishment of the Nazi regime in Germany, causing a series of revolts by Arabs who resented the new arrivals.

Immigration resumed with new intensity after the Second World War as the survivors of the holocaust, as well as other displaced people, sought a new life in Israel. Boatloads of illegal immigrants arrived on the beaches. Militant Zionism had already made an appearance in 1942, when Menachem Begin formed a terrorist group known as Irgun Zvai Leumi to oust the British. War-weary Britain no longer wanted the responsibility for the growing tensions, and after an atrocity in which Israeli terrorists blew up the King David Hotel, killing several British officers, it turned the problem over to the United Nations. In 1947, the latter proposed to partition Palestine into Arab and Jewish states, with Jerusalem being an international zone under its control. In May 1948, a provisional Jewish government unilaterally declared the establishment of such a State, which prompted an attack by Palestinian activists, supported by military forces from Egypt, Lebanon, Syria and Iraq. The United Nations managed to secure an armistice, which set the de-facto frontiers of the new State of Israel. About 600,000 Arabs were forced to flee, becoming refugees in neighbouring countries. World Jewry and the Unites States provided massive economic and military support, intended for defensive purposes. The Israeli Army, supported by Mossad, a highly efficient intelligence service, was thus able to become a first class fighting machine. In October 1956, it marched on Egypt, taking the Gaza Strip and the Sinai Desert, to reach the banks of the Suez Canal, before being persuaded to withdraw under UN mediation.

Sentiment was inflamed by these actions, leading the Arab world under the leadership of Egypt to seek justice and retribution, which in 1967 prompted an Israeli pre-emptive strike simultaneously against Egypt, Jordan and Syria, in what was known as the Six Day War. Israel's superior air power proved decisive. The frontiers of Israel were greatly expanded with the inclusion of the Golan Heights to the North, the Gaza Strip, the Sinai Desert to the South, as well as Arab East Jerusalem and the West Bank of the Jordan. They became known as he Occupied Territories, and were progressively settled by immigrants

The Palestine Liberation Organisation had been formed in 1964 to represent the displaced Arab refugees, being led by Yasir Arafat, partly in exile. In 1973, Egypt and Syria supported its cause by launching a combined attack, known as the Yom Kippur War. These forces were now fighting with better armaments provided by the Soviet Union and funded by Kuwait and Saudi Arabia, which also embargoed exports of oil to the United States, in condemnation of its support for their enemy, as already mentioned. Israel managed to survive but would have faced economic collapse but for a massive infusion of aid and military support from the United States.

Various attempts have since been made to broker a lasting peace settlement, but, given the background outlined above, it is not surprising that they show few signs of success. Left to themselves, the two sides might find, or be forced to find, an accommodation, but neither has a motive to do so, so long as they can call on foreign support for their respective causes.

The tensions have in fact been rising, partly due to the rapidly growing populations of both the Jewish and Arab communities, which now number some six million, not counting the Arab refugees living abysmal lives in camps in neighbouring countries. These six million people are mainly urban dwellers: the earlier image of gentle Jewish settlers transforming the desert into orange groves having substantially passed into history. Certainly, they greatly exceed the carrying capacity of the land where they live. It implies that they must control substantial overseas investments, especially in the United States, which is a further factor impeding progress to peace. The issue is complicated still more by the Arab control of world oil, and the growing US dependence on imports from the region.

In desperation, the Palestinians have found a new weapon in the form of the suicide bomber. It is an effective weapon, and there has been no shortage of young people willing to die for their cause.

September 11th 2001

In the same way as the assassination of Archduke Ferdinand in Sarajevo in 1914 was the catalyst that sparked the First World War, the events of September 11th in New York and Washington may lead inexorably to a Third World War. As in all wars, there is no single cause but rather a nexus of interwoven threads that suddenly finds common expression. Three such threads in this case are: first, the emotional ties between Israel and the influential Jewish community in America, some with dual citizenship being in the government; second, the fact that Israel's enemies control about half the world's remaining oil; and third, the growing dependence of the United States on oil imports. It is profligate user of oil, consuming some 25% of the world's production. Imports, now at some 60%, are rising because its own indigenous supplies are declining due to natural depletion. They are set to reach about 90% within twenty years, assuming even flat demand. These are the ingredients of conflict.

Four airliners were hijacked on the morning of September 11th 2001:

1. American Airlines Flight 11 took off from Boston at 7.59 with 92 passengers, and struck the World Trade Center at 8.45, hitting the 80th Floor;
2. United Flight 93 took off from Newark at 8.42 with 45 passengers, and crashed in Pennsylvania at 10.10;
3. United Flight 175 took off from Boston at 8.14 with 65 passengers; and

struck the World Trade Center at 9.02, hitting the 60th Floor;

4. American Airlines Flight 77 took off from Washington at 8.17 with 64 passengers, and approached the Pentagon at 9.25, where a massive explosion occurred at 9.43

There are many curious aspects of the affair, which have been well documented. Many people are accordingly left wondering what exactly happened, who the real perpetrators were and what their motives might have been. The evidence as presented would hardly satisfy Sherlock Holmes. But whatever the uncertainties and true explanation, the authorities moved with lightening speed to identify the culprits. A Saudi millionaire, by the name of Osama bin Laden, who had previously co-operated with the CIA, having business ties with the Bush family, was immediately identified as the ringleader, being said to control a terrorist network with camps and bunkers in Afghanistan. Before long, images were broadcast of this bearded Christ-like figure in his flowing robes. The actual hijackers were also soon identified as well-educated Saudi and Egyptian nationals, who had been living in Germany. Some of them had been taking flying courses in the United States, displaying a remarkable disinterest in learning how to take off or land.

Not long afterwards, the United States was subject to a curious anthrax scare, as contaminated letters were mailed to senators and others, causing the death of postal workers, all of which was dutifully relayed around with world as impressive television imagery. The scare lasted a few weeks when it ended as mysteriously as it had begun. It later transpired that the particular strain of anthrax is associated with a CIA laboratory.

The Reaction

President Bush soon rallied a nation, which thanks in part to impressive television imagery now felt itself to be under threat by Moslem activists, led by evil men dressed in outlandish robes. The President promptly dispatched an armada to the Middle East, where by remarkable coincidence it encountered most of the British Army on manoeuvres in the Oman. Soon, B52 bombers began pulverising Afghanistan, paving the way for the arrival of special land forces. Thousands of innocent people died in the course of the hostilities that led to the fall of the Taliban government. All of this was dutifully covered on the world's television screens. Some prisoners were strangely conveyed to a US base in Cuba, where they were placed in cages and denied their rights as prisoners of war established under the Geneva Convention. It was evidently important to depict them as terrorists and not soldiers, even though they had been fighting for their government.

It is worth mentioning in this connection that the United States had been taking an interest in the countries around the Caspian for a long time before September 11th with various military missions and contacts. In part, it was interested in the pipeline routes by which to export oil from what were then perceived to be huge resources. One such route for a gas pipeline, sponsored by the American oil company, Unocal, passed through Afghanistan. Much hung on the results of drilling the Kashagan prospect, off Kazakhstan, which, if full, might hold reserves rivalling those of Saudi Arabia. These hopes were dashed when the wells were finally drilled, yielding results far short of expectation. The justification for US intervention in the area has therefore greatly diminished, as it shifts its sights to the Middle East itself.

Within a matter of months, the Afghan campaign ran out of steam, with bin Laden himself eluding capture, such that the country now barely attracts television coverage. But its legacy was a US war on terrorism, which was broadly defined such that anyone who was not for them was deemed to be against them. Israel was not slow to find common cause in the war on terror, emphasising the parallel between itself as a victim of Palestinian suicide bombers and the American public facing Arab hijackers. Its military repression of the Palestinians reached new heights, with the inevitable loss of more innocent lives. It attracted some adverse comment but was now seen to be no more than an element in a wider "war on terror", however subjectively defined.

The new threat to Iraq

As the Afghan War recedes into history, along with the significance of Caspian oil, attention turns to Iraq, which does indeed hold substantial reserves, thanks in part to the UN export embargo that has restricted production over the past few years.

Saddam Hussein was born in 1937, which makes him 65 years of age. He joined the Ba'athist Party in 1957, which was an Arab version of Communist style dictatorship. He helped organise a coup that brought the party to power in 1968, becoming President of the country in 1979. Iraq itself was formed by Britain and France in the carve-up of the Middle East after the First World War. It was a somewhat artificial construction, comprising Kurds, who have long sought independence, in the north; Shi'ites with links to Iran in the south; and Sunni's around Baghdad, the capital, which also housed a substantial Jewish community until it was driven out by popular outrage on the creation of the State of Israel in 1948. The total population exceeds 20 million, most being urban dwellers. Evidently, it takes a strong leader to hold these disparate groups together as a nation.

In 1974, heavy fighting broke out between government forces and Kurdish separatists, who were being backed by Iran, but the dispute was settled when Iran was persuaded to withdraw its support in return, resolution of a long standing boundary dispute, related to the key Sh. Arab estuary. This was important to Iraq as an export outlet, by-passing stranglehold of Kuwait at the mouth the Euphrates River, its main tr. .e route. But tensions with Iran erupted again on the fall of the Shah in 1979 when unrest among the Iranian Kurds spilled over into Iraq. It soon developed into a full-scale war, which dragged on until 1988 with colossal loss of life to both sides. Although nominally neutral, the United States favoured Iraq during this conflict, still smarting from an incident when American citizens were taken hostage in Tehran, following the fall of the Shah. During the late 1980s, the United States supplied Iraq with substantial bank credits and technology to rebuild its military strength.

Meanwhile, President Reagan and Mrs Thatcher had apparently resolved to try to bring down the Soviet regime, ending the policy of co-existence. According to the book, *Victory*, by Peter Schweizer, the first step was to rearm the Afghans to end the Soviet occupation and undermine its military credibility. This was achieved with the help of King Fahd of Saudi Arabia, who funded the covert purchase of arms in Egypt for shipment to none other than bin-Laden, who was backing the Taliban with CIA support.

The next step was to persuade King Fahd to step up oil production to undermine the price of oil. The Soviets relied on oil exports for foreign exchange, which they now needed in greater amounts to buy equipment with which counter the new US "star wars" initiative. It was a successful strategy, which contributed to the fall of the Soviets, but it was achieved at a cost, as the low price of oil was bankrupting King Fahd, along with the Texan oil constituents of Mr Bush Sr.

While all this was in process, Kuwait arbitrarily increased its reported reserves by 50% in 1985 although nothing particular had changed in its reservoirs. It did so in order to raise its OPEC production quota, which was based on reserves. It also began pumping from the South Rumaila field that straddles the ill-defined border with Iraq. The latter complained bitterly both about what amounted to the theft of its oil across the border, and the loss of oil revenue, as prices fell as a result of Kuwait's failure to observe its contractual OPEC agreement.

Now, US strategy moved to strengthen the price of oil. It dispatched Henry Shuyler, an emissary, to encourage its ally, Iraq, to intervene in the councils of OPEC to enforce quota agreements sufficiently to achieve this end. It was recognised that words might not be enough, and that some added inducement could be needed to concentrate the minds of OPEC. Exactly what was proposed is not known, but it seems clear that a border incident to stop Kuwait producing from the southern end of the shared oilfield was contemplated. This interpretation is confirmed by the words of

April Glaspie, the US ambassador to Baghdad, who, on the eve of the invasion of Kuwait, made a statement to the effect that boundary disputes between Arab countries were of no concern to the United States. It was clearly an authorised statement, being released simultaneously in Washington under the signature of James Baker, the Secretary of State.

However, Saddam Hussein, possibly misunderstanding a wink and nod, did not stop with a border incident but mounted a successful full-scale invasion of Kuwait on August 2nd 1990. April Glaspie, on being woken by journalists with the news, reportedly responded "Oh My God, they haven't taken the whole place, have they?"

US policy now changed to condemn its former ally. A series of UN resolutions called for Iraq to withdraw from Kuwait by January 15th 1991, leading to a US aerial bombardment, when it failed to comply. Ground forces under the so-called Desert Storm campaign, led by General Schwarzkopf, crossed into Iraq, killing tens of thousands of Iraqis and destroying most of its military capability, before being ordered to halt at the gates of Baghdad when a cease-fire was agreed. The dissident Shi'ites in the south and the Kurds in the north, saw this as their moment to rise, but were successfully suppressed by remnant government forces. Hundreds of thousands of Kurdish refugees fled into neighbouring Turkey and Iran, where they were not exactly welcome.

The United Nations then imposed a trade embargo on Iraq to prevent it from rebuilding its military strength, making it effectively swing oil producer of last recourse, which provided a useful mechanism for stabilising the world price of oil at no cost to anyone but Iraq. In common with many countries, Iraq had made certain progress in developing modern nuclear, chemical and biological weapons, which were perceived to pose a particular threat. But by 1998, UN inspectors had reported that virtually all such facilities had been destroyed. The embargo remained, however, although it was subsequently partly relaxed for "humanitarian" reasons when the price of oil rose to uncomfortable levels.

The United States is now moving against Iraq on the grounds that it might be re-developing its weapons such that they could pose a threat against unspecified targets in the future. Little concrete evidence has yet been furnished, but the United States claims to be entitled to move on the so-called precautionary principle in this generalised "war on terror".

However, a recent British cabinet minister, Mo Mowlam, sees a different objective in United States' policy. Writing in The Guardian Newspaper on September 5th of 2002, she argues that the administration cannot be so blind as to fail to grasp that a move against Iraq would lead to widespread popular uprisings throughout the Middle East. That being the case, she concludes that the real motive is indeed to promote such uprisings leading to the fall of the Saud regime, which would give the United States the pretext to take its oilfields, a much easier military target. It is not the first time that this hidden agenda for the creation of "The Republic of Aramco" out of the ashes of Saudi Arabia has been mooted, but it is remarkable that it should now be endorsed by no less than a British cabinet minister. It may stretch credulity that President Bush himself would be capable of such a devious strategy and win the support of Mr Blair of Britain. But perhaps, behind the scenes, there are political and military strategists with their own agendas, who successfully implement various indirect steps creating situations, to which their political masters react in predictable ways.

Turning Point

The world is accordingly at a turning point, which may have been passed by the time these words appear in print. We may speculate about two alternative short-term outlooks.

Non-Restraint

If the United States cannot be restrained, it is likely that military attacks against Iraq will commence, no doubt starting with intensive bombing and the inevitable attendant loss of both military and civilian life. That itself may be sufficient to trigger popular outrage, in which the governments of Saudi Arabia, Kuwait, Bahrain, and possibly Egypt could fall, as predicted by Mo Mowlam. A state of extreme chaos would then reign throughout the region, and oil prices will likely surge far above $50 a barrel. US special forces, possibly backed by British troops, would then take the Saudi fields, which would be deemed to be under UN protection. Since there are virtually no local inhabitants in the desert, there will be minimal political issues to resolve beyond erecting a strong fence. Order may return to the cities under new puppet regimes or war lords.

It is hard to know if a subsequent invasion of Iraq would prove possible or desirable, following the collapse of the neighbouring countries. In the event that military strikes or covert actions should succeed in toppling Saddam Hussein, the country would break up with the secession of the Shi'ites in the south, and the Kurds north, the latter controlling the important Kirkuk oilfield. The emergence of a new Kurdish state would in turn trigger uprisings in the Kurdish regions of Iran and Turkey, which might eventually lead to a Kurdish Federation, after a great deal of infighting between factions. It might bring Turkey into the conflict with serious strategic implications.

Iran might emerge relatively unscathed, its Islamic government having been strengthened by events.

These conditions, which are not be easily reversible, would likely lead to a decade of regional instability. It may be imagined that oil production from the five Middle East main countries, which is now running at about 15 Mb/d (million barrels a day) would fall to about 9 Mb/d, coming about equally from Iran, Saudi Arabia and the other countries combined. The 3 Mb/d from Saudi Arabia would be imported by the United States, at nominal cost insofar as it would be free of royalty or tax, having been taken by right of conquest.

It is hard to imagine the creation of a puppet regime in Iraq to rapidly bring about the re-creation of the Iraq Petroleum Company of British and American oil interests. It is even harder to imagine that conditions would calm sufficiently to allow them to step up production.

The world price of oil would accordingly remain very high, imposing deep global recession. Many smaller producing countries would likely restrict exports to conserve supplies for their own use, exacerbating the world shortage. The Far East economy would suffer further decline, and there would be sporadic outbreaks of violence throughout the world as privation dug deeper and people's aspirations were thwarted.

The United States government would manage to use this situation, in domestic political terms, to radically increase the cost of fuel by raised taxes, such that demand would be reined in. The stock market would continue to collapse until a point at which the value of shares reflected their current real and projected earnings in deepening recession. The American people would use their well-known spirit of ingenuity and initiative to rise to the occasion. Immigration would be radically reduced, and the many illegal immigrants already in would likely be expelled. The individual States would assume a greater role in government, and the people might adopt a new frontier mentality, building strong new communities. Urban conditions, already bad enough in the poor ethnic ghettos, could hardly fail but deteriorate further in the face of the new austerity. The government might emerge strengthened, being able to blame the change on world events beyond its control.

Europe would also head into deep recession, but would cement new ties with Russia to supply its needs of oil and gas, as its own North Sea production continued to decline due to natural depletion. Russian exports could increase, in part making good the anomalous fall consequent upon the fall of the Soviets. The French could be invited to launch a crash campaign building nuclear power stations throughout the union. Britain would somehow disentangle itself from its military support for the United States, and become a more enthusiastic member of Europe, belatedly embracing the Euro.

Other countries would react as well as they could within their own abilities and resources in ways that cannot be foreseen.

Although this picture, which stretches the imagination, represents a devastating turn of events upending the established order, it does carry some positive elements. Oil consumption would be cut, so that the supplies would last longer. Recession would cut deep, leading to starvation, conflict and disease, which would reduce the population to a more sustainable level, albeit in the most appalling ways. The forests and fish stocks might begin to recover, and the climate would stabilise as a consequence.

Restraint

Alternatively, there is just a chance that restraint might be exercised through intense diplomacy. Most countries express objections to unilateral action against Iraq by the United States and its ever-faithful ally, Britain, pointing out that international law demands that any move against a sovereign state has to be sanctified by the United Nations. In the face of this pressure, the United States reluctantly retreated, giving the United Nations a chance to pass appropriate resolutions, providing for new inspection.

The Security Council of the United Nations has fifteen members. The United States, Russia, Great Britain, France and China are so-called permanent members with the right to veto a resolution. Britain and the USA would likely vote for military action on any pretext, but Russia, France and China would need much convincing. Russia has had long close ties with Iraq, providing loans and military equipment, in addition to importing much of its oil. Russian and Chinese oil companies are already at work in the country, with the Chinese having a particular need for oil imports to offset the natural decline of its own production.

Russia and China may have to weigh up the relative merits of alternative strategies. On the one hand, they might be tempted to support the US invasion on condition that they received a fair share of the spoils. Or, on the other, they could try to frustrate it in the hope that they could strengthen their position as allies of the Middle East people, and thereby secure a greater share of the region's oil in the longer term. Russia has no good reason to further inflame its large Muslim communities in its own and neighbouring territories. Wider geopolitical considerations may lead them to prefer the latter option, allowing them to strangely emerge victorious in a postscript to the Cold War.

Accordingly, Russia and China, with the support of many other countries, may press the United Nations to await the results of the search by the famous inspectors for offensive weapons. Months might pass, for, as in the case of exploring for oil, it is easier to look than to find, especially if there is in fact little to find.

President Bush, for his part, having strengthened his grip in the elections of November 2002 on the strength of a certain war fever, might be able to take a more relaxed attitude, reacting to the growing opposition to the war in his own country.

So, the World might somehow step back at the edge of the abyss of a new war of devastating world proportions. This in turn would force it to address the deeper real issues of economic decline, environmental degradation, excessive population, and oil and gas depletion.

Tensions in the Middle East itself might diminish if Israel were no longer receiving massive overseas financial support in the face of world recession. It might be finally persuaded to withdraw to earlier frontiers, which would pave the way for the creation of a viable new Palestine state, that would receive subsidies from the oil producing Arab states.

The turning point for a new order could come if the United Nations were to convene a conference of the major countries of the world to sign an Oil Depletion Protocol, providing that:

- no country shall produce oil at a rate above its depletion rate; and
- no country shall import infringements

In effect the importing governments would be required to ration imports where demand should exceed depletion. As a consequence, the revenue from oil would shift from the exporters to the importing governments, transferring control of oil supply, including management of the growing shortage consequent upon natural depletion, from the Middle East to the consuming countries. The price of oil to the consumer would remain high but would be matched by corresponding reductions in other taxes. This would lead to effective energy-saving measures throughout the world. The Middle East would cease to be a tinderbox to world conflict, and could begin to resolve its own destiny, freed from the distortion of excessive un-earned oil revenue. The demand for oil would be curbed in this way, such that the peak would be delayed and the rate of subsequent decline lessened.

Communities everywhere would look inward to find local markets for their produce, discovering new more sustainable ways of life. The days of rampant affluence would be over, and the adoption of changed economic principles would bring about a longer-term view of the future. Immigration would be stopped, such that the populations of ageing societies would fall naturally to levels that were sustainable within their own resources. The end of globalism would mean that the Chinese would make shoes for themselves, retaining the proceeds rather than funding a host of foreign intermediaries with many partly hidden indirect financial benefits through credit, debt and usury.

Discovery of Democracy

Another possibility might be the discovery of democracy. At the present time, most so-called democratic countries operate political parties, representing very small proportions of the total population. These parties not only select the candidates for election, but have their parliamentary representatives vote on central instruction rather than conscience or independent judgement. Denied the secret ballot, they are kept under firm control by the party elites. The result is a form of dictatorship hiding beneath a cloak of democratic imagery. There is little possibility for natural change, as the party leaders have a strong vested interest in maintaining the present system that delivers their power. The funding of these parties is a particularly murky element, based on various degrees of "votes for cash", however subtly implemented.

But some progress might be imposed upon the system by the emerging real-world pressures that call for strong and intelligent government backed by popular support. The "open market" that has freed these governments from real responsibility, would by then have been discredited. Electronic communications might facilitate a more democratic system whereby the issues are placed before the people with their reactions being duly recorded as a basis for decision. A secret ballot for elected representatives would help break the stranglehold of the party machines.

The final chapter

The path of history is at the crossroads, and it is impossible to know which route it will follow. There may well be other tracks through the undergrowth. But twenty or thirty years on, the roads will come together again as the World comes to terms with its new situation whatever that may be. The gradual decline of oil and gas will then be evident to all, and appropriate measures to deal with it will be in place. While the production of renewable energies of all sorts, including nuclear energy, will be stepped up, it will fail to be any real substitute for cheap oil and gas, as known in the past. Instead, the way of the world will have to change so as to be less dependent on abundant cheap energy.

Here, we may take comfort from the words of Mrs Coughlan, the elderly post-mistress of Ballydehob, a village in Ireland. When asked if she thought that the world had got better over the past fifty years, she replied, unhesitatingly, that yes of course it had. But when asked if it had been bad before, she reflected before answering. "Not at all", she said, "although we were as poor as church mice, us girls would happily darn our stockings in the evenings, and think nothing of walking ten miles to a dance. People with fresh eggs would share them. Everyone would give a hand. We'd meet for music and fun in people's houses"

The message is that affluence does not necessarily bring happiness. Perhaps the decline of oil will give a better world in which people will live in better harmony with themselves, each other and their environment.

End Point

The Age of Oil began about 150 years ago, and we have now produced almost half of all that is there to produce from traditional sources. Production must accordingly soon peak and begin to decline, eventually running out in the next century. But in the same way as oil had a minor impact during its infancy, so too will its dying years mean little. The crunch comes around peak, as growth gives way to decline. The perception of this fundamental change is as important as the onset of physical decline itself. The violent days in which we live are symptomatic of the start of the transition.

Part 6

Bibliography

A

AAPG, 1991, *The Arabian Plate – producing fields and undeveloped hydrocarbon discoveries*; map published by Amer. Assoc. Petro. Geol.[#31]

AAPG. Explorer, 2001, *The headline discoveries of 2000*; AAPG [#1210]

AAPG Explorer, 2001, *West Africa's time to shine arrives*; AAPG [#1211]

AAPG Explorer, 2001, *Trends*; AAPG [#1212]

AAPG, 2001, Discussions and Reply re *"Energy Resources – cornucopia or empty barrel"*; AAPG Bulletin, Vol. 85, No. 6, June 2001 [#1333]

AAPG, 1999, *Climate Change*, AAPG Explorer, p 6/8, December, 1999 [#1547]

AAPG, 2000, *World Oil Resource Forecast Increases*; AAPG Explorer, p 24/25, June, 2000 [#1556]

AAPG, 2001, *USGS Assessment Study Methods Endorsed*; AAPG Explorer, November,2001 [#1711]

Abelson A., 1996, *Crude awakening*; Barrons 1.5.96 [#431]

Abernathy, V.D., 2001, *Carrying capacity: the tradition and policy implications of limits*; Science & Environmental Politics

Abernathy, V D., 2001, *Population Dynamics: Poverty, Inequality, and Self-Regulating Fertility Rates*; For John Cobb conference, ???, [#1655]

Aburish S.K., 1994, *The rise, corruption and coming fall of the House of Saud*; 226p Bloomsbury, London [#B15]

Abraham K S., 1994, *Low oil prices killed the USSR*; Pet. Eng. Int., Sept. 1994 [#278]

Abraham K S., 1996, *Sifting the stew of Iraqi sales, US politics and future supplies*; World Oil March 1996 [#427]

Abraham K S., 1997, *Venezuela bets on heavy crude for long term*; Oil & Gas Journ. Jan. [#542]

Abraham K S., 1998, *Weak demand, future reserve concerns color global picture;* World Oil Dec [#931]

Abraham K S., 1998, *Debate grows over accuracy of industry statistics;* World Oil May [#983]

Abraham, K S., 2001, *Brazil sustains period of expansion;* World Oil, p 129 – 134, June 2001 [#1305]

Abraham, S., 2001, *We Can Solve Our Energy Problems;* The Wall Street Journal, Friday, May 18, 2001 [#1591]

Adams T., 1991, *Middle East Reserves;* Oilfield Review p7–9 [#27]

Adelman M.A., 1989, *Problems in modeling world oil supply;* Energy Modeling Forum, MIT Energy Lab Working paper MITEL – 89-010WP [#104]

Adelman M.A., 199?, *Modelling world oil supply;* The Energy Journal 14/1[#327]

Adelman M.A., 1995, *Letter to Petroleum Economist;* Petroleum Economist, July 1995

Adelman M.A., 1995, *The genie is out of the bottle: world oil since 1970;* MIT ISBN 0-262-01151-4

Adelman M.A. & M.C.Lynch, 1997, *Fixed view of resource limits creates undue pessimism;* Oil & Gas Journ. April 7th 1997 [#615-2]

Adelman M.A. and M.C.Lynch, 1997, *More reserves growth;* Oil & Gas Journ. June 9th [629]

Adelman M.A. 1997, *Oilfield developments times and future oil supply;* IEA submission [#716]

Adelman, M A., 2001, *Oil-Use Predictions Wrong Since 1974;* The Wall Street Journal, March 6, 2001 [#1772]

Adelman M.A, 1993 *Economics of Petroleum Supply* [book]

Ahlbrandt, T S. & Charpentier, R R. & Klett, T R. & Schmoker, J W. & Schenk, C J. & Ulmishek, G F., 2000, *Future Oil and Gas Resources of the World;* Geotimes, June 2000 [#1503]

Ahlbrandt, T., 2000, *USGS World Petroleum Assessment 2000,* USGS [#1577]

Ahmed N.M.,2002, *The war on freedom;* Media Messenger pp.398

Ahuja, A. & Bennett, V. & Henderson, M., 2001, *Kyoto – Why care ? Cover Story;* The Times, p 2 – 7, Monday, July 9, 2001 [#1580]

Al-Fathi S.A., 1994, *Opec oil supply outlook to the year 2010;* Wld Petrol. Congr., Stavanger [#236]

Al-Jarri A.S. & R.A. Startzman, 1997, *Worldwide supply and demand of petroleum liquids;* SPE 38782 [#699]

Alahakkone R.R., 1990, *Conventional supply of oil will slow down;* Oil & Gas Journ., Feb 5 1990 [#81]

Alberta Energy & Utilities Board., 1998, *Non-conventional oil production for Alberta* [#708]

Aleklett K.,2001, *Brannpunkt;* Svenska Dagbladet 23 March

Aleklett K, 2001, *Oljebrist hotar var civilisation;* Brannpunkt [#1208]

Alezard N., J.H.Laherrere & A. Perrodon, 1992, *Réserves et resources de pétrole et de gaz des pays Méditerranées*; Revue de l'Energie 441, Sept. 1992 [#140]

Alhajji, A F., 2001, *Middle East: Investment levels rise higher*; World Oil, August, 2001 [#1488]

Ali.M, 1999, *Oil Crisis*; Al Arab, 19 November [1020]

Allen R.V., 1966, *The man who changed the game plan*; The National Interest 44 summer 1996.

Allen R., 1998, *Rafsanjani leads Iranian bid to make new friends*; Financial Times 23/2 [#685]

Alvarez C.G., 2000, *Economía y politica petrolera*: published by Ecopetrol

Amenson H., 1988, *Oil discovery principles, exploration strategy*; Oil & Gas Journ. Oct 17 1988 [#89]

Amenson H., 1992, *World developments since mid 1990*. AAPG Explorer, Nov. 1992.

Amenson H., 1993, *U.S. drilling: up is the only direction left*; World Oil, Feb. 1993.

Amenson H., 1993, *Look for a moderate rise in U.S. exploration*; World Oil, Feb. 1993

Amenson H., 1993, *U.S. oil and gas reserves resume downward trends*; World Oil, Feb 1993.

Amenson H 1993, *Drilling remains strong in many world regions*; World Oil, Feb 1993.

Amuzwgar J.,1999, *Managing the oil wealth*; Tauris 266pp

Anderson, B., 2001, *The creation of a Palestinian state is essential for the war against terrorism*; The Spectator, October 20, 2001 [#1671]

Anderson F. S., 2000, *Industrial Management: Energy – "There Is Not Enough Gas Around"*; Business Week, September 18, 2000 [#1388]

Anderson F. S., 2000, *Unnatural Demand For Natural Gas*; Business Week, April 3, 2000 [#1644]

Anderson R.B., 1999, *Gas tanks on full; supplied heading for empty*; The Oregonian April 10 [#1046]

Annin P., 1998 , *Power on the prairie*; Newsweek, 26 Oct. [#860]

Anon., 1992, *New minister reports on UK oil and gas development*; First Break 10/6 June 1992 [#40]

Anon., 1992, *Certain outlook: uncertainty*; AAPG Explorer March 1992 [#125]

Anon., 1993, *Look for a moderate rise in U.S. exploration*; World Oil Feb. 1993 [#136]

Anon., 1993, *U.S. oil and gas reserves resume downward trends*; World Oil Feb 1993 [#137]

Anon., 1993, *Drilling remains strong in many world regions*; World Oil Feb 1993 [#138]

ARAMCO, 1980, *Aramco and its world: Arabia and the Middle East*; (Ed. Nawwab I.I., P.C.Speers, & P.F.Hoye) Aramco. ISBN 0-9601164-2-7.

Arnold R., G.A. Macready & T.W. Barrington, 1960, *The first big oil hunt: Venezuela 1911–1916*; Vantage Press, New York 353p [#B3]

Arnold, W., 2001, *A Gas Pipeline to World Outside*; The New York Times, October 26, 2001 [#1727]

Ashby T.,2000, *Venezuela not interested in spare capacity*; Reuters [#1058]

ASTM, 1982, *Standard for metric practice*; ASTM E-380-82 [#341]

Attanasi, E D. & Mast, R F. & Root, D H. & USGS, 1999, *Oil, gas field growth projections: wishful thinking or reality ?*; Oil & Gas Journal, April 5, 1999 [#1596]

Ayres R.U. & P. Frankl, 1998, *Toward a non-polluting energy system*; Environ. Sci. & Tech. Sept 1. [#831]

B

Bahree B., 1994, *World oil demand in 4th quarter to top earlier forecast, agency says*; Wall St. Journ July 8 1994 [#243]

Bahree B., 1998, *Oil demand policies distort policies in the Gulf*; Wall St Journ. Feb 23 [#677]

Bahree B., 1999, *Global demand for oil is set to grow in 2000 at faster pace*; Wall St. Journ Aug 11 [#1004]

Bahree, B. & Barrionuevo, A., *OPEC Warns of Price Collapse Next Year*; The Wall Street Journal, Tuesday, October 30, 2001 [#1714]

Baker Institute Study, 2000, *Japanese Energy Security And Changing Global Energy Markets: An Analysis of Northeast Asian Energy Cooperation and Japan`s Evolving Leadership Role in the Region*; James Baker III Institute For Public Policy Of Rice University, No.13, May, 2000 [#1437]

Bakhtiari A.M.S., 1999, *The price of crude oil*; OPEC Review 13/1 March

Bakhtiari A.M.S., 2000, 1999, *IEA, OPEC oil supply forecasts challenged: Oil & Gas Journ, April 30*

Bakhtiari, A.M.S., Shahbudaghlou, F., 2001, *IEA, OPEC oil supply forecasts challenged*; Oil & Gas Journal, Apr 30, 2001 [#1295]

Bakhtiari, A.M.S., 1999, *The price of crude oil*; OPEC Review, Vol. 23, No. 1, March 1999 [#1296]

Bakhtiari, A.M.S., 2001, *2002 to see birth of New World Energy Order*; Oil & Gas Journ, Jan 7, 2002 [#1763]

Bakhtiari Oil Consultants, 2001, *OPEC Member Countries: Forecasts For Oil Production Capacities (2001 – 2020)*; October, 2001 [#1680]

Baldauf S., 1998 *World's oil supply may soon run low*: Christian Science Monitor 23 Sept [#840]

Ball, J., 2002, *Bush Will Shift Gears on Auto Research, Favoring Fuel Cells Over 80 MPG Sedan*; The Wall Street Journal, January 9, 2002 [#1747]

Banerjee N and Kapner S, 2001, *Drilling in Alaska presents hard choices to BP*; New York Times [#1197]

Banerjee, N., 2001, *Fears, Again, of Oil Supplies at Risk*; The New York Times, October 14, 2001 [#1723]

Banerjee, N., 2001, *The High, Hidden Cost of Saudi Arabian Oil*; The New York Times, Oct 21, 2001 [#1725]

Banks H., 1998, *Cheap oil: enjoy it while it lasts*; Forbes June 15 [#748]

Barber A.H., C.J.George, L.H.Stiles and B.B.Thompson, 1983, *Infill drilling to increase reserves – actual experience in nine fields in Texas, Oklahoma and Illinois*; Journ. Petrol. Technol.[#67]

Barker T.,1995, *Taxing pollution instead of employment: greenhouse gas abatement through fiscal policy in the UK*; Energy & Environment 6/1 [#471]

Barker, R., 2001, *The Oil Spigot May Be Closing*; Business Week, November 12, 2001 [#1715]

Barkeshli F., 1996, *Oil prospects in the Middle East and the future of the oil market*; Oxford Energy Forum, 26 August 1996 10–11 [#483-1, #493

Barley B.,1999, *Deepwater problems around the world,* Leading Edge, April [#1098]

Barrett M.E., 1994, *Amoco corporation;* American Graduate School of International Management. [#499]

Barrett W.J., 1992, *Calling all bottom fishers*; Forbes May 11th 1992 [#57]

Barrioneuvo A, 2001, *Oil services firms take a bath on deep-water drilling;* Industry Focus [#1225]

Barry, D., 2000, *Americans are just a few dinosaurs short of a full tank*; The Register Guard, Sunday, April 16, 2000 [#1645]

Barry J., 2002, *Pipeline brigade*; Newsweek, April 8 [#1831]

Barry R.A., 1993, *The management of international oil operations.* PennWell Books,Tulsa.

Bartlett A.A., 1978, *Forgotten fundamentals of the energy crisis*; Am. J.Phys. 46/9 Sept [#681]

Bartlett A.A., 1986, *A management program for non-renewable resources*; Am. J. Physics 54/5 [#734]

Bartlett A.A., 1990, *A world full of oil*; The Physics Teacher Nov.[#696]

Bartlett A.A., 1994, *Reflections on sustainability, population growth and the environment*; Population & Environment 16/1 [#682]

Bartlett A.A., 1996, *The exponential function, XI: the new flat earth society;* The Physics Teacher 34 [#679]

Bartlett A.A., 1997, *Is there a population problem?* Wild Earth Fall 1997 [#714]

Bartlett A.A., 1997, *An analysis of US and world production patterns using Hubbert curves*; draft [#678]

Bartlett A.A., 1998, *Reflections on sustainability, population growth and the environment – revisited*; Renewable resources Journal Winter 1997/8 [# 697]

Bartlett A.A., 1998, *CNES Hearings*; Letter to DOE [#745]

Bartlett A.A., 1998, *An analysis of US and world oil production patterns;* draft [#784]

Bartlett A.A, 1998, *Reflections in 1998 on the twentieth anniversary of the publication of the paper: "Forgotten fundamentals of the energy crisis"*, Negative Population Growth [#1240]

Battle J., 1997, *From waste to potential*; Petrol. Review [#581]

Bauquis, P-R., 2001, *A Reappraisal of Energy Supply and Demand in 2050*; Oil & Gas Science and Technology, Vol.56, No.4, p 389 – 402, 2001 [#1754]

Bayless F., 2000, *Sky high oil prices freeze the northeast*; USA Today 10 Feb [#1080]

BBC, 2000,*The last oil shock:* The Money Programme Nov.8, 2000

Beaumont P. and J. Hooper, 1998, *Energy apocalypse looms as the world runs out of oil*; Observer Newspaper 26.7.98 [#792]

Beardsley T., 1994, *Turning Green; Shell International projects a renewable energy future*; Scientific American 271/3 Sept. 1994 [#260]

Becker, B., 1991, *U.S. Conspiracy to Initiate the War Against Iraq*; The Commission of Inquiry

for the International War Crimes Tribunal, 1992 [#1310]

Bee A.C.,1991, *Long-run industry performance in exploration outside N.America*; Abstract Amer. Assoc. Petrol. Geol. 75/8, 1405 preprint [#41]

Beeby-Thompson A.,1961, *Oil pioneer*; Sidgwick & Jackson, London

Bell Helicopter Textron, 1989, *World Energy Update*; Bell Helicopter Study [#95]

Bell S., 1994, *Distribution of resources knowledge helps locate undiscovered reserves*; Petrol. Eng/Int.[#327]

Beller M., A. Chauvel and P. Simandoux, 1999, *The Challenge of North Sea oil and gas*; Revue de l'IFP 54/1 [1022]

Benoit G., 2000, *Energizing America;* The New American March 27 [#1162]

Bentley R.W., 1997, *The future of oil*; Seminar briefing, Reading University [#604]

Bentley R.W., 1997, *Oil shock imminent if heavies are slow or expensive to produce*; Energy World 250 20–22 [# 617-3]

Bentley R.W et al.,2000, *Perspectives on the future of oil*, Energy Exploration & Exploitation 18/2–3

Bentley R.W., 2001, *Global oil and gas depletion*; own paper for EU Conference: what energy options for Europe in 2020? [#1247]

Bentley, R W. & Whitfield, G R., 2001, *Advice on Nuclear Energy Issues Relevant to NERC Science*; The University of Reading, Version 2, August 28, 2001 [#1357]

Bentley, R W. & Booth, R H. & Burton, J D. & Coleman, M L. & Sellwood, B W. & Whitfield, G R., 2000, *World Oil Supply: Near & Medium Term–The Approach Peak In Conventional Oil Production*; The University of Reading, August 22, 2000 [#1413]

Bentley, R W. & Booth, R H. & Burton, J D. & Coleman, M L. & Sellwood, B W. & Whitfield, G R., 2000, *Perspectives on the Future Oil*; Reprinted from Energy Exploration & Exploitation, Vol.18, Nos.2&3, 2000 [#1414]

Bentley, R W., 2000, *Letter to G R Davis re: Topics related to the "Oil Depletion and reporting;* Royal Society of Chemistry; April 19, 2000 [#1639]

Bentley, R W., 2001, *Submission to the Cabinet Office Energy Review*; Draft, The Oil Depletion Analysis Centre, September 6, 2001 [#1653]

Bentley, R W., 2001, *Letter to The Economist re: "Sunset for the oil business?"* (#1756); The Economist, p 22, November 24, 2001 [#1757]

Bentley, R W., 2002, *Global oil & gas depletion: an overview*; Energy Policy 30, p 189 – 205, 2002 [#1789]

Bentley, R.W., 2002 *Oil forecasts, past and present*; ASPO Workshop, Uppsala [#1839]

Berger W, 2000, *Ol sprudelt nicht mehr lang*; Wirtschaft [#1245]

Bianco, A., 2001, *Cover Story: Exxon Unleashed*; Business Week, April 9, 2001 [#1630]

Binney, G., 2001, *The Petro-Population Parallel*; The Ecologist, Vol 31, No.6, July/August, 2001 [#1677]

BGR, 1994, *World Energy – a changing scene*; BGR Conference abstracts [#275

BGR, 1995, *Reserven, Ressourcen und Verfügbarkeit von Energierohstoffen* 1995; BGR Hannover 498p [#B19]

Bilkadi Z., 1996, *Babylon to Baku*; Stanhope-Seta ISBN 0952-881608 230p.

Billo S.M., 1990, *Complex geology discussed by noted Arabian scientist*: Oil & Gas Journ, Jan 29 199 [#82]

Bird K.J., F.Cole, D.G.Howell & B. Leslie, 1995, *The future of oil and gas in northern Alaska*; Amer. Assoc. Petrol. Geol., 79/4 p.579 [#390]

Bishop R.S., 1998, *"Out of Oil" question demands an answer*; AAPG Explorer Nov [#900]

Bjørlykke K., 1995; *From black shale to black gold*; Science Spectra 2, 1995 44–49

Blakey E.S., 1985, *Oil on their shoes*; Amer.Assoc. Petrol. Geol.192p [#B5]

Blakey E.S.. 1991, *To the waters and the wild*; Amer.Assoc. Petrol. Geol. [#B6]

Bloomberg, L. P., 2000, *Crude Oil Inventories Fall to 24-Year Low in US; Prices Rise*; Bloomberg, L. P., August 8, 2000 [#1371]

Bloomberg, L. P.,2001, *Total Says Oil, Gas Flowing From Second Kashagan Test Well*; Bloomberg, L. P., May 3, 2001 [#1442]

Bloomberg, 2000, *Oil Demand Growth to Outpace New Output Capacity, BP Amoco Says*; Bloomberg, June 21, 2000 [#1518]

Bloomberg 2002, *If we really have the oil*, http://www.bloomberg.com/wealth/0902/sep.ft.crude.pdf

Boeckman O.C. and others, 1990, *Agriculture and fertilizers*; Norsk Hydro 245pp

Bohnet M., 1996, *Promotion of conventional and renewable sources of energy in developing countries*; in Kürsten M.(Ed) *World Energy – Charging Scene*: ISBN 3-510-65170-7

Bookout J.F., 1989, *Two centuries of fossil fuel energy*; Episodes 12/4 [#331]

Boostein D.J.,1985, *The discoverers*; Vintage 745pp

Boulton G., 1998, *World may have more oil than it will ever need*; Salt Lake City Tribune May 19 [#749]

Bourdaire J.M. *et al.* 1985, *Reserves assessment under uncertainty – a new approach*; Oil & Gas Journ. June 10. 135–140.

Bourdaire J.M, 1993, *Le pertrole dans l'economie mondiale*; Energie Universale, Sept. 1993 [#180]

Bourdaire J.M, 1998, World energy prospects to 2020; Brit. Inst. Energy Economics, 2 July [#770]

Bourdaire J.M, 2000, *Petrole – Les fondaments de l'economie petroliere*, Encyclopaedia Universalis, France [#1234]

Bourdaire, J M., 2000, *Energy and Economics: "Moving Backwards To The Future"*; July 2000 [#1361]

Bowen J.M., 1991, *25 years of UK North Sea exploration*; in Abbotts J United Kingdom Oil & Gas fields; Geol. Soc Mem14 [#913]

Bowlin M.R, 1999, *Clean Energy*; CERA conf. [#1051]

Box W., 1999, IEA ; Seatrade [#921]

Boyes R, 2001, *Russian power and Gazprom*; The Times (UK) [#1194]

Bradbury J, 2000, *Reinventing the North Sea*; Hart's E&P [#1253]

Bradley R.L.,1999, *The growing abundance of fossil fuels*; The Freeman [#1130]

Brammer R.,1995, *Oil Change*: Barron's March 6th 1995 4–4 [#333]

Brauer B, 2001, *A Waning Honeymoon in Kazakhstan*; New York Times [#1195]

Breton T.R. & J.C.Blaney, 1991, *Production rise, consumption fall may turn Soviet oil exports higher*; Oil & Gas Journ. Nov 18 1991 110–114 [#7]

British Petroleum Co., *BP Statistical Review of World Energy*; Published annually by BP, London.

British Petroleum Co, 1959, *Fifty years in pictures*; publ. BP, London 159 p.

British Petroleum Co, 1979, *Oil crisis – again*; BP paper [#590]

British Petroleum Co, 1998 *Presentation of Statistical Review of World Energy*; BP [#759]

Bronner M.E., *US pins high hopes on Caspian oil*; Washington Times 22.11.96 [#655]

Brown D.,1998, *How much oil is really there ?* ; AAPG Explorer April [#796]

Brown, D., 2000, *So, Are We Running Out Of Oil ?*; AAPG Explorer, p 46, 47, April, 2000 [#1540]

Brown, D., 2000, *Bulls and Bears Duel Over Supply*; AAPG Explorer, p 12 – 15, May, 2000 [#1541]

Brown L.R., 2001, *The state of the world*; W.E.Norton & Co.,268pp

Brown L.R., 2001, *Eco-economy*; W.E.Norton & Co.,333pp

Browne E.J.P., 1991, *Upstream oil in the 1990s: the prospects for a new world order*; Oxford Energy Seminar, Sept. 1991. Publ. The British Petroleum Company, London [#42]

Browne E.J.P., 1991 *The way ahead – hydrocarbons for the 1990s*: Amer. Assoc. Petrol. Geol. London Conference. preprint [#24]

Brown T., 1991 *Mexico oil assets called inflated*; The Arizona Republic Dec. 10 1991 [#13]

Buderi R., 1992, *Oil's downhill skid may be ending*; Business Week Feb. 1992 [#8]

Bundesministerium für Wirtschaft, 1993, *Securing Germany's economic future*; Report 338 [#481-1]

Bundesministerium für Wirtschaft, 1995, *Sources of energy in 1995: reserves, resources and availability*; Report 383 [#483-1]

Bureau of Resource Sciences, 199?, *Sufficiency of Australia's petroleum resources* [#537]

Burns, S., 2001, *World running out of cheap oil that fuels economies*; Houston Chronicle,, July 16, 2001 [#1322]

Burruss R.C., 1995, *USGS oil and gas CD hits the charts*; Geotimes July 1995 [#384]

Business Week, 2000, *Is Big Oil Getting Too Big?*; Business Week, October 30, 2000 [#1415]

Business Week, 1993, *The scramble for oil's last frontier*; Business Week Jan 11 1993 [#150]

Business Week, 1997,*The new economics of oil* Nov 3 [#939]

Business Week, 1999, *Sheik Yamani on the coming oil price crisis;* April 14 [#1011]

Bryant, A., 2000, *The New War Over Oil;* Newsweek, October 2, 2000 [#1404]

Byman D., 1996, *Let Iraq collapse!*; The National Interest 45 48–60.

C

Cambridge Energy Research Associates, 1991, *The capacity race: the long term future of world oil supply*; James Capel Report 1991, [#46]

Cameron, N. & Bate, R. & Clure, V. & Benton, J., 1999, *Oil and Gas Habitats of the South Atlantic: Introduction*; In: The Oil and Gas Habitats of the South Atlantic, Cameron et al, Special Publication, 153, The Geological Society, London,1999 [#1526]

Campbell C.J. 1965, *The Santa Marta wrench fault of Colombia and its regional setting*; 4th Carib. Geol. Conf

Campbell C.J. & H. Bürgl, 1965, *Section through the Eastern Cordillera of Colombia, South America*; Geol. .Soc. Amer. 76 567–590

Campbell C.J., 1974 *Structural classification of northern South America*; verhnd. Naturf Ges. Basel

Campbell C.J., 1974, *Colombian Andes*; in Spencer A.M. (Ed) *Mesozoic-Cenozoic Orogenic Belts*; Mem. Geol. Soc., London

Campbell C.J., 1974, *Ecuadorian Andes*; in Spencer A.M. (Ed) *Mesozoic-Cenozoic Orogenic Belts*; Mem. Geol. Soc., London

Campbell C.J., and E.Ormaasen, 1987, *The discovery of oil and gas in Norway: an historical synopsis*; in Spencer A.M. (Ed) Geology of the Norwegian oil and gas fields, Norwegian Petrol. Soc.493p.

Campbell C.J., 1991, *The golden century of oil 1950–2050: the depletion of a resource*; Kluwer Academic Publishers, Dordrecht, Netherlands; 345p.

Campbell C.J., 1992, *The depletion of oil* ; Marine & Petrol. Geol., v.9 Dec. 1992, 666–671

Campbell C.J., 1993, *Assumptions*, AAPG Explorer, Readers Forum, Sept. 1993 [#176]

Campbell C.J., 1993, *The Depletion of the world's oil*; Petrole et technique. No.383. 5–12 Paris

Campbell C.J., 1994, *World oil – Reserves, production, politics and prices*; Proc. 1993 NPF Conference, Stavanger, Norway *Quantification and prediction of hydrocarbon resources*. NPF [#495]

Campbell C.J., 1994, *An Oil Depletion Model: a resource constrained yardstick for production forecasting*; Rept., Petroconsultants S.A., Geneva.

Campbell C.J., 1994, *The imminent end of cheap oil-based energy*; SunWorld 18/4 17–19

Campbell C.J., 1995, *Taking stock*; SunWorld 19/1 16–19

Campbell C.J., 1995, *The Coming Crisis*; SunWorld 19/2 16–19

Campbell C.J., 1995, *The next oil price shock: the world's remaining oil and its depletion*; Energy Expl. & Exploit., 13/1 19–46

Campbell C.J., 1995, *Proving the unprovable*; Petroleum Economist, May 1995

Campbell C.J., 1995, *Cassandra or prophet*; Petroleum Economist Oct. 1995 [#362]

Campbell C.J., 1996, *The resource constraints to oil production : the spectre of a pending chronic supply shortfall*; in Kürsten M. (Ed) *World energy – a changing scene*; Proc.7th Int.Symposium BGR, Hannover, E.Schweizerbart'sche Verlagsbuchhandlung, Stuttgart 227p.

Campbell C.J., 1996, *The status of world oil depletion at the end of 1995*; Energy Exploration and Exploitation; March 1996

Campbell C.J., 1996, *Oil Shock*; Energy World, June 1996.

Campbell C.J., & J.H.Laherrère, 1995, *The world's supply of oil 1930–2050*; Report Petroconsultants S.A., Geneva

Campbell C.J., 1997, *Better understanding urged for rapidly depleting reserves*: Oil & Gas Journ. April 7th 1997. [#615-2]

Campbell C.J., 1997, *The Coming Oil Crisis*; Multi-Science Publishing Co. & Petroconsultants 210p.

Campbell C.J., 1997, *Oil shock – now the production peak*; Energy World 250 p.23 [#616-3]

Campbell C.J., 1997, *How to calculate depletion of hydrocarbon reserves*; Petroleum Economist Sept. 111–112

Campbell C.J., 1997, *As depletion increases, energy demand rises*; Petroleum Economist Sept 113–116

Campbell C.J., 1997, *Depletion survey: Syria and Oman*; Petroleum Economist, Oct.

Campbell C.J., 1997, *Egypt's greater emphasis on gas*; Petroleum Economist Dec. p33.

Campbell C.J., 1997, *Depletion patterns show change due for production of conventional oil*; Oil & Gas Journ., Dec 29 33–37

Campbell C.J.,1998, *Remote areas remain underexplored*; Petroleum Economist Jan.p. 44

Campbell C.J., 1998, *How secure is our oil supply?*; Science Spectra 12 18–24 [#636]

Campbell C.J., and J.H. Laherrère, 1998, *The end of cheap oil*; Scientific American March 80–86

Campbell C.J., 1998, *Running out of gas: this time the wolf is coming*; The National Interest, spring [#661]

Campbell C.J., 1998, *Have all the elephants been found?* Petroleum Review March 24–26 [#700]

Campbell C.J., 1998, *Major exporter nearing peak*; Petroleum Economist April 7–8

Campbell C.J., 1998, *Reserves controversy*; Oil & Gas Journal April 20

Campbell C.J., 1998, *The future of oil*; Energy Exploration & Exploitation 16/ 2–3

Campbell C.J., 1998, *Explorers turning to the deepwater Atlantic margin*; Petroleum Economist July [#794]

Campbell C.J., 1998, *The enigma of oil prices in times of pending oil shortage*; World Oil Prices: oil supply/demand dynamics to the year 2020; Centre for Global Energy Studies, 208–225

Campbell C.J., 1998, *Maybe a third peak for Nigeria*; Petroleum Economist Dec.[#883]

Campbell C.J., 1988, *Oil: a case of short-sighted vision*; Energy Day, Dec 17 1998 [#885]

Campbell C.J., 1998, *L'avenir de l'homme de l'age des hydrocarbures*; Geopolitique 63, Oct

Campbell C.J.,and J.H. Laherrère, 1998 '"La fin du petrole bon marche" Pour la Science Mai p30–36

Campbell C.J., 1998, *How soon will oil production peak ?* Petroleum Review Oct [#890]

Campbell C.J., 1999, *Oil madness*; Geopolitics of Energy; January

Campbell C.J., 1999, *A new mission for geologists as oil production peaks*; Swiss Association of Petroleum Geologists and Engineers, 4/1 19–34

Campbell C.J., 1999, *The dating of reserve revisions*; Petroleum Review. July 44–45

Campbell C.J., 1999, *The extinction of Hydrocarbon Man ?* Times Higher Educat. Supplement, Nov 12

Campbell C.J.,1999, *The imminent peak of oil production;* Presentation at House of Commons, oilcrisis.com [#1101]

Campbell C.J.,1999, *Norway profile: looking for the peak*; Tomorrow's Oil 1/1 May

Campbell C.J.,1999, *Definitions*; Tomorrow's Oil 1/2 August

Campbell C.J.,1999, *Venezuela–Conventional Oil*; Tomorrow's Oil 2/1 December

Campbell C.J.,1999, *Study – Non-conventional oil*; Tomorrow's Oil 1/2 August

Campbell C.J.,2000, *Peru and Nigeria*; Tomorrow's Oil 2/2 February

Campbell C.J.,2000, *World's oil endowment and its depletion* Tomorrow's Oil 2/2 February

Campbell C.J.,2000, *Italy* Tomorrow's Oil 2/3 April

Campbell C.J.,2000, *Vietnam* Tomorrow's Oil 2/3 April

Campbell C.J.,2000, *Myth of spare capacity setting the stage for another oil shock*; Oil & Gas Journ. Mar 20 [#1144]

Campbell C.J.,2000, *Oil depletion in the US Lower 48*: Tomorrow's Oil 2/6

Campbell C.J.,2000, *Champion of national sovereignty in OPEC*; Tomorrow's Oil 2/6

Campbell C.J.,2000, *What do the USGS numbers really mean ?* Tomorrow's Oil 2/7

Campbell C.J.,2000, *Depletion: the Democratic and Popular Republic of Algeria*; Tomorrow's Oil 2/7

Campbell C.J.,2000, *Deep water heroism keeps oil imports at bay*; Tomorrow's Oil 2/8

Campbell C.J.,2000, *The myths of oil*; Petroleum Review, Oct.

Campbell C.J.,2000, *Opec's spare capacity and strategy*; Tomorrow's Oil 2/9.

Campbell C.J.,2000, *Analysis: the challenge of reserve evaluation*; Tomorrow's Oil 2/10 Oct

Campbell C.J., 2000 *Indonesia*; Tomorrow's Oil 2/12 Dec

Campbell C.J., 2000 *Depletion and denial: the final years of oil supplies* USA-Today Nov

Campbell C.J.,2000 *The imminent peak in world oil production*; Middle East Economic Databook 101–10 Campbell C.J, 2000, *A new energy crisis: when will we ever learn?*; The Progress Report [#1185] 9

Campbell C.J., 2000, *Die Ara niedriger Olpreise ist zu Ende*; Gastkommentar [#1259]

Campbell C.J., 2000 *The imminent oil crisis* in CMDC-WSEC Blueprint for the Clean Sustainable Energy Age, Proc. Millennium Conference, Geneva

Campbell, C J., 2000, *Depletion And Denial – The Final Years Of Oil Supply*; ? , [#1375]

Campbell, C J., 2000, *Myth of spare capacity setting the stage for another oil shock*; Oil & Gas Journal, p 20,21, March 20, 2000 [#1544]

Campbell, C J., 2000, *Peak Oil: an Outlook on Crude Oil Depletion*; MBendi Information for Africa, Website, February, 2002 [#1773]

Campbell C.J, 2001, *Peak Oil: A turning for mankind*; Hubbert Center Newsletter 2001/2–1 [#1183]

Campbell C J. and Zagar J, 2001, *Oilfields – Maintenance expenses*; Hubbert Center Newsletter 2001/2–2 [1184]

Campbell C.J., 2001 *The four lives of Trinidad*; Tomorrow's Oil 3/1 Jan

Campbell C.J., 2001 *The imminent oil crisis – the missing element in the environment debate*; Tomorrow's Oil 3/2 Feb

Campbell C.J., 2001 *Peak Oil : a turning point for Mankind*; Hubbert Center Newsletter 2001/2–1

Campbell C.J., 2001 *Petroleo: o pico que se aproxima*; www://amerlis.pt

Campbell C.J., 2001 *The Oil Peak Oil : turning point*; Solar Today, 15/4 July–Aug

Campbell C.J., 2001 *Peak Oil : a turning point for Mankind*; Climate Change & Energy Conf. Evora, Portugal proc.

Campbell C.J., 2001 *Oil, Gas and Make-believe*; Energy Exploration & Exploitation 19/2–3 117–133

Campbell, C J.& Zagar J J., 2001, *Peak Oil: A Turning For Mankind / Oilfields – Maintenance Expenses*; Hubbert Center Newsletter # 2001/ 2–1 and 2–2 [#1289]

Campbell, C J., 2001, *The Oil Peak: A Turning Point*; Solar Today, p 40–43, July/August 2001 [#1299]

Campbell, C J., 2001, *"Energy Resources – cornucopia or empty barrel ?": Discussion*; March 5, 2001 [#1634]

Campbell, C J., 2002, *The Fuel that Fires Political Hotspots:* The Times Higher Education Supplement, May17th 2002 [#1864]

Campbell, C J., 2002, *Forecasting global oil supply 2000–2050:* M.King Hubbert Center for Petroleum Supply Studies 2002/3

Campbell, C J., 2002, *Facing up to the future:* Norwegian Petroleum Diary 2/2002 [#1866]

Campbell, C J., 2002, *Bell signals death knell for oil:* Times Higher Educational Supplement [#1865]

Campbell C.J., 2002, *The assessment and importance of oil depletion;* ASPO Workshop, Uppsala [#1836]

Campbell, C J., 2002, *Ekonom nonchalerar klotets begransningar: Svenska Dagbladet 20 August*

Carlisle, T., 2001, *It`s Gooey, Unruly, Mythic – and Priceless*; The Wall Street Journal,, August 9, 2001 [#1352]

Carlton, J., 2002, *Drilling Is Threat To Alaska Herd, U.S. Report Says*; The Wall Street Journal, March 29, 2002 [#1803]

Carmalt S.W. and B.St.John, 1986, *Giant oil and gas fields;* in Future petroleum provinces of the world; Amer. Assoc.Petrol. Geol. Mem.40 [#43]

Carpenter D.,2000, *Oil price rockets higher; America barely flinches;* The Register Guard 22 Jan [#1070]

Cavalo A.J., 2002, *Predicting the peak in World oil production:* Natural Resources Research 11/3 [#1939]

Cazier E.C., et al., 1995, *Petroleum geology of the Cusiana Field, Llanos Basin foothills, Colombia*; Amer. Assoc. Petrol. Geol. 79/10 1444–1463 [#403]

Centre for Global Energy Studies, 1992, *The Gulf crisis* 228pp

Centre for Global Energy Studies, 1992, *The long-run price of oil*: CGES study v.3 [#194]

Centre for Global Energy Studies, 1992, *The costs of future North Sea oil production*; CGES Rept. v3[#205]

Centre for Global Energy Studies,, 1993, *Oil production capacity in the Gulf Vol II Saudi Arabia*; Centre for Global Energy Studies [#284]

Centre for Global Energy Studies, 1994 *Monthly oil report;* 3/5 [#240]

Centre for Global Energy Studies, 1994, *Oil production capacity in the Gulf, Vol.II t*

Centre for Global Energy Studies,, 1998, *World oil prices*; 316p

CGES, 2001, *The future of oil: meeting the growth in world demand*; Global Oil Report, Volume 12, Issue 1 [#1190]

CGES, 2001, *Non-OPEC Production: The Going Is Getting Tougher*; Global Oil Report, Vol.12, Issue 6, November / December, 2001 [#1755]

Chazan G., 2002, *Caspian summit fails to resolve energy claims*; Wall St Journ 25/Apr [#1826]

Chevron Corp., 1990, *Overview, energy outlook to the year 2000*; [#90]

Chevron Corp., 1993, *Saudi memories*; Chevron World winter/spring 1993 p.14–18

Chevron Corp., 19??, *Whatever happened to Standard Oil ?* The First Hundred Years of Chevron [#511]

Chevron Corp., 1996 *The great Arabian discovery*; http://chevron. Com/chevron root/arab 50 [#523]

Chicago Tribune, 1999, *US hails deal for pipeline in Caspian region*, Nov 28 [#1105]

Clack R.W., 2000, *The world needs more oil*; Oil & Gas Journ Feb 21 [#1088]

Clark P, P. Coene and D. Logan, 1981, *A comparison of ten U.S. oil and gas supply models*; Federal Reserves Institute, Washington [#103]

Cleveland C.J., 1992, *Yield per effort for additions to crude oil reserves in the Lower 48 United States 1946–1989*; Amer. Assoc. Petrol. Geol. 76/7 948–958

Cleveland C.J. & R.K.Kaufmann, *Forecasting ultimate oil recovery and its rate of production: incorporating economic forces into the models of M. King Hubbert*; The Energy Journal 12/2 [#26]

Clifford M.L., 2000, *Can this giant fly ?* ; Business Week, Feb 7th [#1072]

Clifford, D., 2000, *The Oil Age*; First published S.A. Humanist Post, June 2000, www.users.on.net , March, 2002 [#1786]

Clifford, M L., 2001, *An Oil Giant Stirs (China)*; Business Week, May 7, 2001 [#1660]

Clover, C., 2001, *The Balance Of Power*; The Daily Telegraph, February 10, 2001 [#1624]

Cocks, D., 2001, *How to energise the clever county*; The Australian Financial Review, Friday, July 20, 2001 [#1702]

Cocks, D., 2001, *There will be life – but not as we know it*; The Australian Financial Review, Friday, July 27, 2001 [#1703]

Cocks, D., 2001, *Coming to terms with complexity*; The Australia Financial Review, Friday, August 3, 2001 [#1704]

Cohn, L. & Crock, S., 2001, *What To Do About Oil*; Business Week, October 29, 2001 [#1693]

Coleman J.L., 1995, *The American whale oil industry: a look back to the future of the American petroleum industry;* Non-renewable Resources 4/3 273–288

Commonwealth of Australia, 1996, *Sustainable energy policy for Australia.* [#538]

Cook, L J., 2001, *Safe For Now*; Forbes, November 26, 2001 [#1739]

Cook W.J., 1993, *Why Opec doesn't matter anymore*; US News & World Report 13/12/93 [#299]

Cope G., 1998, *Have all the elephants been found?* Pet. Review 52/614 24–26

Cope G., 1998, *Have all the elephants been found?* Part 2 Pet. Review 52/615 33–35

Cope G., 1998, *Will improved oil recovery avert an oil crisis*; Pet. Review June 52/616 22–23

Cooper, C., 2001, *A Wheezing Iraqi Oil Refinery Tells All*; The Wall Street Journal, May 14, 2001 [#1622]

Cooper, C. & Herrick, T., 2001, *Oil Giants Struggle to Spend Profits Amid Shortage of Exploration Sites*; The Wall Street Journal Online, July 30, 2001 [#1313]

Cooper, C. & Herrick, T., 2001, *Pumping Money – Major Oil Companies Struggle to Spend Huge Hoards of Cash*; The Wall Street Journal, July 30, 2001 [#1678]

Cooper M., 1998, *Oil production in the 21st Century*; CQ Researcher 8/29 [#788]

Cornelius C. D., 1987, *Classification of natural bitumen: a physical and chemical approach*; in Meyer R.F. (Ed) *Exploration for heavy crude oil and bitumen*; AAPG Studies in geology #25

Corzine R., 1993, *Warning over Saudi output*; Financial Times, 4/11/93 [#282]

Corzine R., 1999, *Saudi diplomats lead first-aid effort*; Financial Times, April 15 [#970]

Coward, M P. & Purdy, E G. & Ries, A C. & Smith, D G., 1999, *The distribution of petroleum reserves in basins of the South Atlantic margins*; In: The Oil and Gas Habitats of the South Atlantic, Cameron et al, Geological Society, London, Special Publications, 153, p 101–131, 1999 [#1524]

CPN., 1994, *A forecast of China's petroleum resources*; China Pet. Newsletter 1/9 April 27 1994 [#253]

Crandell J.D. *et al.*, 1993, *'93 E&P spending: shifting back to the U.S.*, World Oil Feb. 1993 [#134]

Crandell J.D. *et al*, 1994, *Depends on oil prices*; World Oil, Feb 1994[#200]

Creswell J., 1996, *Global demand fuels need for more drilling*; Press & Journal Aberdeen [#404]

Creswell J., 1999, *How big oil is slashing*; Press & Journal Aberdeen [#920]

Crilley & Co., 1969, *Stock Market Action Report: Oil In The Alaskan Arctic – The Prudhoe Bay Discovery*; August, 1969 [#1770]

Crock, S., 2001, *Rogue States: Why Washington May Ease Sanctions*; Business Week, May 7 2001 [#1661]

Crow P., 1991, *Oil and bullets*; Oil & Gas Journ., Oct 21. 1991[#183]

Curnow, J., 2000, *Australia`s Carrying Capacity*; Sustainable Population Australia Inc., April 26, 2000 [#1636]

Curran R., 1998 *Uncertainty clouds fundamental strength of Canada's industry*; World oil [#672]

D

Dallas Morning News, 1993, *Arco says Alaskan fields could increase reserves 60%*; Dallas Morning News April 14 1993 [#185]

Daly M.C., M.S. Bell & P.J. Smith, 1996, *The remaining resource of the UK North Sea and its future development*; in Glennie K & A Hurst NW Europe Hydrocarbon Industry; Geol. Soc. [#912]

Davies P., 1994, *Oil supply and demand in the 1990s*; World Petrol. Congr. Topic 16 [#235]

Davidson J.D. & W. Rees-Mogg, 1988, *Blood in the streets – investment profits in a world gone mad*; Sidgwick & Jackson, London 385p

Davidson J.D. & W. Rees-Mogg, 1994, *The great reckoning*; Pan Books 602pp

Davidson J.K., 1995, *Globally synchronous compressional pulses in extensional basins: implications for hydrocarbon exploration*: APEA Journ. 1995 169–88 [#369]

Davidson K., 1998, *Some experts say oil demand will soon exceed supply again*; Sta Barbara News 2 Sept. [#839]

Davidson K., 1998, *Petrol Pessimists*; Rocky Mountain News 13 Sept [#841]

Davis, G., 1999, *Foreseeing a refracted future*; ???, April, 1999 [#1696]

Deffeyes K.S., 2001, *Hubbert's peak – the impending world oil shortage*; Princeton University Press 208pp

Derakhshan M., 1999, *Iran's oil and gas: background and policy challenges*; International Bureau for Energy Studies.

Deitzman W.D. *et al.* 1983, *The petroleum resources of the Middle East*; U.S.Dept. of Energy / Energy Information Administration Report. DOE/EIA-0395 May 19 83 169p.

Dell P.R., 1994, *Global energy market : future supply potentials*; Energy Exploration and Exploitation 12/1 59–72 [#377]

Demaison G. and B.J. Huizinga, 1991, *Genetic classification of petroleum systems*; Amer.Assoc. Petrol. Geol.. 75 1626–43

Demaison G. and A. Perrodon, 1994, *Petroleum systems and exploration strategy*; AAPG Course, Mexico. Oct. 1994 [#258]

Department of Energy, , 1995, *US crude oil, natural gas and natural gas liquids reserves*; 1994 Annual Rept. [#429]

DeSorcy G.J., *et al.*, 1993, *Definitions and guidelines for classification of oil and gas reserves*; Journ. Canadian Petrol. Technology 32/5, 0–21 [#338]

Deutsche Bank, 2000, *Prescription for the OPEC Meeting: A Dose of Prozac*; Deutsche Bank Global Oil & Gas Research, Energy Wire, June 15, 2000 [#1498]

Deutsche Bank, 2001, *Oil Team Conference Call – Energy Rebound in Sight: Looking through the near-term murk*; Deutsche Bank Alex. Brown, , [#1750]

Dieoff.com, *Economics error: Illusion that money can replace physically exhausted resources*; dieoff.com, p 185, [#1342]

Dismal Scientist, 2000, *Oil and Gas Inventories, Analysis and Business Implications*; Released April 26, 2000 [#1574]

Djurasek, S., 1998, *The coming oil crisis*; NAFTA 49/6 167–168 [#884]

Dorsey, J M., 2001, *Saudi Leaders Warns U.S. of "Separate Interests"*; The Wall Street Journal, Monday, October 29, 2001 [#1708]

Dowthwaite R., 1999, *The ecology of money*; Green books 78pp

Douthwaite, R., 2001, *Essay: When should we have stopped?*; The Irish Times, Saturday, December 29, 2001 [#1782]

Dromgoole P. & R. Speers, 1997, *Geoscore: a method for quantifying uncertainty in field reserve estimates*; Petroleum Geoscientist 3 1–12.

Duncan R.C., 1995, *The energy depletion arch: new constraints enhance the Hubbert model*: Public Interest Environmental Law Conf., Univ. of Oregon [#400]

Duncan R.C. and W. Youngquist, 1998, *The world petroleum life-cycle*; PTTC Workshop, Petroleum Engineering Program, Univ. of S. California 22 Oct 1998 [#875]

Duncan R.C. and W. Youngquist, 1998 *Is oil running out?* Science 282 [#882]

Duncan R.C., 1999, *Oil production per capita*; Oil & Gas Journal 92/20 May 17 [#1009]

Duncan R.C, 2000, *The peak of world oil production and the road to the Olduvai gorge*; Pardee Keynote Symposia Geological Society of America [#1256]

Duncan, R C., 2001, *World Energy Production, Population Growth, and the Road to the Olduvai Gorge*; Population and Environment, Vol. 22, No. 5, May 2001, [#1307]

Duncan, R C., 2000, *Crude Oil Production And Prices: A Look Ahead At OPEC Decision Making Process*; Presented at the PTTC Workshop, September 22, 2000 [#1405]

Duncan, R C. & Youngquist, W., 1999, *Encircling the Peak of World Oil Production*; Natural Resources Research, Vol. 8, No.3, p 219–232, 1999 [#1462]

E

Easterbrook, G., 2000, *Hooray for expensive oil! – Opportunity Cost*; The New Republic, p 21–25, May 15, 2000 [#1548]

Eberhardt W., 2000, *Am Tropf der Scheichs*; Focus 200/9 [#1134]

Economist The, 1993, *A shocking speculation about the price of oil*; The Economist, Sept.15 1993. 87–88. [#189]

Economist The, 1994, *Power to the people – a survey of energy*; June 18th 1994 [#242]

Economist The, 1995, *The future of energy*; Oct.7 1995[#379]

Economist The, 1996, *From major to minor*; May 18th [#437]

Economist The, 1996, *Pipe dreams in central Asia*; May 4th [#472]

Economist The, 1966, *Kurdistan, which one do you means?* Aug.10th

Economist The, 1996, *The oil buccaneer*; Nov. [#521]

Economist The, 1997, *Connections needed*; 15.3.97 [#563]

Economist The, 1998, *Colombia on the brink*; 8/8/98 [#815]

Economist The, 2000, *In praise of Big Oil*; The Economist, October 21, 2000 [#1425]

Economist The, 1994, *Russia's Caucasian cauldron*; The Economist, August 6, 1994, [#1457]

Economist The, 2001, *Oil Depletion – Sunset for the oil business?*; The Economist, p 97/98, November 3, 2001 [#1756]

Edwards Economic Research, 1999, *Into the Millennium*; HSBC Middle East Economic Bulletin, [#987]

Edwards J.D., 1997, *Crude oil and alternate energy production forecasts for the twenty-first century: the end of the hydrocarbon era*; Amer. Assoc. Petrol .Geol. 81/8 1292–1305 [#631]

Edwards J.D., 1998, *"Optimist" predicts world oil demand will outstrip production in 2020*; Univ of Colorado [#937]

Edwards R.H., 1999, *The imminent peak in world oil production;* Gulf Business Economic Databook [#1148]

Edwards R.H., 2002, *The prospects for sustainable development in the Middle East.* HSBC M.East Economic Bulletin 2 qtr [#1820]

Edwards R., 1998, *High level risk*; New Scientist [#768]

Edwards W.R., 1997, *Energy supply*; Oil & Gas Journ. July 9 [#629]

Ehrenfeld D., 1999, *The coming collapse of the age of technology;* Tikkun 14/1 [#966]

Ehrenfeld T., 2002, *Iraq: It's the oil, stupid – what if this is the beginning of an oil war?* Newsweek Sept 30 [#1935]

Ehrlich P.R. & A.H. Ehrlich, 1990, *The population explosion*; Simon & Schuster

EIA, 2000, *Long Term World Oil Supply (A Resource Base / Production Path Analysis)*; ???, [#1768]

Einhorn C.S., 1994, *Well oiled*; Barron's June 27 [#244]

Ellis-Jones P., 1988, *Oil: a practical guide to the economics of world petroleum*; Woodhead-Faulkner 347 pp

Emerson T., 2002 *The thirst for oil;* Newsweek, April 8 [#1831]

Energy Economist, 1993, *The 1996 oil shock?*; Energy Economists, May 1993, 139/17 [#339]

Energy Economist, 1993, *Saudi Arabian sands*; Energy Econ. 145/14 Nov. 1993 [#281]

Energy Information Administration, 1985, *International energy outlook 1985*; DOE/EIA 0484(85) [#18]

Energy Information Administration, 1988, *Annual energy review*; DOE/EIA 0384 (88) [#19]

Energy Information Administration, 1990, *U.S. oil and gas reserves by year of field discovery*; DOE/EIA 0534 [#611-2]

Energy Information Administration, 1993, *International oil and gas exploration and development 1991*; U.S.Department of Energy, Washington

Energy Information Administration, 2000, *First quarter 2000 oil production and capacity estimates*; [#1147]

Energy Exploration & Exploitation, 1998, *The world's non-conventional oil and gas resources – review* [#793]

Energy Information Administration, 1986, *International energy outlook 1986*; DOE/EIA 0484(96) [#609-2].

Energy Information Administration, 1999, *Projections of oil production capacity*; [#951]

Energy Information Administration, 2001, *World Oil Market and Oil Price Chronologies 1970–2000 (3 pages of 35)* [#1193]

Esau I., 1992, *Azeri initiative*; Offshore Eng. Aug. 1992 [#116]

Esser R., 2000, *Discoveries of the 1990s–were they significant?*; CERA [#1151]

Eureka, 1991, *Total – Colombian discovery*; Eureka July 1991 [#4]

European Commission, 1995, *For a European Union Energy Policy*; Green Paper [#605-1]

European Commission The, 1999, *European Union Energy Outlook To 2020*; Special Issue, November 1999 [#1511]

European Commission, 2000, *The European Union's oil supply*; Oct. 4th Report

ExxonMobil, 1999, *Tomorrow's energy needs*; Wall Street Journal [#1196]

ExxonMobil, 2001, *Exxon Mobil Contribution To The Debate On The Green Paper – Towards a European Strategy for the Security of Energy Supply*; Report, September 28, 2001 [#1774]

F

Feldstein, M., 2001, *Vouchers Can Free Us From Foreign Oil*; The Wall Street Journal, December 27, 2001 [#1752]

Ferguson A.R.B, 2001, *Planning for the demise of cheap energy*; Optimum Population Trust [#1209]

Ferguson, A R B., 2001, *Renewable Liquid Sunshine:its scope and limits*; The Optimum Population Trust (U.K.), 30 May, 2001 [#1301]

Ferguson, A R B., 2001, *Depletion of Oil and Gas – globally and in the US*; The Optimum Population Trust (U.K.), 8 July, 2001 [#1339]

Ferguson, A R B., 1999, *Judging the Oil Experts*; Optimum Population Trust (U.K.), 1 December, 1999 [#1468]

Ferguson, A R B., 2001, *Natural gas resources and US population growth*; Optimum Population Trust (U.K.), 17 June, 2001 [#1663]

Ferguson, A R B., 2001, *The US Population Explosion*; Optimum Population Trust (U.K.), 29 July, 2001 [#1664]

Ferrier R.W., 1982, *The history of the British Petroleum Company: Volume 1, the developing years 1901–1932*; Cambridge University Press 801p [#B7]

Ferriter J.P., 1994, *International energy cooperation*; Petrole et Technique 389 [#264]

Fesharaki, F. & Varzi, M., 2000, *Investment Opportunities Starting To Open Up In Iran's Petroleum Sector*; Oil & Gas Journal, February 14, 2000 [#1474]

Financial Times, 1998, *Realignment in the Gulf*; 26.2.98 [#684]

Financial Times, *Expecting the Improbable – Middle East capacity and global demand to 2020*; Energy

Economist Briefings [#1186]

Flavin C., 2000, *Energy for a new century*; Worldwatch Mar/Apr [#1129]

Fleay B., 1995, *The decline in the age of oil*; Pluto Press 152p.

Fleay B., & J.H. Laherrère, 1997, *Sustainable energy policy for Australia; submission to the Department of Primary Industry and Energy Green Paper 1996*; Paper 1/97 Institute for Science and Technology Policy, Murdoch University, W.Australia [#606-1]

Fleay B, 1998, *Chemical dependency grows*; West Australian 26/6/98 [#814]

Fleay B, 1998, *Its time to refuel debate on oil* ; West Australian 18/4/98 [#815]

Fleay B, 1998, *The Greens WA – Energy Platform*; [#835]

Fleay B., 1998, *Climaxing oil: how will transport adapt*; Chartered Inst., Transport Symposium Tasmania 6–7 Nov., [#868]

Fleay B., 1999, *Beyond oil*; Chartered Inst of Transport in Australia conf. [#977]

Fleming D., 1999, *The spectre of OPEC*; Sunday Telegraph March 21 [#953]

Fleming, D., 1999, *The Next Oil Shock?*; Prospect, p 12 – 15, April, 1999 [#1549]

Flores G., 1987, *Arc of the sun*; Lion & Thorne, Tulsa. 229p [#B4]

Flower A.R., 1978, *World oil production*; Scientific American 238/3 [#720]

Forbes, 1992, *The next oil crisis?* Forbes Feb 17 1992 [#159]

Forbes S., 1999, *Unending oil?* Forbes 14 June [#993]

Foreman N.E, 1996, *Opec influence grows with world output in next decade*; World Oil Feb 1996 [#412]

Foreman N.E, 1997, *The bear awakens: resurgence of FSU oil and gas*; World Oil Feb [#557]

Fowler, R M. & Burgess, C J. & Otto, S C. & Harris, J P. & Bastow, M A., 2000, *World Conventional Hydrocarbon Resources: How Much Remains To Be Discovered, Where Is It?*; Robertson Research International Limited, 2000, [#1384]

Franco A., 1997, *Latin America activity soars to new heights*; Hart's Int. Petrol. Eng., July [#625]

Friedman A., 1990, *Opec needs $60 bn extra for productivity increase*; Financial Times Feb 8th 1990 [#76]

Friedman T.L., 1994, *Opec's lonely at the tap, but China's getting thirsty*; New York Times Apr 3 1994 [#222]

Fromkin D., 1989, *A peace to end all peace*; Avon Books 635p ISBN 0-380-71300-4.

Fuerbringer, J., 2000, *Market Place – With oil prices increasing and the election season at hand, the cry goes up for OPEC to raise output*; The New York Times, August 16, 2000 [#1396]

Fuller J.G.C., 1971, *The geological attitude*; Amer. Assoc. Petrol. Geol 55/11 [#141]

Fuller J.G.C., 1993, *The oil industry today*; Brit. Assoc. Advancement of Sci. Published as *The British Association Lectures 1993* by The Geological Society, London .[#53]

Futurist, 1997, *Permanent oil shortage impossible*; May–June [#583]

Futurist, 1997, *The world is not running out of oil*; May–June [#1812]

G

Garb F.A., 1985, *Oil and gas reserve classification, estimation and evaluation*; Journ. Petrol. Technol. March 373–390

Gardner T., 2000, *US ups estimate of non-US recoverable oil by 20 pc*; Reuters [#1132]

Gauthier D.L. et al. 1995, *1995 national assessment of United States oil and gas resources – results, methodology and supporting data*: U.S. Geological Survey CD-ROM

George D., 1993 *Caspian Sea reserves could rival those of Saudi Arabia*; Offshore/Oilman July 1993 [#172]

George R., 1997, *Canada's oil sands: the unconventional alternatives*; Hart's Pet. Eng. Int., 29/5/97 [#597]

Georgescu-Roegen, N., 1995, *La decroissance*; Sang de Terre, Paris 254pp

Geoscientist, 1996, *World oil supplies*, 6/1 [#440]

Gever J., R. Kaufmann & D. Skole, 1997, *Beyond Oil*; http://dieoff.org.[#644]

Ghadhban T.A. et al., 1995, *Iraq oil industry: present conditions and future prospects*; Report Iraq Oil Ministry [#344]

Giampietro M., S. Ulgiati and D. Pimental., 1997, *Feasibility of large-scale biofuel production*; Bioscience 47-9 587–598 [#704]

Giddens P.H., 1955, *Standard Oil Company (Indiana) – oil pioneer in the Middle West*; Appleton-Century-Crofts, New York

Glain, S J., 1999, *Iraq Says It Can Far Exceed Pre-Embargo Oil Output*; The Wall Street Journal, p 20, November 29, 1999 [#1641]

Gold D.H, 2000, *Oil price surge brings fear of crisis, but also hope for more crude output*; Investor's Business Daily [#1219]

Gold T., 1988, *Origin of petroleum; two opposing theories and a test in Sweden*; Geojournal Library 9 85–92.

Goldberg J., 1998, *Getting crude in Baku*; New York Times 4/Oct. [#851]

Goldman, I., 2000, *Natural Gas: The Five Stages to Market Panic*; SolarQuest Net NewsService, August 10, 2000 [#1377]

Goldman Sachs, 1998, *Reserve estimates*; report [#669]

Goldman Sachs, 1998, *Canadian data* [# 775]

Goldman Sachs, 1998, *Oil data* [# 787]

Goldsmith J, 1994, *The trap*; Macmillan 216p [#B17]

Goodman M. & N.C. Chriss, 1979, *Mexico oil estimates inflated, experts say*; Los Angeles Times |May 18 1979 [#14]

Goodwin S., 1980, *Hubbert's Curve*; Country Journal, Nov 1980 [#164]

Gore A., 1998, *Finding a third way*; *Newsweek*, 23 Nov. [#858]

Grace J.D., R.H. Caldwell and D.I. Heather, 1993, *Comparative reserve definitions: U.S.A., Europe and the former Soviet Union*; Journ. Petrol. Technol. Sept.1993 866–872 [#174]

Graham H., 1990., *US oil plot fuelled Saddam*; Observer. 21 Oct 1990 [#721]

Graham H., 1991., *The wolf bites back*; South [#736]

Gray D., 1989, *North Sea outlook and its sensitivity to price*; Petrol. Revue. Jan 1989 [#35]

Greenwald J., 1994 *Black Gold*, Time June 20 [#239]

Grob G., 1999, *New total approach to energy statistics and forecasting*; CMDC-WSEC [#1043][1819]

Grollman N.G., 1997, *Environmentally sustainable energy for the East Asia/Pacific Region*; APPEA Journ 722 [#752]

Grollman N.G., 1998, *Pipelines, politics and prosperity*; APPEA Journ. 815 [#t53]

Gruy H.J., 1998, *Natural gas hydrates and the mystery of the Bermuda triangle*; Hart's Petrol. Eng. Int. March [#717]

Gulbenkian N., 1965, *Portrait in oil*; Simon & Schuster, New York [#B9]

Gunther F., 2000, *Vulnerability in agriculture: energy use, structure and energy futures*: INES Conf., Stockholm

Gurney J., 1997, *Migration or replenishment in the Gulf*; Petrol. Review May [#593]

H

Hage, G., 2001, *The incredible shrinking society*; The Australian Financial Review, Friday, September 7, 2001 [#1705]

Haines L., 1994, *Oil price recovery : for real or a dream*; NewsWell[#272]

Halbouty M.T., 1970, *Geology of giant petroleum fields*; Amer. Assoc. Petrol. Geol. Mem.14

Haldorsen. H.H., 1996, *Choosing between rocks, hard places and a lot more: the economic interface*; in Doré A.G and R. Sinding Larsen (Eds), *Quantification and prediction of hydrocarbon resources*; NPF Sp. pub. 6., Elsevier ISBN 0-444-82496-0.

Hall, C A S. & Cleveland, C J. & Kaufman, R., 1992, *Oil And Gas Availability: A History of Federal Government Overestimation*; In: Energy And Resource Quality, by all et al, p 343 – 349, 1992, dieoff.com, [#1552]

Hammer A., 1988, *Hammer, witness to history*; Coronet Books 752pp

Hanson J., 1997, *Fossilgate – the biggest coverup in history;* http://dieoff.org [#645]

Hanson, J., 2000, *[energyresources] draft newsletter*; , April 28, 2000 [#1563]

Hanson, J., 2000, *Energy Synopsis*; dieoff.com/synopsis.htm , December 30, 2000 [#1579]

Hardman, R.F.P., 1998, *The future of Britain's oil and gas industry*; Inst.Mining Eng. March [#711]

Hardman, R.F.P., 2001, *New Petroleum Provinces of the 21st Century?*; 20 July, 2001 [#1341]

Harigal G.G., 1998, *Energy in a changing world*; Pugwash Meeting 243 [#946]

Harper, F G., 1999, *Ultimate Hydrocarbon Resources in the 21ˢᵗ Century*; BP Amoco, AAPG Birmingham 1999 [#1399]

Harris R., 1994, *Production Analysis shows growth in worldwide oil and gas reserves*: World Watch, Petroconsultants [#308]

Harts Petroleum Engineer International, 1996, *New paradigm: mining for oil*; Sept. [#514]

Hart's, *OPEC's ability to manage oil prices shaky, but group's long-term prospects still bright*; OPEC and oil prices Hart's 12/10/00 [#1257]

Hatfield C.B., 1995, *Will an oil shortage return soon?* Geotimes Nov. 1995 [#406]

Hatfield C.B., 1997, *Oil back on the global agenda*; Nature 367 121 [#589]

Hatfield C.B., 1997, *A permanent decline in oil production*; Nature 388 [#632]

Hatley A.G., 1995, *The oil finders*; Centex 267pp

Haun J.D. (Ed), 1975, *Methods of estimating the volume of undiscovered oil and gas resources*; Amer. Assoc. Petrol. Geol. Studies in Geology No.1.

Hawken P., 1993, *The Ecology of Commerce – a declaration of sustainability*; Publ. Harper Business, New York 250p.

Hawkes N., 1999, *Huge reserve of gas will fuel 21ˢᵗ Century*, The Times [#897]

Hebert H.J., 2000, *Clinton preaches patience on oil*; Assoc. Press. [#1141]

Henry J.C., 1996, *Managing oil and gas companies*; Oil & Gas Journ. Nov 4. [#524]

Herald Sun, 1997, *Power Cut*; 31ˢᵗ. May [#618-3]

Herkstroter C., 1997, *Contributing to a sustainable future — the Royal Dutch/ Shell Group in the global economy;* Shell paper [#588]

Herrick, T. & Bahree, B., 2001, *OPEC Won`t Curb Oil Until Others Do*; The Wall Street Journal, November 15, 2001 [#1730]

Herrick, T 2002, *Oil prices get a lift from decline in inventories*; Wall St Journ 2 Apr [#1827]

Hersh, S M., 2001, *King`s Ransom – How vulnerable are the Saudi royals?*; The New Yorker, , October 24, 2001 [#1682]

Herwijnen T. van & M. Groeneveld, 1999, *Global energy perspectives for power generation*; 3 IERE Conf. Kobe, proc.[971]

Hewins R., 19??, *Mr Five Per Cent: the biograph of Calouste Gulbenkian*; Hutchinson

Higgins, A. & Hutzler, C., 2001, *China Pursues a Great Game of Its Own*; The Wall Street Journal, June 14, 2001 [#1398]

Higgins G.E., 1996, *A history of Trinidad Oil*; Trinidad Express Newspapers, 498pp

Hill P., 2000, *Gulf states will likely agree to increase oil production*; Washington Times [#1081]

Hiller K., 1996, *Depletion midpoint and the consequences for oil supplies*; pre-print WPC [#433]

Hiller K. 1997, *Future world oil supplies – possibilities and constraints*; Erdol Erdgas Kohle 113/9 349–352 [#572]

Hiller K., 1997, *Kohnwasserstoff-projekte der BGR von 1970 bis 1995*; Z. angew. Geol. 43/1 [#573]

Hiller K., 1998, *Depletion midpoint and the consequences for oil supplies*; WPC 15. [#780]

Hiller K., 1999, *Verfugbarkeit von Erdol*; Erdol, Erdgas, Kohle Jahrgand *115/2 [#918]*

Hoagland J., 1994, *The US-Saudi line is off the hook*; Int. Herald Tribune 31.1.94 [#206]

Hobbs G.W., 1995, *Oil, gas, coal, uranium, tar sand resource trends on rise*; O&GJ Sept 4 1995 [#374]

Hobson G.D., 1980, *Musing on migration*; Journ. Petrol. Geol. 3/2 [#311]

Hobson G.D., 1991, *Field size distribution – an exercise in doodling*? Journ. Petrol Geol 14/1 [#312]

Hogan L., et al., *Australia's oil and gas resources*; ABARE report [# 534]

Hogarty T.F., 1999, *Gasoline: still powering cars in 2050*; The Futurist March [#1000]

Holley D., 1997, *China's thirst for oil fuels competition*; Los Angeles Times July 29 [#635]

Hollis R., 1996, *Stability in the Middle East – three scenarios*; Pet. Review May p 205 [# 442]

Horton S. & N. Mamedov, 1996, *Investment in Azerbaijan's upstream requires attention to legal details*, World Oil April 1996 [#441]

Hotellet R., 1998, *Tangled web of an oil pipeline*; Christian Science Monitor May 1st [#740]

Hotelling H., 1931, *The Economics of exhaustible Resources*; Journ. Politic. Economy 1931 [#190]

Hovey, H H., 2001, *Matt Simmons: Energy Crisis is "Extremely Serious"*; Dow Jones News, April 6 2001 [#1287]

Howard K., 2000, *Running on Empty*; Autocar 23. Feb.

Hubback, A., 2000, *10 years on, Iraq and West still poles apart*; Hart`s E&P, August 2000 [#1368]

Huber P & M.Mills, 1998, *King Fahd and the tide of technology*; Forbes Nov 16 [#942]

Hubbert M.K., 1949, *Energy from fossil fuels*; Science 109, 103–109

Hubbert M.K., 1956, *Nuclear energy and the fossil fuels*; Amer. Petrol. Inst. Drilling & Production Practice. Proc. Spring Meeting, San Antonio, Texas. 7–25.[#187]

Hubbert M.K., 1962, *Energy resources, a report to the Committee on Natural Resources*; Nat. Acad. Sci. Publ. 1000D

Hubbert M.K., 1969, *Energy resources*; in Cloud P. (Ed) *Resources and Man*; W.H.Freeman

Hubbert M.K., 1971, *Energy resources of the Earth*; in *Energy and Power*, W.H.Freeman

Hubbert M.K., 1976, *Exponential growth as a transient phenomenon in human history*; in Strom Ed., Scientific Viewpoints, American Inst. of Physics 1976 [#952]

Hubbert M.K., 1980, *Oil and gas supply modeling*; in Gass S.I., ed. proceedings of symposium, U.S. Dept. of Commerce June 18–20, 1980 [#492]

Hubbert M.K., 1981, *The world's evolving energy system*; Amer. J. of Physics 49/11 1007–1029

Hubbert M.K., 1982, *Technique of prediction as applied to the production of oil & gas*; in NBS Special Publication 631. U.S. Dept.Commerce/ National Bureau of Standards, 16–141.

Hubbert M.K., 199? *Senate committee hearings of efforts to suppress study*; [#849]

Hubbert M.K., 1998 *Exponential growth as a transient phenomenon in human history*; Focus [#902]

Huber P & M. Mills, 1998, *King Faisal and the tide of technology*; Forbes Nov.16 [#878]

Hupe R., 2000, *Entzug der Öldroge*; Die Woche, 22.12.00

I

IFP, 1996, *Panorama*, Inst. Francais du Petrole

IHS Energy Group, 2000, *1999 A Star Year For Oil And Gas Discoveries...*; Press Release, June 13, 2000 [#1514]

Imbert P., J.L. Pittion and A.K. Yeates, 1996, *Heavier hydrocarbons, cooler environment found in deepwater*; Offshore April [#446]

Independent Petroleum Association of America, 1993, *The promise of Oil and Gas in America*; final report of IPAA Potential Resources Task Force [#225]

Ibrahim Y.M., 1990, *Widespread unrest threatens world oil supply, experts say*; Int. Herald Tribune 6/3/90 [#73]

Institute of Petroleum, 1993, *Valuable Saudi upstream data published*; Pet. Review Dec 1993 [#283]

Institute of Petroleum, 1995, *The UK continental shelf in 2010: is this the shape of the future?* Report [#481-1]

International Center for Technology Assessment, 2000, *The real price of gas* [#1136]

International Energy Agency, 1993, *Oil Market Report*, 7 Sept. 1993 [#175]

International Energy Agency, 1994, *World Energy Outlook* 1994 Edition [#241]

International Energy Agency, 1995, *World Energy Outlook* 1995 Edition [#357]

International Energy Agency, 1998, *World Energy Prospects to 2020*; Report to G8 Energy Ministers, March 31 (www.iea.org/g8/world/oilsup.htm)

International Energy Agency, 1998, *World Energy Outlook* 1998 Edition

International Energy Agency, 1999, *World Energy Outlook and impact of Economic turmoil in Asia on oil prospects*; Report June [#998]

International Energy Agency, 1999, *Meeting of the governing board*; Press Release [#1094]

International Energy Agency, 1999, *World Energy Outlook* 224pp

International Energy Agency, 2001, *World Energy Outlook* 226pp

Investors Chronicle, 1994, *Driven by supply*; Investors Chronicle 23/12/94 [#326]

Ion D.C., 1980, *The availability of world energy resources;* Graham & Trotman 345pp

Iritani E, ?, *Bush pushes energy plan in Mexico;* Los Angeles Times [#1203]

Ismail I.A.H, 1994, *Untapped reserves, world demand spur production expansion;* Oil & Gas Journ. May 2, 1994 95–102 [#224]

Ismail I.A.H, 1994, *The world oil production perspective; the future role of Opec and Non-Opec;* Preprint, APS Conf. [#285]

Ismail I.A.H, 1994, *Future growth in OPEC oil production capacity and the impact of environmental measures;* Energy Exploration and Exploitation 12/1 17–58 [#378]

Ivanhoe L.F., 1976, *Evaluating prospective basins;* in three parts – Oil & Gas Journ. Dec 13. 1976 [#108]

Ivanhoe L.F., 1980, *World's prospective petroleum areas;* Oil & Gas Journ. April 28 1980 [#91]

Ivanhoe L.F., 1984, *Oil discovery indices and projected discoveries;* Oil & Gas Journ. 11/19/84

Ivanhoe L.F., 1985, *Potential of world's significant oil provinces;* Oil & Gas Journ. 18/11/85 [#144]

Ivanhoe L.F., 1886, *Oil discovery index rates and projected discoveries of the free world;* in Oil & Gas Assessment. Amer. Assoc. Petrol. Geol. Studies in Geology #21, 159–178

Ivanhoe L.F., 1987, *The decline of giant oilfield discoveries;* draft [#19]

Ivanhoe L.F., 1987, *Permanent oil shock;* AAPG 71/5 [#309]

Ivanhoe L.F., 1988, *Future crude oil supply and prices;* Oil & Gas Journ. July 25 111–112 [#97]

Ivanhoe L.F., 1990, *Liquid fuels fill vital part of US Economy;* Oil & Gas Journ. Apr.23 106–109 [#15]

Ivanhoe L.F., 1990, *Competition increases to obtain oil imports;* Oil & Gas Journ Oct 29. 1990 [#49]

Ivanhoe L.F., 1991, *Oil, gas dominant sources of energy in U.S.;* Oil & Gas Journ. Sept 30 1991 #387]

Ivanhoe L.F., 1995, *Future world oil supplies: there is a finite limit;* World Oil Oct.1995 [#381]

Ivanhoe L.F., & G.G. Leckie, 1991, *Data on field size useful to supply planners;* Oil & Gas Journ, April 29 1991 [#30]

Ivanhoe L.F., 1995, *Oil reserves and semantics;* M.King Hubbert Center for Petrol. Supply studies [#405]

Ivanhoe L.F., 1996, *World oil supply;* personal comm. [#439]

Ivanhoe L.F., 1997, *Updated Hubbert curves analyze world oil supply;* World Oil Nov. [#508]

Ivanhoe L.F., 1997, *Get ready for another oil shock;* Futurist Jan–Feb 1997

Ivanhoe L.F., 2000, *Petroleum positions of the United Kingdom and Norway;* M.King Hubbert Center for Petrol. Supply studies 200/1 [#1166]

Ivanhoe L.F., 2000, *Oil reserve revisions: Major OPEC and Communist countries 1979–99*; M.King Hubbert Center for Petrol. Supply studies 200/1 [#1167]

Ivanhoe L.F, 2001, *re critique of "Would opening up ANWR really make a difference?"*; Fax to MacKenzie, World Resources Institute [#1221]

Ivanhoe L.F., 2001, *Hubbert Center Newsletter 2001/1, Petroleum positions of Saudi Arabia, Iran, Iraq, Kuwait, UAE Middle East Region* [#1232]

Ivanhoe L.F., 2000, *graph depicting OPEC production (Millions barrels per day) against time* [#1248]

Ivanhoe, L.F., 2001, *Petroleum Positions Of Egypt, Libya, Algeria, Nigeria, Angola, Africa Region*; Hubbert Center Newsletter # 2001/3 [#1326]

Ivanhoe, L F. & Riva, J P., 2000, *Exports – The Critical Part Of Global Oil Supplies /Petrophobia*; Hubbert Center Newsletter # 2000 / 4-1 and 4-2, October 2000 [#1412]

Ivanhoe, L F., 2000, *World Oil Supply – Production, Reserves, And EOR*; Hubbert Center Newsletter # 2000 / 1-1, January 2000 [#1427]

Ivanhoe, L F., 1999, *Petroleum Positions Of Brazil And Venezuela*; Hubbert Center Newsletter # 99 / 3, July, 1999 [#1464]

Ivanhoe, L F., 2001, *Petroleum Positions Of China, India, Indonesia, Malaysia, Australia, Far East & Oceania Region*; Hubbert Center Newsletter # 2001 / 4, October, 2001 [#1694]

Ivanhoe, L F., 1986, *Limitations of Geological Consensus Estimates of Undiscovered Petroleum Resources*; AAPG Studies in Geology #21, September, 1986 [#1799]

Ivanhoe, L F., 2002, *Canada's future oil production,* Hubbert Center Newsletter 2002/2 [#1818]

Ivanovich, D., 1999, *World may learn to wean itself from oil*; Houston Chronicle, Sunday, October 24, 1999 [#1546]

J

Jacque M. 1994, *Reserves mondiales de petrole*; geochronique 49. [#257]

Jaffe A.M. and R.A. Manning, 1999, *The Shocks of a World of Cheap Oil*; Foreign Affairs 79/1[1031]

Jaffe A.M. and R.A. Manning, 1999, *The myth of the Caspian great game: the real geopolitics of energy;* [#1123]

Jaffe, A., 2001, *ExxonMobil, Others Say Needed Refineries Unlikely to Be Built*; Bloomberg.com : energy news, June 21, 2001 [#1315]

Jaffe, A., 2000, *The Outlook For OPEC and International Oil Markets*; James A. Baker III Institute For Public Policy Energy Forum, ???, [#1438]

James, M., 1953, *The Texaco Story – the first fifty years 1902–1952*: Publ. The Texas Company . 115p

Jefferson M., 1994, *World energy prospects to 2010*; Petrole et Techn. 389 [# 262]

Jenkins D.A.L., 1987, *An undetected major province is unlikely*; Petrol. Revue Dec 1987 p 16 [#336]

Jenkins S., 1997, *Exploding the myth*; The Times 12.Nov [#653]

Jenkins S., 1999, *Will they never learn*; Times 9 April [#949]

Jenkins S., 1999, *Weep for poor Orissa*; [1025]

Jennings J.S., 1996, *The millennium and beyond*; Energy World 240 June.

Jones, N., 2002, *Strike it rich*; New Scientist, February 2, 2002 [#1765]

Journal de Geneve, 1998, *Un eminent geologue predit la fin du petrole a bon marche des 2008*; April 2 [#705]

K

Kahn J., 2000, *Surging prices show that oil's far from irrelevant in US economy*; Minneapolis Star Tribune Feb 21 [#1085]

Kahn J., 2000, *Surge in oil prices is raising specter of inflation spike*; N.Y. Times 21 Feb [#1090]

Kaletsky A., 1996, *Time has come to review the demand side disaster*; Times 25.1.96 [#411]

Kassler, P., 1994, *Two global energy scenarios for the next thirty years and beyond*; World Petrol. Congr. Stavanger.[#233]

Kaufmann R., 1991, *Oil production in the Lower 48 States* : Res. & Energy 13 [#96]

Kaufmann R. & C.J. Cleveland, 1991, *Policies to increase US oil production likely to fail, damage the economy, and damage the environment*; Ann.Rev.Energy Environ.1991 [#210]

Kaufmann R., W. Gruen & R. Montesi, 19 9?, *Drilling Rates and expected oil prices: the own price elasticity of US oil supply*; Centre for Energy & Environmental studies, paper [#211]

Keegan W., 1985, *Britain without oil*; Penguin 128pp

Kelley A, 2000, USA:Experts duel over future oil supply scenarios; Reuters [#1178]

Kemp A.G & L. Stephen, 1996, *UKCS future beyond 2000 depends on oil price and reserve trends*; World Oil Oct [#540].

Kempf H., 2000, *Le petrole et la planete*; Le Monde, 4th Sept.2000

Kennedy P., 1994, *Preparing for the 21st century*; Fontana 428pp

Kenney J.F., 1996, *Impeding shortages of petroleum re-evaluated*; Energy World 250 June 1992.

Kerr J., 1998, *The next oil crisis looms large – and perhaps close*; Science 281 [#790]

Khalimov E.M., 1993, *Classification of oil reserves and resources in the former Soviet Union*; Amer. Assoc. Petrol. Geol. 77/9 1636 (abstract)

Khan H.K., 1989, *Exploration promotion in India*; UN Seminar, Policy and Management of Petroleum Resources, Oslo [#101]

Kinzer S., 1998, *On piping out Caspian oil*; New York Times 8/11/98 [#855]

Kinzer S., 1998, *US bid to build a pipeline in Caucasus appears to fail*; New York Times 11/10/98 [#881]

Kjaergaard T., 1994, *The Danish revolution 1500–1800*; Cambridge 314pp

Klare M.T., 2002, *Resource wars : the new landscape of global conflict*; Owl Books. pp.289

Klare M.T., 2002 *Bush's master plan;* www.alternet.org [#1822]

Klebnikov P., 1998, *Opec: the cowardly lion*; Forbes April 6 [#719]

Klemme H.D., 1983, *Field size distribution related to basin characteristics*; Oil & Gas Journ.Dec. 25. 1983 169–176

Klemme H.D. & Ulmishek G.F., 1991, *Effective petroleum source rocks of the world: stratigraphic, distribution and controlling deposition factors*; Amer. Assoc. Petrol. Geol 75/12, 1908–185

Knoepfel H., 1986, *Energy 2000*; Gordon & Breach 181pp

Knott D., 1994, *Opec, once all-powerful, faces a cloudy tomorrow*; Oil & Gas Journ. Aug 22 1994 [#255]

Knott D., 1996, *Reserves debate*; Oil & Gas Journ. Jan 29. 40 [#526]

Knott D., 1998, *Oil orthodoxies*; O&GJ 21 Sept [#837]

Koen A. 1996, *Day of reckoning;* Upstream news 23.12.96 [#529]

Kroenig, J., 1999, *Am Tropf der Scheichs*; Die Zeit, No.40, September 30, 1999 [#1452]

Krauss C., 1997, *Mexican data suggest 30% overstatement of oil reserves*; New York Times 18th March [#577]

Kvint V., 1990, *Eastern Siberia could become another Saudi Arabia*; Forbes Sept 17 1990 [#388]

L

Labibidi M.M. al, 2000, *Depletion of petroleum resources;* Pres. Clean Energy 2000 Conf. [#1064]

LaGesse, D., 2001, *A low-gas warning*; U.S. News & World Report, September 17, 2001 [#1691]

Laherrère J.H., 1990, *Hydrocarbon classification rules proposed*; Oil & Gas Journ. Aug.13. p.62

Laherrère J.H., 1990, *Les Reserves d'hydrocarbures*; BIP 6629 [#167]

Laherrère J.H., 1992 *Reserves mondiales restantes et a decouvrir* :ATFP Conf. Paris 18.4.91 Revue de Presse TEP No3 20.1.92 [#168]

Laherrère J.H., 1993, *Le petrole, une ressource sure, des reserves incertaines*; Petrol et Technique, 383, October 1993 [#208]

Laherrère J.H., A. Perrodon and G. Demaison, 1993, *Undiscovered petroleum potential: a new approach based on distribution of ultimate resources*; Rept. Petroconsultants S.A., Geneva

Laherrère J.H., 1994, *Published figures and political reserves*; World Oil, Jan 1994 p. 33.[#207]

Laherrère J.H., 1994, *Study charts US reserves yet to be discovered*; American Oil & Gas Reporter 37/9 99–104.

Laherrère J.H., 1994, *Nouvelle approche des reserves ultimate – application aux reserves de gaz des Etas-Unis*; Petrole et Technique, Paris 392. 29–33

Laherrère J.H., 1994, *Reserves mondiales de petrole: quel chiffre croire?*; Bull. Inform. Petrol. 7727, 7728, 7729

Laherrère J.H., 1995, *World oil reserves: which number to believe?*; OPEC bull. Feb draft.[#256] Final [#346]

Laherrère J.H., 1995, *An integrated deterministic/probabilistic approach to reserve estimations – discussion*; draft for Journ. Pet. Technol. [#332]

Laherrère J.H., 1996, *Distributions de type "fractal parabolique" dans la Nature*: C.R.Acad. Sci. Paris 322 II

Laherrère J.H., 1996, *Que valent les chiffres de réserves publiées?* ; draft paper to Commissiariat au Plan Energie 2010–2020, Paris [#447]

Laherrère J.H., 1996, *Distributions de type <fractal parabolique> dans la nature*; C.R.Acad. Sci. Paris 322 IIa 535–541 [#449]

Laherrère J.H., & A. Perrodon, 1996, *Technologie et réserves*; Draft Petrole et Technologie [#496]

Laherrère J.H., 1996, *Upstream potential of the Middle East in a World context*; Proc. Oil& Gas project finance in the Middle East IBC Dubai Conf. May 1996 [#565]

Laherrère J.H., A. Perrodon, and C.J.Campbell, 1996, *The world's gas potential*, Report, Petroconsultants

Laherrère J.H., 1997, *Production decline and peak reveal true reserve figures*; World Oil Dec. 77– [#658]

Laherrère J.H., 1997, *Evolution of development lag and development ratio*; IEA submission [#715]

Laherrère J.H., 1998, *Development ratio evolves as true measure of exploitation*; World Oil, [#673]

Laherrère J.H., 1998, *Modeles et realite*; Assoc. Nat. Dir. Finance, France (in press) [#810]

Laherrère J.H., 1999, *Erratic reserve reporting*; Petroleum Review Feb [#914]

Laherrère J.H., 1999, *World oil supply – what goes up must come down – but when will it peak ?*; Oil & Gas Journ. Feb 1 57–64 [#968]

Laherrère J.H., 1999, *Reserve growth: technological progress, or bad reporting and bad arithmetic* Geopolitics of Energy [#1065]

Laherrère J.H., 2000, *Oil reserves and potential of the FSU*, Tomorrow's Oil 2/7

Laherrère J.H, J H. & Campbell, C J. & Duncan, R C. & McCabe, P J., 2001, Discussions & Reply acc. to *"Energy Resources– cornucopia or empty barrels?"* in AAPG Bulletin, V. 85, No. 6 (June 2001), p 1083–1097, 2001[#1306]

Laherrère J.H., 2002, *Forecasting future oil production*; Presentation for the BGR, January, 2002 [#1764]

Laherrère J.H., 2002, *Is FSU oil growth sustainable?*; Petroleum Review April, 2002 [#1810]

Lamar L., 1992, *World energy statistics*; Shale Shaker, May/June 1992 [#114]

Lanier, D., 1998, *Heavy Oil – A Major Energy Source for the 21ˢᵗ Century*; UNITAR Centre for Heavy Crude and Tar Sands, No. 1998.039, 1998 [#1697]

Leach, G., 2001, *The coming decline of oil*; Tiempo, Issue 42, December, 2001 [#1766]

Leckie G.G., 1993, *Hydrocarbon reserves and discoveries 1952 to 1991*; Energy Exploration & Exploitation, 11/1, 1993 [#214]

Leggett, K., 2000, *China Is Likely to Establish A Strategic Oil Reserve*; The Wall Street Journal, August 28, 2000 [#1380]

Lenzner R & J.M. Clash, 1994, *Wrong again*; Forbes June 6 1994 [#231]

Lenzner R., 1994, *The case for hard assets*; Forbes, June 20 1994 [#237]

Leonard R.C., 1984, *Generation and migration of hydrocarbons on southern Norwegian shelf*; AAPG. 68 796 [#531]

Leonard R.C., 1993, *Distribution of subsurface pressure in the Norwegian Central Graben and applications for exploration*; Petrol. Geol. of NW Europe Proc. 4th Conf. [#298]

Leonard R.C., 1996, *Caspian Sea regional hydrocarbon development: opportunities and challenges*; 4ᵗʰ Kazakhstan Int. Oil & Gas projects conf. [# 539]

Leonard, R., 2001, *The Current Oil Crisis And The Coming Age Of Natural Gas*; February, 2001 [#1633]

Leonard, R., 2002, *Russian Oil And Gas: A Realistic Assessment*;, ASPO Workshop, Uppsala, Sweden, May, 2002 [#1796]

Lewis, B., 2002, *What Went Wrong?*; The Atlantic Monthly, January, 2002 [#1742]

Lewis M., 1999, *Oil price uncertainty: restructuring and the implications for NOCs*; CWC conf [#1005]

Licking E., 1998, *The world's next power surge*; Business Week, 14 Dec. [#873]

Liesman S., 2000, *OPEC oil cuts raise threat of shortage*; Wall St Journ., 20ᵗʰ Jan [#1065]

Liesman S., 1999, *Texaco's strategy: produce less oil more profitably*; Wall St Journ., Oct 27 [#1019]

Lieven, A., 2001, *The Search For Strategy*, The Australian Financial Review, Friday, September 28, 2001 [#1698]

Lifsher, M., 2001, *Oil Producers Balk as Venezuela Tighten Terms On Investment*; The Wall Street Journal, November 15, 2001 [#1731]

Lippman T.W., 1990. *Saudis come up with major oil find*; Washington Post Oct.15 [#52]

Lippman T.W., 1992, *Saudis and US teamed on oil issues*; Fort Worth Star Telegram, July 24 1992 [#34]

List F., 2000, *Die Ölförderung droht ab dem Jahr 2005 zu sinken; Technik & Wirtschaft*; 52 Dez 29 [#1243]

Littell G.S., 1999, *World crude production: bad statistics produce poor conclusions*; World Oil June [#986]

Littell G.S., 1999, *Bad data distorts industry, market perceptions*; World Oil April [#1172]

Longhurst H., 1959, *Adventure in oil: the story of British Petroleum*; Sidgwick and Jackson, London 286p.[B#8]

Los Angeles Times, 1991, *Mexico lied about Proven Oil Reserves, report says*; Dec.10 1991 [#12]

Lotter C and S. Peters, 1996, *The changing European security environment*; Bohlau 335pp

Lovelock J., 1988. *The ages of Gaia*; Oxford University Press 252pp.

Lugar R.G. and R.J. Woolsey, 1999, *The new petroleum*; Foreign Affairs Jan–Feb [#899]

Lynch M.C., 1992, *The fog of commerce: the failure of long-term oil market forecasting*; MIT Center for Int. Studies Sept. 92

Lynch M.C., 1998, *Imminent peak challenged*; Oil & Gas Journal 28.1.98 p.6 [#657]

Lynch M.C., 1998, *Crying Wolf: warnings about oil supply*; MIT Center for International Studies [#728]

Lynch M.C., 1998, *Farce this time*; Geopolitics of Energy; December–January [#925]

Lynch M.C., 1999, *The debate over oil supply: science or religion*; Geopolitics of Energy, Aug. [#1133]

Lynch M.C., 2001, *Oil prices enter a new era*; Oil and Gas Journal [#1214]

Lynch M.C., 2001, *Forecasting Oil Supply: Theory and Practice*; DRI-WEFA, 2001, [#1311]

Lynch M.C., 2000, *Michael Lynch, the economist, responds to Colin Campbell*; from , June 17, 2000 [#1516]

Lynch M.C., 2001, *Closed Coffin: Ending the Debate on "The End of Cheap Oil" – A commentary*; M C Lynch, Chief Energy Economist, DRI-WEFA, Inc., September, 2001 [#1675]

M

Mabro R., 1996, *The world's oil supply 1930–2050 – a review article*; Journ. of Energy Literature II.1.96 [#469]

MacArthur C.E., 2002, *The adjustment*; 208pp

Macgregor D.S., 1996, *Factors controlling the destruction or preservation of giant light oilfields*; Petroleum Geoscience 2. 197–217

Mack T., 1991, *Are the big oil stocks a buy now?*; Forbes Feb 18 1991 [#47]

Mack T., 1992, *The last frontier*; Forbes, May 8th 1992. [#56]

Mack T., 1994, *History is full of giants that failed to adapt*; Forbes 28 Feb 1994 [#195]

Mack T., 1998 *Venezuela is changing the balance of power among the world's oil producers*; Forbes Aug.1 [#695]

Mackenzie A.S., 2000, *Energy, petrochemicals and progress*; Clean Energy 2000 [#1124]

MacKenzie J.J., 1994, *Transportation in the People's Republic of China: beginning the transition to sustainability*; preprint World Resources Inst. [#316]

MacKenzie J.J., 1995, *Oil as a finite resource: the impending decline in global oil production*: World Resources Inst [# 394 436]

MacKenzie, J J., 1996, *Oil as a finite resource: When is global production likely to peak?*; World Research Institute, Updated March 2000, [#1519]

MacKenzie J.J., and K. Courrier, 1996, *Cutting gas tax will make things worse*; Los Angeles Times May 8 [#438]

Mackenzie, W., 1999, *Maturing Gracefully – Overview of the 1999 Probable Developments*; UK Upstream Report, No. 316, September 1999 [#1376]

Mackenzie, W., 2000, *Production*; UK Upstream Report, No.329, October 2000 [#1423]

Mackenzie, W., 2000, *Make Hay While The Sun Shines?*; UK Upstream Report, No.328, September 2000 [#1426]

Mackenzie, W., 2001, *2001 UK Oil Production Review etc.* ; UK Upstream Report, No.339, August, 2001 [#1477]

Macleay J., 1996, *Running on empty*; The Australian 24.10.96 [#510]

MacLeod, F., 1999, *The real motives for industry consolidation*; Deutsche Bank, November, 1999 [#1466]

Magoon, L.B, 2000, *Are We Running Out Of Oil?*, [#1529]

Maiello M., 2000, *Gas up and go*; Forbes, Mar 7 [#1083]

Main, A., 2000, *"Perspective" – Old drivers for the new oil crisis*; The Australian Financial Review, Saturday, September 9, 2000 [#1387]

Malone A., 1996, *Revenge of the Saudi exile*; Sunday Times, 7 Jan 96 [#417]

Manor, R. (Chicago Tribune), 2002, *Non-members may benefit as OPEC cuts*; San Jose Mercury News, Wednesday, January 2, 2002 [#1736]

Mansfield P., 1992, *A history of the Middle East*; Penguin Books 373p.[#B15]

Marine and Petroleum Geology, 1993, *Book Review of "The Golden Century of Oil"*; Marine and Petroleum Geology, Vol 10, p 182, April 1993 [#1455]

Martell H., 1989, *Exploration and Resources, Venezuela*; UN Seminar Policy and Management of Petroleum Resources, Oslo [#98]

Martin A.J., 1985, *Prediction of strategic reserves in prospect for the world oil industry*; Eds.T. Niblock & R.Lawless. Univ. of Durham 16–39

Martin A.J. & P.B. Lapworth, 1998, *Norman Leslie Falcon*; R.Soc. Lond. 44 [#933]

Martin, P., 2002, *Oil and "conspiracy theories" a reply to a liberal apologist for the US war in Afghanistan*; World Socialist Web Site [#1936]

Martinez A.R. et al. 1987, *Study group report: classification and nomenclature system for petroleum and petroleum reserves*; Wld Petrol. Congr. 325–342.

Martoccia D, 1997, *Permanent oil shortage impossible*; Futurist May–June [#935]

Mast R.F. *et al.*, 1989, *Estimates of undiscovered conventional oil and gas resources in the United States – a part of the nation's energy endowment*; U.S.Dept of Interior [#94]

Masters C.D., 1987, *Global oil assessments and the search for non-OPEC oil*; OPEC Review, Summer 1987, 153–169 [#92][#1687]

Masters C.D., 1991, *World resources of crude oil and natural gas*; Review and Forecast Paper, Topic 25, p.1–14. Proc. Wld. Petrol. Congr., Buenos Aires 1991 [#113]

Masters C.D. D.H. Root & E.D. Attanasi, 1991, *Resource constraints in petroleum production potential*; Science 253.[#28]

Masters C.D., 1993, *U.S.Geological Survey petroleum resource assessment procedures*; Amer. Assoc. Petrol. Geol. 77/3 452–453 (with other relevant references).

Masters C.D., 1994, *World Petroleum analysis and assessment*; Wld. Petrol. Congr. Stavanger [#226]

Masters C.D., 1994 *Bibliography of the world energy resources program*; USGS Open File 94–556 [#328]

Masters C.D., D.H. Root, and R.M. Turner, 1997, *World resource statistics geared for electronic access*; Oil&Gas Journ, 13.Oct. 98–104 [#660]

Masuda, T., 2000, *World Oil Supply Outlook to 2010*; For the 5th Annual Asia Oil & Gas Conference, May 28–30, 2000 [#1521]

Maxwell C.T, 2000, *Integrated oil companies' liquids and natural gas (equivalent barrels) reserves ranking by size*; maxwell@weeden.com [#1268]

Maxwell, C T., *Oil Stocks Are Pressured By Crude Price Forecasts Into The Teens.But, Does The Evidence Support Such A Drop?*; maxwell@weeden, vol. 2 / i.13, July 6, 2001 [#1312]

Maxwell, C T., 2001, *Enough is Enough. Many Oil-Related Companies Are Now Undervalued And Will Confirm This In Their Future Earnings Streams*; maxwell@weeden, vol. 2 / i.14, July 20, 2001[#1324]

Maxwell, C T., 2000, *"Shape and Timing of the Next E & P Capital Cycle: Are We Doing Enough Now?"*; Letter, August 9, 2000 [#1419]

Maxwell, C T., 2000, *The Seven Sisters Reborn?*; maxwell@weeden, vol. 1 / i.25, October 17, 2000 [#1420]

Maxwell, C T., 2001, *Integrated Oil Companies` Liquids And Natural Gas (Equivalent Barrels) Reserves Ranking By Size And Value To Shareholders*; maxwell@weeden, vol. 2 / i. 17, September 5, 2001 [#1481]

Maxwell, C T., 2001, *America Will Strike Back*; maxwell@weeden, vol. 2 / I.19, September 17, 2001 [#1482]

Maxwell, C T., 2001, *Heightened Terrorism: Potential Effects On The Oil Industry*; maxwell@weeden, vol. 2 / i.20, September 19, 2001 [#1483]

Maxwell, C T., 2001, *La Creme de la Creme (Revised)*; maxwell@weeden, September 4, 2001 [#1485]

Maxwell, C T., 2000, *Natural Gas: Stage Five Commences*; maxwell@weeden, December 28, 2000 [#1585]

Maxwell, C T., 2001, *OPEC`s Oil Pricing Leadership Will Bend, But Not Break*; maxwell@weeden, vol.2 / i.21, September 27, 2001 [#1690]

Maxwell, C T., 2001, *U.S. Natural Gas Prices Have Fallen So Far, So Fast, Wildcatters Are Rejoicing*; maxwell@weeden, vol.2 / i.23, October 15, 2001 [#1718]

Maxwell, C T., 2001, *Presentation on Oil and Gas Stocks*; ???, October 31, 2001 [#1719]

Maxwell, C T., 2002, *Suncor Energy (SU – $30)*; maxwell@weeden, vol.3 / i.1, January 24, 2002 [#1780]

Maxwell, C T., 2001, *OPEC vs. Non-OPEC Producers: The Outcome Will Be Lower Oil Prices For A While*; maxwell@weeden, vol.2 / i.25, November 15, 2001 [#1781]

Maybury, R., 2000, *Richard Maybury`s U.S. & World Early Warning Report For Investors*; August 2000 [#1379]

McCabe P.J., et al., 1993, *The future of energy gases*; USGS Circ.1115 [#477-1]

McCabe P.J., 1998, *Energy resources – cornucopia or empty barrel*; Amer.Assoc. Petrol. Geol. 82/11 2110–2134

McCormack M., 1999, *Twenty-first century energy resources: avoiding crisis in electricity and transportation*; Amer. Assoc. Advancement of Science., [#905]

McCrone A.W., 2001, *Looking beyond the petroleum age*; San Francisco Chronicle, Mar 4th [#1191]

McCutcheon, H. & Osbon, R. & Mackenzie, W., 2001, *Risks temper Caspian rewards potential*; Oil & Gas Journal, p 22–28, December 24, 2001 [#1761]

McKenzie A., 2000, *The quality of life – a shared concern*; Presentation to Clean Energy 2000 [#1060]

McKibben B., 1998, *A special moment in history*; Atlantic Monthly, May [#738]

McMahon, P., 2001, *Power suppliers run dry at worst time*; USA Today, Tuesday, March 27, 2001 [#1626]

McMullen, Porgam Murray, 1976, *Energy resources and supply;* Wiley [# 692]

McRae H., 1994, *The world in 2020: power, culture and prosperity, a vision of the future*; Harper Collins 302p [#B16]

McRay H., *Oil looks slippery*; Independent 7/5/93 [#161]

McRay H, *No end of cheap oil*; Independent 11/11/93 [#191]

McRay H, *The real question is why the price of oil is not even higher*; Independent 07/04/02 [#1821]

Meadows D.H. and others, 1972. *The limits to growth*; Potomac 205 pp

Megill R.E., 1993 *Discoveries lag oil consumption*; AAPG Explorer Aug.1993 [#171]

Megill R.E., 1994, *Another look at finding cost*; AAPG Explorer July 1994 [#250]

Melloen G., 1996, *The schism between Islam and the west deepens*; Wall St Journ 16.9.96 [#502]

Miller R.G., 1992, *The global oil system: the relationship between oil generation, loss half-life and the world crude oil resource*; Amer. Assoc. Petrol. Geol. 76/4 489–500.[#302]

Milling M.E., 2000, *Oil in the Caspian Sea*; Geotimes [#1173]

Mineral Management Service, 1995, *Gulf of Mexico OCS Region* ; Internet[#955]

Mineral Management Service, 1995, *Gulf of Mexico*; Internet [#1076]

Mineral Resources of Canada, 1998, *Canada Energy Resources*; [907]

Minnear M.P., 1998, *Forecasting the permanent decline in global petroleum production*; Thesis, Univ. of Toledo

Mitchell J., 1995, *The geopolitics of energy*; EU Report [#371]

Mitchell J., 1996, *The new geopolitics of energy*; Royal. Inst. Int. Affairs April 1996 [#475-1]

Mitchell J., 1997, *Renewing the geopolitics of energy*; Energy World 245 Jan. [#555]

Mohammed A.H., 1989, *Crude oil production, refining, petrochemicals and research activities in Iraq*; UN Seminar: Policy and Management of Petroleum Resources, Oslo [#102]

Monastersky R., *Geologists anticipate an oil crisis soon*; Science News 154/18 31 Oct [#865]

Monty M., 1999, *Use data mining to discover odds of making a billion bbl discovery*; World Oil March [#1063]

Moody-Stuart M, 2000, *Realising the value of scientific knowledge – geosciences in energy industries*; Sir Peter Kent lecture, Geological Society [#1278]

Morehouse D.F., 1997, *The intricate puzzle of oil and gas "reserve growth"*; EIA, Nat. Gas Monthly [#1103]

Morrison D.R.O., 1999; *World energy and climate in the next century*; [#961-F1]

Morrison D.R.O., 2000, *Energy in Europe – comparison with other regions*; Clean Energy 2000 [#1112]

Morse, E L. & Richard, J., 2002, *The Battle for Energy Dominance*; Foreign Affairs Magazine, March/April, 2002 [#1792]

Mortished C., 1995, *Shell thinks, then does the unthinkable*; Times 31/3/95 [#334]

Mortished C., 1996, *Iraq oil could be back on sale soon*; Times March 11 [#421]

Mortished C., 1997, *Tempus – BP*, 6.2.97 [#551]

Mortished C., 1998, *BP shrugs off influence of oil prices with aid of self help*; The Times 11.2.98 [#666]

Mortished C., 1998 *Set-back for Western fuel long term gains*; The Times 19.2.98 [#667]

Mosley L., 1973, *Power play: oil in the Middle East*; Random House [#B10]

Mowlam M., 2002, *The real goal is the seizure of Saudi oil*; The Guardian 5 Sept [#1906]

Myers Jaffe, A. & Manning, R A., 2000, *The Shocks of a World of Cheap Oil*; Foreign Affairs, Vol. 79, No.1, p 16–29, Jan/Feb, 2000 [#1434]

N

Narimanov A. and A. Palaz, 1994, *Baku region rich with oil, history*; AAPG Explorer Oct. 1994 p.40 [#315]

Nasmith J., 1996, *A story of oil price reporting 1945–1985*, Pipeline 13 May [#608-1]

Nation L., 1993?, *DOE revises resource estimates*; AAPG Explorer ref. unknown [#118]

Nation L., 1993, *Delegates told oil prices must rise*; AAPG Explorer [#124]

Nation L., 1995, *Hodel sees looming energy crisis*; AAPG Explorer May 1995 [#358]

N.E.R.A., 1993, *Oil and natural gas price outlook*; Energy Outlook, Nat.Econ.Research. Assoc. Feb. 15 1993 [#119]

Nehring R, 1978, *Giant oil fields and world oil resources*; CIA report R-2284-CIA [#16]

Nehring R, 1979, *The outlook for conventional petroleum resources*; Paper P-6413 Rand Corp.21p.

Nehring R, 2000, *America's bigger resource*; Hart's E&P, Jan [#1071]

Newswell, 1991, *Higher oil prices to spur 1991 economic growth in southwest.* Newswell [#58]

New York Times, 2001, *Renewable Energy: today`s basics*; The New York Times, May 3 2001 [#1314]

New York Times, 2001, *Iran Is Accused of Threatening Research Vessel in Caspian Sea*; The New York Times, July 25, 2001 [#1344]

New York Times, 2001, *Reconsidering Saudi Arabia*; The New York Times, October 14, 2001 [#1722]

New York Times, 2001, *Osama bin Laden`s Wildfire Threatens Grip of the Saudi Royal Family*; The New York Times, November 6, 2001 [#1729]

Niiler, E., 2000, *Awash in Oil*; Scientific American, September 2000 [#1370]

Nikiforuk A, 2000, *Running on empty, when Canada's natural gas reserves hit the crisis point, who will be left out in the cold?;* Canadian Business Magazine [#1255]

Norwegian Petroleum Directorate, 1993, *Improved oil recovery*; [#482-1]

Norwegian Petroleum Directorate, 1997, *The petroleum resources of the Norwegian continental shelf* ; report [#620-3]

Norwegian Petroleum Directorate, 1997, *Discoveries on the Norwegian continental shelf*; report [#621-3]

Norwegian Petroleum Directorate, 1997, *Trends in petroleum resource estimates*; reprt [#659]

Norwegian Petroleum Directorate, 1997, *Classification of petroleum resources on the Norwegian continental shelf;* Report [#664]

Norwegian Petroleum Directorate, 1999, *Undiscovered Petroleum Resources*; website [#974]

Norwegian Petroleum Directorate, 2000, *Two-thirds left to go*; NPD Diary [#1154]

Norwegian Petroleum Directorate, 2000, *Production Figures From The Norwegian Continental Shelf*; Updated: February 3, 2000 [#1443]

Norwegian Petroleum Directorate, 2000, *Norwegian Production Of Stabilized Oil and Gas*; 2000, [#1445]

O

Obaid N.E., 2000, *The oil kingdom at 100*; Washington Inst. For Near East Policy 136pp

O'Dell S., 1994, *Prospects for non-opec oil supply*; 13th CERI Conf., [#314]

O'Connor T.E., 199?, *The international development banking view of petroleum exploration and production in developing countries*; World Bank [#932]

Odell P.R., 1994, *World Oil resources, reserves and production*; The Energy Journal v. 15 89–113

Odell P.R., 1996, *Middle East domination or regionalisation*; Erdol, Erdgas, Kohle Heft4 [#435]

Odell P.R., 1996, *Britain's North Sea oil and gas production – a critical review*; Energy Exploration & Exploitation 14/1/

Odell P.R., 1997, *Oil shock – a rejoinder*; Energy World 245 [#554]

Odell P.R., 1997 *Oil reserves: much more than meets the eye*; Petroleum Economist Nov. [#656]

Odell P.R., 1998, *Fossil fuel resources in the 21st Century; Presentation*, Int. Atomic Energy Agency [#869]

Odell P.R., 1999, *Oil and gas reserves: retrospect and prospect*; Geopolitics of Energy Jan [#956]

Odell P.R., 1999, *Predicting the future: what lies ahead for fossil fuels?* Horizon June [#1096]

Odell P.R., 1999, *Fossil fuel resources in the 21st Century;* Financial Times Energy [#1279]

Odum H. and E., Odum, 1981, *Energy basis for man and nature*; McGraw Hill, New York

Offshore, 1989, *Non-opec producers limited in ability to take advantage of rising oil demand*; Offshore June 1989 [#21]

Offshore, 1989, *Poor US performances underlie reserve buys, international move*; Offshore June 1989 [#22]

Offshore, 1996, *Improved recovery grow Norwegian reserves*; Offshore April 1996

Offshore, 1989, *Cat and mouse: the story continues*; April [#725]

Oil & Gas Journal, *World Production Reports*; December each year.

Oil & Gas Journal, 1959, *Where and when oil was discovered throughout the world*; map [#188]

Oil & Gas Journal, 1989, *IPAA, US oil production headed for biggest slide sine the 1970s*; Oil & Gas Journ., Nov 6th 1989 [#77]

Oil & Gas Journal, 1989, *Crude/condensate production slips in Soviet Union*; Oil & Gas Journ. Nov. 27 1989 [#78]

Oil & Gas Journal, 1994, *Steady rise in oil, gas demand ahead*; Oil & Gas Journ June 6 1994 [#238]

Oil & Gas Journal, 1994, *Opec draws praise for restraint*; Oil & Gas Journ. Aug.18 1994 [#212]

Oil & Gas Journal, 1994, *Worldwide oil flow up, reserves steady in 1994*; Oil & Gas Journ 26/12/94 [#322]

Oil & Gas Journal, 1995, *OGJ Newsletter*; Oil & Gas Journ 2/1/95 [#325] Oil & Gas Journal, 1996, *New life in US oil*; Jan 29 [#527]

Oilman The, 1996 ; *UK reserves and potential*; The Oilman 18 March p 3–5 [#430]

OPEC Bulletin, 1997, *An early touch of millennium fever: the coming oil crisis?* Oct .p.3

Orphanos A., 1995, *Looking for oil prices to gush*; Fortune 15/5/95 [#359]

Oswald, A., 2001, *Oil price puts skids under growth*; September 2, 2001 [#1648]

P

Parade Magazine, 2002, *Chinese Oil Fills Foes With Fear*; Parade Magazine, p 14, January 27, 2002 [#1776]

Parent L., 1989, *Natural gas: life after the bubble*; World Oil Feb. 1989 [#397]

Parsons Tony, 1992, *Thick on the ground, thick in the head*; Times Saturday Revue Oct 3 1992 [#71]

Patricelli J.A. & C. L. McMichael, 1995, *An integrated deterministic/probabilistic approach to reserve estimations*; Journ. Petrol. Technol. 47/1 49–53

Pauwels J-P and F. Possemiers, 1996, *Oil supply and demand in the XXIst century*; Revue de l'energie, 477. [#458]

Pearl, D., 2001, *Suddenly, Pipeline Project Sparks Interest*; The Wall Street Journal, June 21, 2001 [#1332]

Pearce F., 1999, *Dry Future*; New Scientist; 10 July [#1007]

Pearce F., 1999, *Iceland's power game;* New Scientist 1 May [#1053]

Pearson J.C., 1997, *Estimating oil reserves in Russia*, Petroleum Engineer Int., Sept. [#650]

Perrodon A., 1988, *Hydrocarbons* in Beaumont E.A. and N.H.Foster (Eds) *Geochemistry.* Treatise of petroleum geology, Amer. Assoc. Petrol. Geol. Reprint Series No.8 3–26

Perrodon A., and J. Zabek., 1990, *Paris Basin*: in Leighton M.W. *et al.* eds. *Interior cratonic basins*. Amer. Assoc. Petrol. Geol. Mem 51 819p [#106]

Perrodon A., 1991, *Vers les reserves ultimes*; Centres Rech.Explor.-Prod. Elf-Aquitaine 15/2 253–369. [#36]

Perrodon A., 1992, *Petroleum systems, models and applications*; Journ. Petrol. Geol.15/3, 319–326.[#38]

Perrodon A., *Crise petroliere?*; Geologues 94 [#37]

Perrodon A., 1993, *Historique des recherches petrolieres en Algerie*; 118 Congr. Nat. des soc. hist et scient, Pau, 323–340 [#395]

Perrodon A., 1995, *Petroleum systems and global tectonics*: Journ. petrol. Geol. 18/4 471–476 [#396]

Perrodon A., J.H.Laherrere and C.J.Campbell, 1998, *The world's non-conventional oil and gas*; Pet. Economist [#683]

Perrodon A., 1998, *Production: les premices du declin*; Petrol. Int. 1735 [#870]

Perrodon A., 1999, *Vers un changement de decor sur la scene petroliere*; Pet. Informations 1738 [#1095]

Perrodon A., 1999, *Quel pétrole demain*, Technip, Paris 94p

Peters K.E., 2000, *Review of the Deep Hot Biosphere by Thomas Gold*; AAPG bull. 84/1 Jan

Peters S, 1999, *The West against the Rest: Geopolitics after the end of the cold war*; Frank Cass, Journal offprint from Geopolitics [#1235]

Petrie Parkman, 1992, *Llanos foothills Trend/Cusiana Field's potential reserves and production economics assessed*; Petroleum Research v IV EPPO1 Feb 22 1992 [#80]

Petrodata, 2000, *Similar Paths – Three Oil Price Cycles*; Petrodata, February, 2000 [#1568]

Petroconsultants S.A., 1993, *World Production & Reserve Statistics; oil and gas 1992*; Petroconsultants, London

Petroconsultants S.A., 1993, *Strategic petroleum insights*; Report [#192]

Petroconsultants S.A., 1994, *Oil production forecast*; Report [#340]

Petroconsultants S.A., 1996, *World petroleum trends 1996* [#512].

Petroconsultants S.A., 1996, *World exploration – key statistics 1995/1986*, [#525]

Petroconsultants S.A., 1997, *Ten-year decline in new field wildcat drilling reversed*; International oil letter June 28th [#623]

Petroconsultants, 1997 *Listing of heavy oil fields* [#637]

Petroconsultants, 1997 *Oil and Gas Reserves 1997* [#647]

Petroconsultants, 1997, *Venezuelan oil* [#675]

Petroconsultants 1997, *World Petroleum Trends* [#676]

Petroconsultants, 1998, *Deepwater reserves*; report [#689]

Petroconsultants, 1998 *Russian reserves and wildcats* [#707]

Petroconsultants, 1998, *UK reserves* [#722]

Petroconsultants, 1998, *US production* [#731]

Petroconsultants, 1998 *Petroleum Trends* [#760]

Petroconsultants, 1998, *Oil production – discovery gap widens*: news release [#771]

Petroconsultants, 1998, *Oil & gas reserves added 1993–97* ; Petroleum Review Oct [#899]

Petrodata, 1999, *The impact of oil price changes*; Upstream Economics Symposium [#1054]

Petroleum Economist, 1995, *Interview with C J Campbell – "Prophet or Cassandra"*, Petroleum Economist, September, 1995 [#1436]

Petroleum Engineer Int., 1993, *The Norwegian North Sea contains about 75 billion bbl*; Petrol. Eng. Int June 1993 [#162]

Petroleum Engineer Int., 1994, *Worldwide oil and gas activity forecast*; Petrol. Eng. Int Jan 1994 [#203]

Petroleum Engineer International, 1997, *SPE/WPC reserve definitions to provide more accurate consistent estimates*; Sept [#652]

Petroleum Review, 1996, *Report on travel trends and fuel consumption in the United Kingdom*; Pet. Review May 1996 p 202 [#443]

Petroleum Review, 1996, *Ample energy reserves for the future*; Petroleum Review Aug. 1996 [#473]

Petroleum Review, 1998, *Insights from the statistics*; Pet. Review Sept. [#820]

Petroleum Review, 2000, *Discovery still lags production despite good 1999 results*; Petroleum Review, September 2000 [#1366]

PetroMin, 2000, *Report: The looming crisis – Are you ready for this?*; PetroMin, November, 2000 [#1417]

Pettingill, H S., 2001, *Giant field discoveries of the 1990s*; The Leading Edge, July, 2001 [#1657]

Petzet A., 1999, *Decline in world crude reserves is first sense '92*; O&GJ December 22 [1026]

Petzet, A., 2000, *World resource estimate shows more liquids, slightly less gas* and *The oil resource grows*; ???, April 17 & 10, 2000 [#1531]

Phipps S.C., 1993, *Declining oil giants, significant contributors to U.S. production*; Oil & Gas Journ. Oct.4. 1993 [#181]

Picerno J., 2002, *If we really have the oil*; Bloomberg Wealth Manager Sept [#1899]

Pickens T.B., 1987, *Boone*; Houghton Miffin, Boston [#B1]

Pilger, J., 2001, *Inevitable ring to the unimaginable*; September 14, 2001 [#1673]

Pimental D., 1994, *Implications of the limited potential of technology to increase the carrying capacity of our planet*; Human Ecology Review Summer [#904]

PIW, 1998, *The end of cheap oil*; PIW April 6th [#702]

PIW, 1998, *Crying wolf, warnings about oil supply*; PIW April 6th [#723]

PIW, 1998, *Cusiana News*; Sept 14 [#844]

PIW, 1999, *End of sight for North Sea output growth*; March 22 [#956]

PIW, 1999, *Mexico gains credibility by losing reserves*; March 29 [#975]

PIW, 2001, *US Majors` Hot Streak Ends In Second Quarter, But Profits Remain Robust*; PIW, July 30, 2001 [#1323]

Pooley, E., 2000, *Who`s Right About Oil?*; Time, October 2, 2000 [#1403]

Pope H, 1997, *Oil and geopolitics in the Caucasus*; Wall St. Journ, April 25 [#598]

Popular Science, 2001, *Are We Really Running Out Of Oil?/What`s Next: Cars*; Popular Science Special Issue, Summer, 2001 [#1335]

Porter E., 1995, *Are we running out of oil?* API Discussion Paper 081 [#425 & 479-1]

Poruban, S. & Bakhtiari, A M S. & Emerson, S A., 2001, *OPEC`s Evolving Role – Analysts discuss OPEC`s role*; Oil/Gas Journal, July 9, 2001 [#1327]

Potter N., 1997, *Caspian production sharing*; Petroleum review. Feb [#546]

Power M., 1992, *Lognormality in observed size distribution of oil and gas pools as a consequence of sampling bias*; Int. Assoc. of Mathematical Geology 24/8 [#806]

Power M., 1992, *The effects of technology and basin specific learning on the discovery rate*; Journ. Canadian Pet. Technology 31/3 [#807]

Power M and J.D. Fuller, 1991, *Generating and using basin specific discovery and finding costs forecasts*; Energy Exploration & Exploitation 9/6 [#819]

Pratt W., 1952, *Toward a philosophy of oil finding*; Amer.Assoc. Petrol Geol., 26/12 2231–36.

Preusse A., 1966, *Coalbed methane production – an additional utilization of hard coal deposits*; in Kürsten M. (Ed) *World Energy – a changing scene*; E. Schweizerbart'sche Verlagshandlung, Stuttgart ISBN 3-510-65170-7

Priddle, R., 2000, *Opinion: Oil market volatility – what can or should be done?*; June 6, 2000 [#1512]

Protti G.J., 1994, *Canada's upstream petroleum industry*; Journ. Canada. Pet. Tech. May 1994 [#252]

Pursell D., 1999, *Depletion: the forgotten factor in the supply demand equation, Gulf of Mexico analysis:* Simmons & Co rept. [#982]

Pursell, D A. & Eades, C., 2001, *Crude Oil And Natural Gas Price Update – Tough Sledding Near Term… But Optimistic About 2003*, Simmons & Co Internat, Energy Industry Research, March 4, 2001 [#1790]

Q

Quinlan M., 1999, *The oil price factor kicks in*; Pet. Review April [#1102]

R

Radler, M., 2001, *World crude, gas reserves expand as production shrinks*; Oil & Gas Journal, p 125, December 24, 2001 [#1762]

Raeburn, P., 2001, *Commentary: This Clean Oil Deal Is Already Tainted*; Business Week, May 7, 2001 [#1662]

Rahmani B. Mossavar, 1983, *The Opec Multiplier*, Foreign Policy 52 [#223]

Randol W., 1995, *No gushers*; Barron's 6/2/95 [#318]

Rasmusen H.J., 1996, *Bright future for natural gas*; Oil Gas European 2/1/96 [#467]

Rauch, J., 2001, *The New Old Economy: Oil, Computers, and the Reinvention of the Earth*; The Atlantic Monthly, p 3–49, January, 2001 [#1623]

Read, R D., 2001, *The North Sea: Oil Production Has Peaked! The GOM Model Must Come To The North Sea*; Simmons & Company International, October 18, 2001 [#1670]

Reed S., 2000, *Energy; Business Week*, Jan 10 [#1128]

Reed, S., 2001, *A Test For The House Of Saud*; Business Week, November 26, 2001 [#1721]

Reed, S., 2001, *Can The Saudis Step On The Gas?*; Business Week, December 24, 2001 [#1740]

Rees-Mogg W., 1992, *Picnics on Vesuvius: steps towards the millennium*; Sidgwick & Jackson 396p [#B17]

Rees-Mogg W., 1999, *Troubled waters for oil*; The Times 30 Aug. [#991]

Rees-Mogg, W., 2001, *Paying the price for the triumph of illusion*; The Times, September 10, 2001 [#1668]

Reese C., 1998, *We may get another great depression* ; Evansville Press 8/20/98 [#818]

Reich, K., 2001, *Gauging the Global Fuel Tank`s Size*; Los Angeles Times, Monday, June 11, 2001 [#1308]

Rempel H, ?, *Hydrocarbon Potential of Mediterranean Region;* Bundesanstalt fur Geowissenschaften und Rohstoffe [#1260]

Rempel, H., 2000, *Will the hydrocarbon era finish soon?*; BGR, Presentation at the DGMK/BGR event "Geosciences in Exploration…" ; Hannover, May 23, 2000 [#1608]

Reuters, 1990, *Oil stocks in west at low point*; Int. Herald Tribune 7/3/90 [#74]

Riahi M.L., 1998, *Deepwater exploration in the Gulf of Mexico*; Hart's Pet. Eng. Int., April [#737]

Ridley M., 1998, *Only hot air fuelled the petrol crisis*; Daily Telegraph 17 Mar [#710]

Rifkin J, 2002, *The Hydrogen Economy*; Penguin Putnam 294pp.

Riley D and M.McLaughlin, 2001, *Turning the corner: energy solutions for the 21st Century*; Alternative Energy Inst. 385pp

Rist, C., 1999, *Why We'll Never Run Out Of Oil*; Discover, p 80–87, June, 1999 [#1460]

Ritson N., 1998, *Maintaining production in the new millennium*; Petroleum Review, July [#714]

Riva J.P., 1991, *Dominant Middle East oil reserves critically important to world supply*; Oil & Gas Journ., Sept 23 1991 [#25]

Riva J.P., 1992, *Petroleum in the Muslim Republics of the Commonwealth of Independent States: more oil for Opec*; US Congressional Research Report 92-684 SPR [#165]

Riva J.P., 1992, *The domestic oil status and a projection of future production*; US Congressional Research Report 92-826 SPR [#166]

Riva J.P., 1993, *Large oil resource awaits exploitation in former Soviet Union's Muslim republics*; Oil & Gas Journ., Jan 4 1993 [#120]

Riva J.P., 1996, *World production after year 2000: business as usual or apocalypse*; Geopolitics of Energy 18/9 September 2–6. [#503]

Riva J.P., 1997, *U.S.conventional wisdom and natural gas*; M.King Hubbert Center for petroleum supply studies, Newsletter 97/3 [#622]

Riva J.P., 1999, *Is the world's oil barrel half full or half empty?* , M.King Hubbert Center 99/2 [#1052]

Riva J.P., 2002, *Canadian gas, our ace in the hole?* , M.King Hubbert Center 99/2 [#1018]

Roach J.W., 1997, _Reserves growth_; Oil & Gas Journ June 2 [#628]

Roadifer R.E., 1986, _Size distribution of world's largest oil, tar accumulations_; Oil & Gas Journ. Feb.26. 1986 93–98 [#17]

Roberts J., 1992, _Saudi Ambitions_; Petrol. Revue. July 1992 [#39]

Roberts J., 1995, _Visions & Mirages – the Middle East in a new era_; Mainstream ISBN 1-85158-429-3.

Roberts J., 1996, _IEA studies Middle East_; Petrol. Review, Feb. 1996 [#416]

Roberts J., 1999, _Caspian oil and gas flows west but doubts remain_; Petroleum Review Feb [#915].

Robertson J., 1989, _Future Wealth_; Cassell 178p.

Robertson Research International Limited, 2000, _Extra 30 years oil and gas supply_; Press Release, June 12, 2000 [#1515]

Robinson A.B. and Z.W. Robinson, 1997, _Science has spoken: global warming is a myth_; Wall St Journ Dec 4 [#746]

Robinson, B., 2001, _"The Big Rollover": World oil production decline predictions_; CSIRO Sustainability Network Australia, , Updated October 5, 2001 [#1713]

Robinson B., 2002, _Australia's growing oil vulnerability_; [#1859]

Robinson J., 1988, _Yamani – the inside story_; Simon & Schuster 302pp

Robinson, L. & Cary, P., 2002, _Princely Payments – Saudi royalty, it is claimed, make out like bandits on U.S. deals_; U.S. News & World Report, January 14, 2002 [#1741]

Robinson, M., 2002, _Venezuela syncrude challenging Mideast oil in U.S._; Reuters Limited, March, 2002 [#1794]

Rocky Mountain Institute, 1998, _Oil, oil everywhere…_; Newsletter Spring [#733]

Rodenburg E., _The decline of oil_; World Resources Inst., handbook [#343]

Roeber J, 1994, _Oil industry structure and evolving markets_; Energy Journal 15 [#351]

Roeber J, 1995, _Time to end exporters' self-defeating refusal to allow their crudes to be traded_; unknown ref. [#353]

Roger J.V., 1994, _Use and implementation of SPE and WPC petroleum reserve definitions_; Wld.Petrol. Congr., Stavanger [#228]

Rohter, L., 2001, _Energy Crisis in Brazil Brings Dim Lights and Altered Lives;_ The New York Times International, Wednesday, June 6, 2001 [#1302]

Roland K., 1998, _Perceptions of future, often flawed , shape plans and policies_; Oil & Gas Journ Feb 23 [#701]

Root D., E. Attenasi, and R.M. Turner, 1987, _Statistics of petroleum exploration in the non-communist world outside the United States and Canada_; U.S.G.S. Circ. 981 [#110]

Root D., E. Attenasi, and R.M. Turner, 1989, _Data and assumptions for three possible production schedules for non-Opec countries_; Memorandum U.S.Dept. of Interior 26 Sept. 1989 [#100]

Rossant J. & P. Burrows, 1994, _Pain at the pump_; Business weekly July 4 [#246]

Rothenberg M., 1994, *Risk factors in Azerbaijan*; World Oil, Feb 1994 [#202]

Rothschild E.S., 1992, *The roots of Bush's oil policy*; The Texas Observer Feb. 14 1992 [#33]

Rubin D, 2001, *OPEC cuts seen as new strain on economies;* Seattle Times [#1227]

Rubin, J. & Buchanan, P., *Why Oil Prices Will Have To Go Higher*; CIBC World Markets Inc. Occasional Report #28, February 2, 2000 [#1401]

Rubin, J. & Shenfeld, A. & Buchanan, P., 2000, *The Wall/How High Must Oil Prices Rise?*; CIBC World Markets Inc., Monthly Indicators, October 2000 [#1409]

Rubin, J., 2000, *Running On Empty*; CIBC World Markets, October 2000 [#1411]

Rudel, D., 2000, *Das Zeitalter des Erdoels geht zu Ende*; Yahoo! Schlagzeilen, Monday, June 5, 2000 [#1493]

Russian Information Agency; 1998, *Russia's E&P sector undergoes a slow transition*; World Oil June [#757]

S

Salameh M.G., 2000, *Can the oil price remain high?* ; Pet. Review April [#1078]

Salameh M.G., 2002, *Filling the global energy gap in the 21st Century*; Pet. Review Aug [#1905]

Salomon Smith Barney, 2001, *OPEC Analysis*; OPEC Monitor, No.103, July 30, 2001 [#1479]

Salpukas A., 1999, *An oil outsider revives a cartel;* New York Times Oct 24 [#1132]

Samuelson R.J, 2001, *The American energy fantasy;* Newsweek [#1228]

Samuelson, R J., 2001, *The Energy War Within Us*; Newsweek, p 28/29, May 28, 2001 [#1298]

Sander N.J., 1996 *Ibn Saud – King by Conquest;* Hats Off Books pp244.

Sander N.J., 1996 *Nestor Sander*, Calfutures.9/1 [#505]

Sandrea I. & O. al Buraiki., 2002, *Future of deepwater, Middle East hydrocarbon supplies*; Oil & Gas Journ [#1896]

Sampson A. , 1988, *The seven sisters: the great oil companies and the world they created*; Coronet, London. [B#11]

Sauer J.W, 1993, *Crude oil prices: why the experts are baffled*; World Oil Feb. 1993

Schindler, J. & Zittel, W., 2000, *Der Paradigmawechsel vom Oel zur Sonne*; Natur und Kultur, 1/1, p 48–69, 2000 [#1510

Schollnberger, 1996, *A balanced scorecard for petroleum exploration*; Oil Gas European 2/1/96 [#466]

Schollnberger, 199?, *Energievorrate unde mineralische rogrstoffe : wie lange noch*; Osterichische Akad. Wissenshft. Bnd 12.[#834]

Schollnberger, 1996, *Projections of the world's hydrocarbon resources and reserve depletion in the 21 century;* Houston Geol. Soc. Bull Nov. [#861]

Schrempp, J E., 2000, *Energy for the Future*; Speech at the opening of the World Engineer's Convention, Hannover, Germany, June 19, 2000 [#1538]

Schroeder Salomon Smith Barney, 2001, *Differentiating the Barrels*; Global Oil and Gas, August 23, 2001 [#1475]

Schuyler J., 1999, *Probabilistic reserves definitions, practices need further refinement*; O&GJ May [#1042]

Schuler G.H.M., 1991, *A history lesson: oil and munitions are an explosive mix*; Oil & Gas Journ. Nov.18, 1991[#182]

Schweizer P., 1994 *Victory: the Reagan administration's secret strategy that hastened the collapse of the Soviet Union*; Atlantic Monthly Press, New York 284p ISBN 0-87113-567-1

Science et Vie, 2001, *Energie*; No 214 Mars

Sciences et Avenir, 1997, *Petrole : la penurie a partir de 2015?* Aug.[#634]

Sciolino E., 1998, *It's a sea! It's a lake! No Its a pool of oil*; New York Times 21 June [#761]

Scott R.W., 1995, *Bloody fiasco*; World Oil Feb 1995 [#345]

Scott R.W., 1997, *Points to ponder*; World Oil July [#626]

Scott R.W., 1998 *Saudi stuff*; World Oil November [#871]

Seago D, 2001, *This energy crisis isn't going away*; The News Tribune [#1224]

Seal C., 2002, The 9/11 evidence that may hang George W. Bush; www.scoop.co.nz/mason/stories/HL0206/S00071 [#1860]

See M., 1996, *Oil mining field test to start in East Texas*; Oil & Gas Journ Nov. [#541]

Sell G., 1938, *Statistics of petroleum and allied substances*; The Science of Petroleum v1 1938 [#145]

Selley, ? , 2000, *World oil discoveries and production diagram*; Adapted from Selley, Changing Oil: Briefing Paper New Series, 10, January, 2000, The Royal Institute of International Affairs, London, [#1640]

Shafranik Y.K., 1993, *Fuel and energy complex of Russia: modern conditions and perspectives*: printed lecture to Univ. of Leiden May 1993

Shammas P., 1994, *LNG business – is it sustainable beyond the 2010s*; Preprint APS Conf. [#286]

Shell, 1995, *The evolution of the world's energy system 1860–2060*; Shell [#434]

Shell, 1999, *Arguments for solar energy*; Shell [#926]

Shell, 2001, *Energy needs, choices and possibilities*; Shell 69 pp

Shepherd R., 2000, *All eyes on the deepwater prize*; Offshore engineer, [#1061]

Shirley K., 1992, *Colombia finds wow explorers*; AAPG Explorer, Aug. 1992 [#121]

Shirley K., 1997, *Russia's potential still unrealized*; AAPG Explorer Aug.[#653]

Shirley K., 2000, *Caspian ready to face major tests*; AAPG Explorer Feb [#1077]

Shirley K, 2001, *Angola Hottest of the Hot Offshore*, Explorer [#1192]

Shirley, K., 2000, *Discoveries Are Getting Smaller*; AAPG Explorer, p 6, 10, January, 2000 [#1545]

Sierra J., 1994 *European energy supply security*; Petrole et Tech. 389 [#263]

Simienski A, 2000, *Energy Puzzler on Saudi upstream investment*; Deutsche Bank [#1089]

Simienski A, 2000, *Energy Puzzler on $25–30 oil price sustainability*; Deutsche Bank [#1139]

Simon B., 1990, *Oil project approaches last hurdle*; Financial Times Feb 2 1990 [#75]

Simmons M.R., 1994, *It's not like '86*; World Oil Feb. 1994 [#201]

Simmons M.R., 1995, *Strong market indicators*; World Oil, Feb 1995 [#330]

Simmons M.R., 1995, *Despite sloppy prices, fundamentals tighten*; Pet. Eng. Int. Sept 1995 [#392]

Simmons M.R., 1995, *1995 global wellhead review and drilling review; a new era for the oil service industry*; Simmons & Co report.[#402]

Simmons M.R., 1996, *Robust demand strengthens outlook*; World Oil Feb 1996 [#415]

Simmons M.R., 1997, *Are our oil markets too tight?* World Oil Feb. [#558]

Simmons M.R. *et al*, 1997, *Our hydrocarbon system in uncharted waters*; Simmons & Co [#602]

Simmons M.R., 1998, *Facts don't support weakening market*; World Oil, Feb. [#671]

Simmons M.R., 1998, *The impact of Asia's economic crisis on oil and gas now and in the future*; SPE Conf. Perth, Australia 13 Oct. [#857]

Simmons M.R., 1998, *The perils of predicting supply and demand*; Energy conf. N.Orleans [#909]

Simmons M.R., 1999, *1998: a year of infamy*; World Oil, Feb.

Simmons M.R., 1998, *It is not 1986, but could it be worse*; A.Andersen 19th Ann Energy Symp. [#950].

Simmons M.R., 1999, *Why oil prices need to rise: the illusion of oilfield cost and technology:* Oil, Gas & Energy Quarterly [#1034]

Simmons M.R., 2000, *The Earth in balance: has energy capacity maxed out;* Proc. Conf. Bridgewater House, London Nov. 16

Simmons, M R., 2001, *Digging Out of Our Energy Mess: The Need For An Energy Marshall Plan*; AAPG, June 5, 2001 [#1330]

Simmons, M R., 2001, *Investing in Energy: An Exercise Not for the Faint-hearted*; Managed Funds Association Forum 2001, New York City, July 11, 2001 [#1336]

Simmons, M R., 2000, *An Energy White Paper*; April 2000 [#1393]

Simmons, M R., 2000, *Energy in the New Economy: The Limits to Growth*; Energy Institute of the Americas, October 2, 2000 [#1402] and [#1418]

Simmons, M R., 2002, *2001: In like a lion, out like a lamb*; WorldOil.com- Online Magazine, February, 2002 [#1779]

Simmons M.R., 2002, *Depletion and US Energy Policy*; ASPO Workshop, Uppsala [#1853]

Simon, B., 2001, *Embracing Canadian Energy*; The New York Times, October 24, 2001 [#1726]

Sitathan, T., 2001, *Far East: Natural gas use increasing*; World Oil, August, 2001 [#1489]

Skrebowski C., 1998, *Sisters to wed*; Pet. Review Sept. [#822]

Skrebowski C., 1998, *Is this the third oil shock*? Petroleum Review October [#887]

Skrebowski C., 1998, *Iraqi production set to triple* Petroleum Review October [#888]

Skrebowski C., 1999, *Reducing drilling costs – the key to further non-OPEC development*; Pet. Review June [#981]

Skrebowski C., 1999, *Fossil fuel resources in the 21st Century*; Pet. Review [#992]

Skrebowski C., 2000, *How much can OPEC actually produce*; Pet. Review April [#1069]

Skrebowski C., 2000, *A silly game with no winners*; Pet. Review June [#1177]

Skrebowski C., 2000, *The North Sea – a province heading for decline?*; Pet. Review Sept

Skrebowski C., 2000, *The perils of forecasting*; Pet. Review Sept

Skrebowski C., 2000, *Discovery still lags production despite good 1999 results* Pet. Review Sept

Skrebowski, C., 2000, *Asking the wrong question about oil reserves*; Petroleum Review, September 2000 [#1364]

Skrebowski, C., 2000, *The perils of forecasting*; Petroleum Review, September 2000 [#1365]

Slessor M and J. King, 2002, *Not by money alone*; Jon Carpenter 160 pp

Smith M.R., 2001, *Environmentalists can relax. Oil supplies will decline sooner than most geoscientists are prepared to accept*; PESGB, Newsletter Aug.–Sept.

Smith M.R., 2002, *US oil supply vulnerability growing*; Offshore August [#1918]

Smith M.R., 2002, *Energy security in Europe*; Petroleum Review Aug. [#1871]

Snow N., 1995, *Nazar's dismissal does not mean change in Saudi oil policy*; Pet. Eng. Int. Sept. 1995 [#393]

Solomon C., 1993. *The hunt for oil*; Wall Street Journ. Aug.25 1993 [#169]

Sonnen Zeitung, 2000, *Bis zum letzten tropfen*; 4/00

Spencer, J E. & Rauzi, S L.(Arizona Geological Survey), 2001, *Crude Oil Supply and Demand: Long-Term Trends*; Arizona Geology, Vol.31, No.4, Winter, 2001 [#1753]

S.P.E. & W.P.C, 1999, *Petroleum resource definitions*; J. Petr. Geol [#1100]

Speight R, 1998, *Hydrocarbon Man: a threatened subspecies*; Shell International [#930]

Spiegel Der, 2000, *Auswege aus dem Energienotstand*; Der Spiegel, No. 23, 2000 [#1513]

Spring C., 2001, *When the lights go on: understanding energy*; Emerald Resource Solutions 120pp

Stabler F.R., 1998, *The pump will never run dry*; Futurist Nov. [#943]

Srodes J., 1998, *No oil painting*; Spectator 20 Aug. [#789]

Srodes J., 1998, *Here we go again*; Barron's 19 Oct. [#902]

Stanley B., 2002, *Oil supply seen set to fall*; Washington Times 28 May [#1856]

Starling P., 1997, *Oil market outlook*; Petroleum review Feb. [#544]

Starobin, P. & Crock, S., 2001, *Putin's Russia* and *From Evil Empire To Strategic Ally*; Business Week, November 12, 2001 [#1710]

Steakley, L., 2002, *In The Pipeline*; Wired, January, 2002 [#1748]

Steeg H., 1994, *World energy outlook to the year 2010*; 7th Int. Oil & Gas Seminar, Paris [#251]

Steeg H., 1997, *De nouveaux chocs pétroliers nous menacent… mais notre insouciance est totale*; Le Temps Stratigique, Geneve [#642]

Stone, R., 2002, *Caspian Ecology Teeters On the Brink*; Science, Vol. 295, January 18, 2002 [#1775]

Stoneley, R., 1993, *Book Review of "The Golden Century of Oil 1950–2050"*; Cretaceous Research, p 250, No. 14, 1993 [#1453]

Stosur G.J. & R.W. Luhning, 1994, *Worldwide EOR activity in the low price environment*; Pet.Eng.Int. Aug. 1994 p.46.[#267]

Stow A.R., 1996, *Consequences of US oil dependence*; Energy 3 Nov [#649]

Stow D., 1999, *Into the abyss*; Guardian 29 Sept. [#1014]

Stuermlinger, D., 2000, *Die grosse Explosion der Benzinpreise kommt erst*; Hamburger Abendblatt, May 29, 2000 [#1509]

Suddeutsche Zeitung, 2001, *Kassandra-rufe in Ol*; Suddeutsche Zeitung Online, Wissenschaft [#1244]

Sullivan A., 1992, *Iraq isn't expected to resume oil exports soon despite progress in talks with UN*; Wall St. Journ., Jan 13 1992

Sunday Times, 1993, *King Fahd*; Sunday Times 10/10/93 [#310]

Sunday Times, 1994, *Uncle Sam gets heavy with Fahd*; Sunday Times 9/10/94 [#269]

Sustainable Population Australia Inc., 2001, *A Dangerous Gamble – The assessment of Australia's carrying capacity based on use of fossil fuel*; Sustainable Population Australia Inc.(Occasional paper), June 2001 [#1317]

Sydney Morning Herald, 1997, *Report claims not enough oil to meet growth in Asian demand* 12.2.97 [# 550]

Syncrude, 1998, *Mining black gold*; World Oil June 1998 [#758]

Szulc. E., 1998, *Will be run out of gas?* ; Evansville Courier 19.7.98 [#795]

T

Takin M., 1972, *Iranian geology and continental drift in the Middle East*; Nature 235 [#303]

Takin M., 1988, *Energy cycles: can they be avoided*? Opec Bulletin Oct. 1988 [#305]

Takin M., 1989, *The high cost of misunderstanding Opec*; 14th Congr. Wld Energy Conf.[#306]

Takin M., 1990, *Opec, Japan and the Middle East*; Opec Bulletin, April 1990 [#304]

Takin M., 1993, *OPEC, Japan and the Middle East*; OPEC Bull. 4/2 (March–April 1993) 17–34.[#158]

Takin M., 1994, *How much gas is there in the Middle East*; Pet. Review July 1944 [#307]

Takin M., 1996, *Many new ventures in the Middle East focus on old oil, gas fields*; Oil & Gas Journ. May 27 [#457]

Takin M., 1996, *Future oil and gas: can Iran deliver?* World Oil Nov. [#516]

Takin M., 1998, *Prospects after the Riyadh agreement*; Petroleum Review, July [#713]

Takin M., 2001, *OPEC–consumer cooperation needed to ensure adequate future oil supply*; Oil & Gas Journal, December 3, 2001 [#1760]

Tanner J, 1990, *Looming shock: Mideast peace could trigger a sharp drop in crude oil prices*: Wall St. Journ. Dec. 19. [#50]

Tanner J, 1992, *Agency says world oil demand rose in 4th quarter and sees 1% rise in 1992*; Wall St. Journ., Jan 15 1992 [#11]

Tanner J, 1992, *Iran, in need of revenue, lifts oil output*; Wall St. Journ 19/10/92 [#115]

Tanzer, A. & Ghosh, C., 2001, *Insatiable – China and the rest of developing Asia are driving the world oil market*; Forbes, July 23, 2001 [#1321]

Tavernise, S. & Brauer, B., 2001, *Russia Becoming an Oil Ally*; The New York Times, Oct 19, 2001 [#1724]

Tavernise, S., 2001, *Exxon Says Way Is Cleared for Development in Russia*; The New York Times, October 30, 2001 [#1728]

Taylor III, A., 1999, *Oil Forever*; Fortune, p 193/194, November 22, 1999 [#1745]

Taylor B.G.S., 1997, *Towards 2020: a study to assess the potential oil and gas production from the UK offshore*; Petroleum Review, February [#543]

Taylor E., 1996, *Future Chill*; Tyler Morning Telegraph 12 Feb [#468]

Tchuruk S., 1994, *Les relations entre les societes et les gouvernements dans l'industrie mondiale du petrole et du gas*; Petrole et Technique 389, July 1994, [#266]

Teitelbaum R.S., 1995, *Your last big play in oil*; Fortune 30 Oct. 1995 [#382]

Thackery F., 1998, *Opec future rests on Asian tigers' return to growth*; Petroleum Review July [#772]

Thews K., 2000, *Allmahlich lauft die Erde leer*, Die Stern 4.5.2000 [#1369]

Third World Traveller, 2000, Global oil reserves alarmingly overstated; 1999 Censored Foreign Policies News Stories [#1180]

Thomas, M., 2000, *Azerbaijan back on track*; Hart's E&P, August 2000 [#1369]

Thomasson, M R., 2000, *Petroleum Geology: Is There a Future?*; AAPG Explorer, p 3–10, May, 2000 [#1543]

Thomson, B., 2001, *The Coming Oil Crash*; , 2001 [#1654]

Thurow L., 1996, *The future of capitalism*; Nicholas Brealey, 385pp

Tibbs H., 1997, *Global change: a context for transport planning*; Ecostructure Report [#670]

Tickell C., 1996, *Climate & history*; Oxford today 8/2 [#428]

Tissot B and D.H. Welte, 1978, *Petroleum formation and occurrence*; Springer verlag, New York [#B12]

Toal, 1999, *The big picture;* Oil & Gas Investor 1/99 [#866]

Toman M & J. Darmstadter, 1998, *Is oil running oil?* Science 282 2Oct [#882]

Townes H.L., 1993, *The hydrocarbon era, world population growth and oil use – a continuing geological challenge;* Amer. Assoc. Petrol. Geol. 77/5, 723–730.[#157]

Tomitate T., 1994, *World oil perspectives and outlook for supply-demand in Asia-Pacific region;* World Petrol. Congr. Topic 16. [#232]

Total, 1994, *Gas in FSU and European security of Supply;* Company Presentation [#215]

Trainer T., 1997, *The death of the oil economy;* Earth Island Journal Spring [#743]

Traynor, J J. & Sieminski, A. & Cook, C., 2000, *OPEC – Shortage? What Shortage?;* Deutsche Bank, Global Oil & Gas, November 9, 2000 [#1751]

Trout, R., 2001, *Global insight from oil expert – Interview with C J Campbell;* Hobbs News, Sun, August, 2001 [#1665]

Tugendhat C and A. Hamilton, 1968, *Oil – the biggest business;* Eyre Methuen. ISBN 0-413-33290-X.

Tull S., 1997, *Habitat of oil and gas in the Former Soviet Union;* Geoscientist 7/1 [#549]

U

Udall J.R., 1997, *When will the joy ride end;* report CORE [#665]

Udall, J. R. & Andrews, S., ???, *Methane Madness: A Natural Gas Primer;* Community Office for Resource Efficiency, Aspen, Colorado, ???, [#1592]

Udall, S L., 1974, *The energy balloon;* 288pp McGraw Hill

Udall, S L., 1980, *America`s Trip in the Energy Swamp;* The Washington Post, January 6, 1980 [#1395]

Udall, S L., 1980, *Energy: Fifty Years of American euphoria* [#1824]

Udall, S L., 1998, *The Myths of August;* Rutgers University Press 397pp

UKOOA & Nat. History Museum, *Britain's offshore oil and gas;* 56pp

Ulmishek G.F., R.R. Charpentier, and C.C. Barton, 1993, *The global oil system: the relationship between oil generation, loss, half-life and the world crude oil resource: discussion;* Amer. Assoc. Petrol. Geol. 77/5 896–899.

Ulmishek G.F, and C.D. Masters, 1993, *Oil, gas resources estimated in the former Soviet Union;* Oil & Gas Journ, Dec 13. 59–62 [#249]

USGS, 1995, *1995 national assessment of United States oil and gas resources;* USGS circular 1118. [#319]

USGS, 1996, *Ranking of oil basins* Open File [#638]

USGS, 2000, *USGS reassesses potential world petroleum reserves;* Press release [#1135]

USGS, 2000, *USGS Undiscovered Assessment Results;* website [#1161]

USGS, 2000, *World petroleum assessment – Description and results;* DDS-60

USGS, 2000, *USGS Petroleum Assessment 2000 – Description and Results;* ?, [#1362]

USGS, 2002, *2002 petroleum resource assessment of the NPRA, Alaska*; Fact Sheet 045-02 [#1890]

V

ven Koevering and N.J. Sell, 1986, *Energy – a conceptual approach;* Prentice Hall 271pp

Vlierboom F.W., B. Collini and J.E. Zumberge, 1986, *The occurrence of petroleum in sedimentary rocks of the meteor impact crater of Lake Siljan, Sweden*; 12th Europ. Assoc. Organic. Geochem. International meeting, Julich Sept 1985; Org. Geochem 10 153–161

Vogel T.T., 1998, *Oil's weakness sparks a sinking feeling in Venezuela*; Wall St Journ. [#802]

Von Rainer H, 2000, *Entzug der Oldroge*; DieWoohe [#1242]

Von Sternberg, B., 2000, *Energy independence is no longer en vogue*; Star Tribune, Minneapolis, Minnesota, Monday, April 3, 2000 [#1643]

W

Wall Street Journal, 2000, *Iraq may have filled tankers, but oil is still not flowing;* Wall Street Journal [#1254]

Wall Street Journal, 2001, *No Policy, No Win*; The Wall Street Journal, Thursday, August 9, 2001 [#1348]

Wall Street Journal, 2001, *World Demand for Oil To Rise a Million Barrels A Day Despite Slump*; The Wall Street Journal, ?, [#1349]

Walsh, J., 2000, *Canadian Heavy Oil*; Draft, November 11, 2000 [#1599]

Walton, P., 2000, *Case Study: FIOC IN Crisis*; DHP P 214 Strategy, July 26, 2000 [#1363]

Wardt J.P. de, 1996, *Operational realities in the '90s*; World Oil Feb. 1996 [#413]

Warman H.R., 1972, *The future of oil*; Geographical Journ. 138/3 287–297

Warman H.R 1973, *The future availability of oil*; proc. Conf. World Energy Supplies, by Financial Times, London 11p [#1]

Washington Inst., 1995, *Saudi succession uncertain*; Washington Inst. News Release [#385]

Washington Post, 2002 *Russia makes big comeback as oil producer*; 20 March [#1825]

Wattenberg R.A., 1994, *Oil production trends in the CIS*; World Oil June

Weiner J., 1990, *The next hundred years*; Bantam 312pp

Wellmer F-W, 1994, *Rerserven und reservenlebensdauer von energierohstoffen*; Energie Dialog, July 1994 [#274]

Wellmer F-W., 1997, *Factors useful for predicting future mineral commodity supply trends*; Geol. Rundsch 86 311–321 [# 735]

Weyant J.P. and D.M. Kline, 1982, *Opec and the oil glut: outlook for oil export revenues in the 1980s and 1990s*; Opec Review Winter 1982 334–365 [#93]

Wheelwright, T., 1991, *Oil and world politics*; Left Book Club 220pp

Whipple D., 1998, *Expert: world faces new oil crisis in next 12 years*; Caspar Star Tribune 6/11/98 [#850]

Widdershoven, C. & King, G H H. & Pabst, M. & Fuller, D., 2001, *Africa: Major Discoveries abound*; World Oil, August, 2001 [#1486]

Will G., 1996, *Heavy oil – the jewel in Western Canada's oil play*; Petroleum Review Nov. [#513]

Will G., 1997, *Promise from Canada's east coast*; Petrol. Rev. April [#580]

Williams C.J., 2000 *Norway looks beyond oil boom*; Los Angeles Times Feb 26 [#1082]

Williams P., 1998, *Half full or half empty*; Oil & Gas Investor 18/2 Feb [#680]

Williams P., 1997 *Orinoco*; Oil & Gas Investor 17/11 [#686]

Willingham B.J., 1994, *Energy shortage looms again*; The Nat. Times Mag, Oct/Nov [#261]

Wind Energy Weekly, 1996, *New oil price shock seen looming as early as 2000*; Wind Energy Weekly 15 684 12 Feb. [#895]

Winter F. de, 1976, *Solar energy happenings in other countries*; US Energy Res. & Dev. Admin. [#221]

Winter F. de, (?) *Economics and policy aspects of solar energy*; Altas Corp., Santa Cruz [#230]

Wood Mackenzie, 1997, *UKCS technical reserves*; N.Sea Report 295 [#824]

Wood Mackenzie, 1999, *Maturing gracefully*, UK Upstream Report 323 [#1174]

Wood Mackenzie, 1999, *Something new in the pipeline?*, UK Upstream Report 316 [#1062]

Wood Mackenzie, 2001, *UK Oil production forecast 2001*; UK Upstream Report, Feb.2001, Number 333 [#1188]

World Energy Council, 2000, *Energy for Tomorrow's World* 175pp

World Energy Assessment, 2000, *Summary description*; Clean Energy 2000 [#1107]

World Oil, 1998, *Accelerated development of Brazil's huge Roncador Field*; July [#890]

World Oil, 1999, *Outlook 99*; Feb [#960-F1]

World Oil, 2001, *Western Europe: Exploration targets becoming smaller* and *FSU / Eastern Europe: A roaring comeback continues*; World Oil, August, 2001 [#1487]

World Oil, 2001, *OPEC retains control*; World Oil, August, 2001 [#1749]

World Oil, 2002, *Gulf of Mexico Deepwater Map* [#1815]

World Tribune. Com, 2002, *Saudi royal family "in complete panic" during December riots*; Thursday, January 3, 2002 [#1743]

Wright D., 1994, *Small investors can strike tax bonanza in oil industry*; Sunday Times 27/2/94 [#218]

Wright T.R., 1991, *Whither oil prices*; World Oil May [984]

Wright T.R., 1999, *Oil prices could rise still more*; World Oil Oct.[#1018]

Wright T.R., 2001, *A guyser of oil, at Gladys City;* World Oil [#1223]

X

Xiaojie X, 1997, *China reaches crossroads for strategic choices*; World Oil April [#568]

Y

Yamani A.Z., 1995, *Oil's global role – the outlook to 2005*; MEES 38/33 [#355]

Yamani A.Z, 1997, *Containment is too risky;* Petrol Review May [#594]

Yamani H.A.Z., 1997, *Sheik's son urges Saudi Arabia to flood world oil markets*; Reuter Info. Service 25/8/97 [#687]

Yasunaga Y., 1994, *Japan's new energy policy and strategy*; Preprint APS Conf. [#279]

Yergin D., 1991, *The Prize: the epic quest for oil, money and power;* Simon & Schuster, New York, 877p [#B2].

Yergin D., 1988, *Energy security in the 1990s*; Petroleum Review Nov. 1988 [#88]

Yergin D., 2002, *US energy security lies in diversity of supply;* Oil & Gas Journ, 2 Sept 2002 [#1904]

Yergin D., and J. Stanislaw 1989, *The commanding heights*; Simon & Schuster pp 457

Youngquist W., 1997, *Geodestinies: the inevitable control of earth resources over nations and individuals*; Nat. Book Co., Portland 500p.

Youngquist W., 1998, *Spending our great inheritance – then what ?* Geotimes 43 7 24–27 [#791]

Youngquist W., 1998, *Shale Oil – the elusive energy*; Hubbert Center Newsletter 98/4 1–7

Youngquist W., 1999, *Comments about "The Coming Oil Crisis"*; Journal of Geoscience Education, v. 47, p.295, 1999 [#1432]

Youngquist W., 1998, *Shale Oil – The Elusive Energy*; Hubbert Center Newsletter #98 / 4 Oct, 1998 [#1463]

Youngquist W., 1999, *The Post-Petroleum Paradigm – and Population*; Reprinted from Population and Environment: A Journal of Interdisciplinary Studies, Vol.20, No.4, 1999 [#1706]

Youngquist W., 2000, *Alternative Energy Resources*; Pre-Conference Papers – Kansas Geological Survey Open-File Report 2000-51, October 2000 [#1421]

Z

Zach B.A., 1998, *Canada's petroleum industry* [#898]

Zagar J.J., 1999, *World oil depletion: the crisis on our threshold*; IADC Conf. [#958]

Zellner W., 1994, *Steamed about natural gas*; Business Week 10/10/94 [#273]

Ziegler P.A., 1993, *Plate-moving mechanisms: their relative importance*; Journ. Geol. Soc., 150/5

Ziegler, W. H., ???, *Book Review: "The Golden Century of Oil"*; ???, [#1456]

Zischka A., 1933, *La guerre secrète pour le pétrole*; Payot, Paris

Zittel W., 2000, *Fossile Energiereserven (nur Erdol und Erdgas) und mogliche Versorgungpasse aus Europaische Perspektive*; LB-Systemtechnik

Zittel W, 2001, *Analysis of the UK oil production, contribution to ASPO*; own paper [#1218]

Zittel W. & Schindler J., 2001, *Natural Gas – Assessment of the Term Supply Situation in Europe*; L-B-Systemtechnik Gmbh, May 28, 2001 [#1294]

Zittel, W., 2000, *Comment on EIA Presentation: Long Term world oil supply*; 2000 [#1386]

Index